Re-engineering the Enterprise

IFIP – The International Federation for Information Processing

IFIP was founded in 1960 under the auspices of UNESCO, following the First World Computer Congress held in Paris the previous year. An umbrella organization for societies working in information processing, IFIP's aim is two-fold: to support information processing within its member countries and to encourage technology transfer to developing nations. As its mission statement clearly states,

> IFIP's mission is to be the leading, truly international, apolitical organization which encourages and assists in the development, exploitation and application of information technology for the benefit of all people.

IFIP is a non-profitmaking organization, run almost solely by 2500 volunteers. It operates through a number of technical committees, which organize events and publications. IFIP's events range from an international congress to local seminars, but the most important are:

- the IFIP World Computer Congress, held every second year;
- open conferences;
- working conferences.

The flagship event is the IFIP World Computer Congress, at which both invited and contributed papers are presented. Contributed papers are rigorously refereed and the rejection rate is high.

As with the Congress, participation in the open conferences is open to all and papers may be invited or submitted. Again, submitted papers are stringently refereed.

The working conferences are structured differently. They are usually run by a working group and attendance is small and by invitation only. Their purpose is to create an atmosphere conducive to innovation and development. Refereeing is less rigorous and papers are subjected to extensive group discussion.

Publications arising from IFIP events vary. The papers presented at the IFIP World Computer Congress and at open conferences are published as conference proceedings, while the results of the working conferences are often published as collections of selected and edited papers.

Any national society whose primary activity is in information may apply to become a full member of IFIP, although full membership is restricted to one society per country. Full members are entitled to vote at the annual General Assembly, National societies preferring a less committed involvement may apply for associate or corresponding membership. Associate members enjoy the same benefits as full members, but without voting rights. Corresponding members are not represented in IFIP bodies. Affiliated membership is open to non-national societies, and individual and honorary membership schemes are also offered.

Re-engineering the Enterprise

Proceedings of the IFIP TC5/WG5.7 Working Conference on Re-engineering the Enterprise, Galway, Ireland, 1995

Edited by

Jim Browne and David O'Sullivan
CIMRU, University College, Galway
Ireland

Published by Chapman & Hall on behalf of the
International Federation for Information Processing (IFIP)

CHAPMAN & HALL
London · Glasgow · Weinheim · New York · Tokyo · Melbourne · Madras

Published by Chapman & Hall, 2–6 Boundary Row, London SE1 8HN, UK

Chapman & Hall, 2–6 Boundary Row, London SE1 8HN, UK

Blackie Academic & Professional, Wester Cleddens Road, Bishopbriggs, Glasgow G64 2NZ, UK

Chapman & Hall GmbH, Pappelallee 3, 69469 Weinheim, Germany

Chapman & Hall USA, 115 Fifth Avenue, New York, NY 10003, USA

Chapman & Hall Japan, ITP-Japan, Kyowa Building, 3F, 2-2-1 Hirakawacho, Chiyoda-ku, Tokyo 102, Japan

Chapman & Hall Australia, 102 Dodds Street, South Melbourne, Victoria 3205, Australia

Chapman & Hall India, R. Seshadri, 32 Second Main Road, CIT East, Madras 600 035, India

First edition 1995

Printed in Great Britain by TJ Press, Padstow, Cornwall

ISBN 0 412 64260 3

A catalogue record for this book is available from the British Library

Printed on permanent acid-free text paper, manufactured in accordance with ANSI/NISO Z39.48-1992 and ANSI/NISO Z39.48-1984 (Permanence of Paper).

CONTENTS

PREFACE

The business environment throughout the world is currently going through rapid and far reaching change. Companies are re-structuring the way they do business. They are analysing their business processes and scrutinising ways to make their systems more streamlined and competitive. In parallel with this change, developers are putting forward new concepts and new tools which can facilitate companies in carrying out what has become known as 'business process re-engineering' (BPR). Together these developments have yielded a tremendous amount of new knowledge and will continue to offer us new challenges well into the future.

This book brings together a number of leading experts from around the world who have worked extensively in the area of business process re-engineering. Through individual chapters in this book, these consultants, analysts, and researchers put their expert views forward on various aspects of re-engineering the business enterprise. Some authors offer new ideas of the way processes should be re-engineered. Other authors present new tools and new approaches for the effective implementation of process re-engineering. Still other authors present the future issues for change in the business environment and a glimpse of what solutions may be developed in the future.

The book has been loosely structured to allow chapters which address common themes to be grouped together. In these chapters, the reader will learn about all of the key issues currently being addressed in BPR research and practice throughout the world. The reader will discover new approaches to BPR and learn about new software tools which facilitate BPR change. In short, this book presents the latest thinking and solutions to BPR implementation from leading experts around the world.

Each of the chapters in this book are the result of presentations given by the authors at the international conference 'Re-engineering the Enterprise' held in Galway, Ireland in April of 1995. The conference, which was supported by the International Federation of Information Processing, attracted experts from around the world who presented and reviewed the latest thinking and ideas on business process re-engineering. The conference was organised by Working Group 5.7 from within IFIP and hosted by CIMRU of University College Galway.

We would like to thank all of our contributors for taking the time and trouble in preparing and presenting their papers to a very high standard. We would also like to thank members of IFIP Working Group 5.7 for their support in reviewing and selecting each of the papers. Special thanks goes to the organising committee of 'Re-engineering the Enterprise' especially William Kelly, Catherine Hayden, Paul Higgins, Gerry Lyons, Padraic Bradley, Attracta Brennan, Brian Kelly, and Garry Monaghan for their help in the preparation and administration of the conference.

Jimmie Browne
David O'Sullivan

Galway, July 1995

1

Planning centres - new conception for re-engineering activities in non-producing areas

M. Remme, A.-W. Scheer
Institut für Wirtschaftsinformatik IWi (Institute for Information Systems) at the University of the Saarland,
Im Stadtwald, Geb. 14.1, 66123 Saarbrücken, Germany
Tel.: +49/681/302-4297, Fax: +49/681/302-3696
E-Mail: remi@iwi.uni-sb.de

Abstract

The definition of tasks and functions as well as their relationships is one of the main tasks of organising an enterprise. Usually, this is done step by step. This means the complexity of an enterprise's overall task is broken down. The resulting subtasks must be assigned to individual task performers. Only if the task is adequately broken down, task performers will be able to execute it properly. Starting from present-day requirements, this paper explains fundamental organisational patterns to reduce the complexity in enterprises. As point of departure these patterns are used to derive the conception of planning centres. Afterwards, the way down to planning centres—i.e. the enterprise's re-engineering—is lined out and experiences gained in an industrial project are explained.

Keywords

Planning centre, business process re-engineering, complexity of organisational structures, functional-, object-oriented and hybrid paradigm for organising enterprises, networked decentralisation, Architecture of Integrated Information Systems (ARIS), re-usability of organisational knowledge.

1 INTRODUCTION

Over the last decades major improvements in technology and personnel's qualification have effected that those who take advantage of this development in re-designing their organisational structures meet the competition of the market. Catchwords like lean production, decentralisation, business process re-engineering etc. represent paradigms of how to adapt the enterprises to these new challenges.

Four industrial companies assisted by five research institutes and consultancies are investigating the conception of the planning centres which represent the missing component to complete current decentralised and process-oriented approaches to an overall flexible and customer-oriented enterprise organisation called networked decentralisation (Scheer 1994, p.585). Figure 1 illustrates the structure of the project "Funktionsintegration in Planungsinseln" (Integration of functions in planning centres) which is funded by the German Ministry for Education, Science, Research and Technology (BMBF). It defines an architecture of decentralised organisational units in non-producing areas of enterprises and how this architecture is applied. There are four autonomous partial projects which in each case consist of one industrial company and one or two supporting partners. All the partial projects have to implement a pilot planning centre but they differ in the main functionality of the planning centre and in the enterprise-specific preconditions. While e.g. fischerwerke manufactures dowels in mass production Fortschritt manufactures machines for harvesting in small serials. The project is co-ordinated by the Institut für Wirtschaftsinformatik IWi (Institute for Information Systems), which thus, has a double-function beside being a supporting partner.

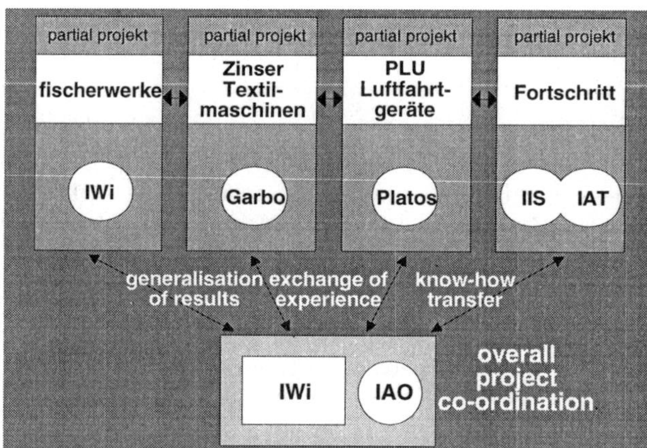

Figure 1 Structure of project referred to.

2 BREAKING DOWN THE COMPLEXITY OF ORGANISATIONAL STRUCTURES

In industrial organisations, when a problem's complexity is defined by the effort a human being has to expend in order to solve it, this complexity is the result of the multitude of functions to be processed and the large number of goods and services to be scheduled (materials, component parts, assemblies, all the way to end products). (Grossmann 1992, pp.18, Flood/Jackson 1991, pp.33) Figure 2 compares these two dimensions, with the size of the area representing the overall complexity. (Scheer 1994, p.7)

Dividing this complexity area into individual sections serves to reduce the overall complexity. The predominant form of structuring is function-oriented. In a functional structure, one organisational unit (e.g. staff, department) is given responsibility for all areas and products. This principle is illustrated in Figure 2 by horizontal lines. This approach takes advantage of specialisation effects of the employees. A disadvantage of a highly departmentalised functional structure is the high communication and co-ordination effort. Transferring data between all these departments causes high transferring, waiting and working-in times.

Since this structural principle was the prevailing organisational form in the early days of electronic data processing, DP structures reflect horizontal compartmentalisation schemes. Usually, these function-oriented information systems contain their own databases, as illustrated in Figure 2. A closer look, however, shows that these business functions are linked by decision and process relationships: When objects (or outputs) are processed, they typically pass through several functions; for example, an order passes from Sales through Production and Purchasing to Accounts Receivable and is processed further in the Production Controlling and Sales/Marketing. If each function manages its own data, the object relationship of processes will result in redundancy, because data that belong to one object are stored by several functions. Although this leads to increased physical storage requirements, the primary problem is one of logical data consistency, because the data definitions would not necessarily have to be consistent if each function defined its data according to its own requirements.

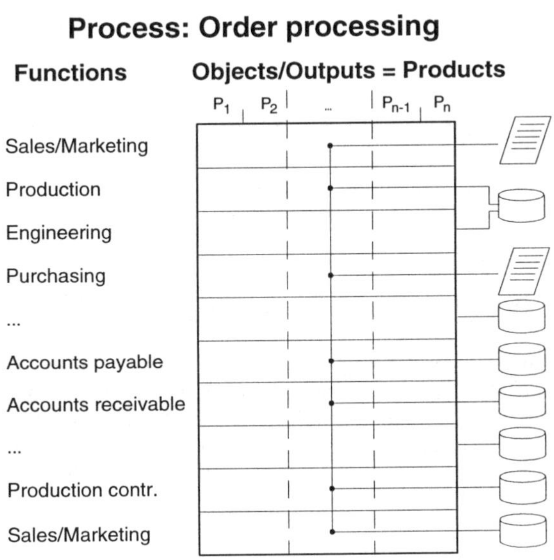

Figure 2 Reduction of complexity.

As a result, functions tend to become integrated. This process is accompanied by dividing the complexity of an enterprise on the basis of other criteria, such as territories, locations or products. (Neis 1994, pp.98-101) In a purely object-oriented structure (vertical lines in Figure 2) as many functions as possible that are responsible for an object class are integrated within a group, which processes them. This makes it possible to decrease co-ordination and communication efforts between the individual processing functions. The most important disadvantage of this organisational form is that synergy effects that arise between the product groups are not exploitable to that extent.

This is the reason for hybrid organisational forms (thick line in Figure 3). Figure 3 gives an idea about it. The functions “supplier selection” and “terms” are performed centrally for all purchases and thus conform to a functional structure, while “planning,” “ordering” and “checking and auditing invoices” are decentralised according to object-oriented structural criteria. (Scheer 1994, p.26) The object-oriented units can plan and order the components tailored to their special needs quickly and in a process-oriented manner, while the central functions “supplier selection” and “terms” reflect the synergy effects of the overall company vis-à-vis the suppliers.

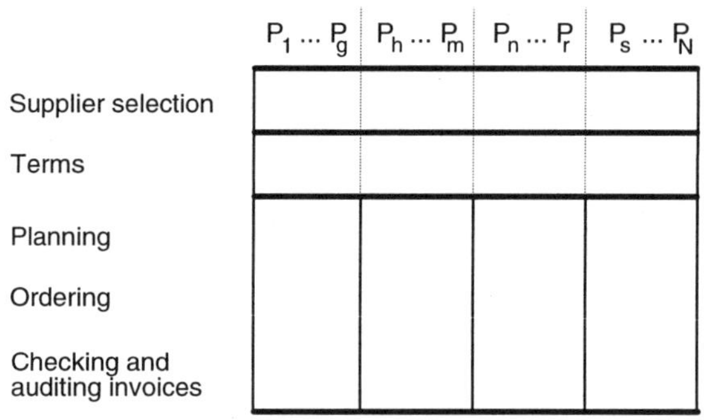

Figure 3 Hybrid organisational structure.

3 CONCEPTION OF THE PLANNING CENTRE

A comprehensive structure which consequently bases on these hybrid forms is called "networked decentralisation", because independently operating organisational units are supported that are both decentralised and networked with each other. (Scheer 1994, p.585) Operatively, the processes are handled in the decentralised units. The relationships between these decentralised units are treated at the superordinate co-ordination level. As a rule, these problems are not as closely linked in terms of time as the processes within the process-oriented organisational unit. Further planning units can be formed that encompass several co-ordination units. The planning functions can also be co-ordinated at a superordinate level.

There are already existing approaches of hybrid organisational forms at the shop floor level. The respective organisational units are called leitstands or production islands (see Figure 4). A leitstand is an autonomous production system which is usually controlled by a central system like a production planning and control system. (Scheer/Hars 1990, pp.377-380) A consequent use of hybrid organisational form leads to the conception of the planning centre.

A planning centre represents an organisational paradigm that pays special attention to the needs of human beings working in it. It is a permanent organisational unit consisting of a group of staff. This group is responsible for the execution of a wide range of closely linked planning, controlling and implementation functions. (Kruse/Scheer 1994) The planning centre is installed in non-producing areas like sales, materials management, research and development etc. The planning centre acts to a large extent autonomously. Once the centre is built one essential task is—besides the daily work—to organise the own structure and the personnel's further qualification.

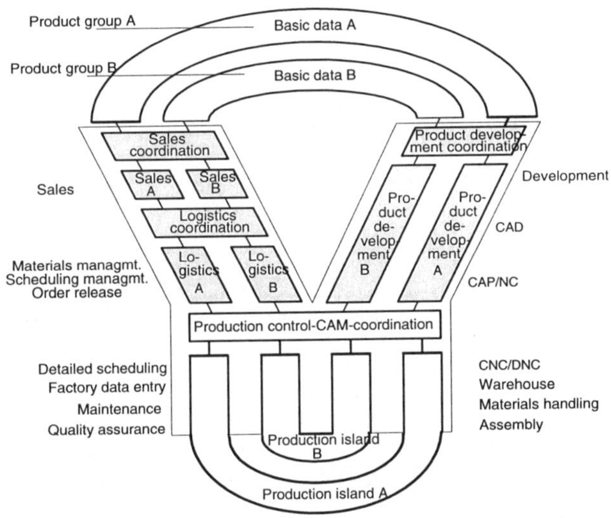

Figure 4 Planning centres.

Figure 4 illustrates the philosophy of the planning centre. Comprehensive processes are broken down into planning centres. Intra-actions within these centres are high and interactions between them low. Again there are the two dimensions of separation: processes and products. The result of re-engineering an enterprise may be an organisation, as shown in Figure 4, consisting of planning centres, two in each, for sales, logistics and product development. Special regard must be given to the co-ordination of planning centres to meet the requirements arisen by the fact that there are relationships between the planning centres. Therefore, planning centres have to be co-ordinated by central units. They also perform additional functions in order to take advantage of synergy effects like mentioned before.

Other types of planning centres may have to process, for example, a wider range of functions while there is a refined specialisation on objects. As an extreme case, you can think of one planning centre for each existing product processing the complete functional spectrum from marketing and order processing up to production and dispatch. All the planning centres are co-ordinated by a superordinated instance. This extreme idea is very close to the one of profit centres. (Keller 1993, pp.304-306)

Table 1 Characteristics and benefits of planning centres.

Characteristics	Consequence	Benefit
autonomy (concerning determination of operative processes, design of PC-internal organisation, right of disposal of resources)	decision processes are quicker	faster reactions, more flexibility, higher satisfied customers
	decision of those, which have the best know-how	"best" decision, "problem-near solutions
	organisational weaknesses can be eliminated immediately when appearing	continuous improvement process
	working environment is designed according to the requirements of staff	motivation, reduction of absence times of staff
	relocation of former management tasks into the PC	unburding of management, reduction of number of hierarchical levels
process-orientation	staff activities cover comprehensive functionality in process chain	less communication and co-ordination effort, less waiting and transferring times
overlapping qualification	higher flexibility for employing staff	higher process reliability, higher use of capacity, easier disposition of staff
primarily internal qualification	steadily growing qualification of staff with only low occupation of capacities	improving quality
no superior in the group	all members of PC have equal rights	higher motivation, innovation, less absence
easy to survey location and tasks	communication and co-operation without formal pre-definition	reduction of overhead functions, less management necessary
multifarious work	working means learning	continuous increasing of staff qualification, motivation, absence

Table 1 gives an overview of the most important characteristics of planning centres and derived benefits from the perspective of the enterprise goals.

4 CONDUCTION OF RE-ENGINEERING

In a first phase data about the current organisation of an enterprise has been collected and modelled by using the ARIS-Toolset of the IDS Prof. Scheer GmbH, Germany, which is based on the Architecture of Integrated Information Systems (ARIS) (Scheer 1992). The ARIS approach pre-defines four descriptive views and 3 life-cycle phases as shown in Figure 5 and thus, comprises 12 components. For each component a set of different suitable and integrated description methods is offered.

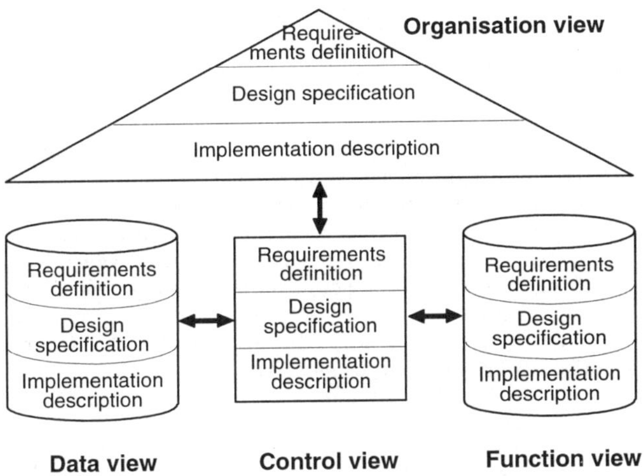

Figure 5 Architecture of Integrated Information Systems (ARIS).

The languages used for modelling the enterprise are entity-relationship models (ERM) for modelling the data view, hierarchy diagrams for the functions and organisations view and extended process chain diagrams (eEPK) (Keller, Nüttgens, Scheer 1992) for the process view. These models are analysed. Weaknesses like time-delays, unnecessary changes of information carriers, redundant execution of functions etc. are identified and eliminated. As a result, there are basic process structures. Parallel to this, principles and rules for distributing processes to organisational units are defined. Hierarchical and sequential completeness will be one of these principles, for example. This means, that functions are grouped together and processed by a planning centre, if they are directly sequential or if the output of a certain function represents the plan for conducting another one.

All the experiences about designing and implementing planning centres gained in the four partial projects are generalised by the IWi. As a result, reference models are defined that give generally accepted proposals (Hars 1993, p.15) for organising decentralised enterprise structures and their logical dependencies on important characteristics of the consortium's enterprises. The basic idea of a reference model is the re-usability of organisational know-how. Several projects with industrial partners have shown that in every project there is a large amount of knowledge involved that could be re-used for other projects. Thus, knowledge that is required for successfully carrying out re-organisation projects, is divided into three categories: (Remme/Allweyer/Scheer 1994, p.234)

1. General knowledge which is independent from a certain company or a certain industrial branch. This includes general knowledge about organisation, business processes, and information technology.
2. Knowledge specific to important characteristics. This may be e.g. a certain industrial branch. This knowledge includes branch-specific requirements, typical processes and structures, common problems etc.
3. Knowledge specific to a certain company. This includes those characteristics of the company that are different from other companies of the same characteristics.

Instead of starting from scratch in every company, a re-organisation project based on reference models created in this project has already the knowledge covered by point 1 and 2 and only has to adapt the reference model to the company's very specific situation. It is therefore not necessary to "re-invent the wheel" in every project.

The ARIS-Toolset offers the functionality for storing reference models and their logical relationships to enterprise characteristics and for configuration of these reference models according to specific enterprise characteristics. Figure 6 describes the use of ARIS-Toolset and reference models. A certain company records models of the current organisation and a list of relevant characteristics has to be filled in. From this information which is based on the reference models and the logical rules, first proposals for the planning centre target specification, for the re-engineering process and for necessary staff qualification are configured automatically. Usually, these models have to be adapted to very individual characteristics of the enterprise, which cannot be covered by reference models.

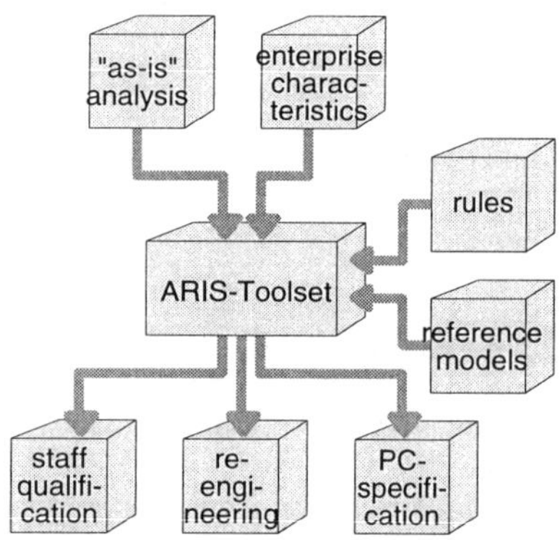

Figure 6 EDP-support of re-engineering.

5 FINAL STATEMENT

Often, re-engineering projects fail because they are single-focused. The approach, presented here, focuses on organisation, human resources, and information and communication technology. All of these aspects are of equal importance. Re-engineering has to balance these to achieve an "organisational fit". The storing of experiences in tools makes re-engineering knowledge automaticable, and what's more important: transportable.

6 REFERENCES

Flood, R.L., Jackson M.C.: *Creative Problem Solving - Total Systems Intervention*, Chichester 1991.

Grossmann, C.: *Komplexitätsbewältigung im Management, Anleitungen, integrierte Methodik und Anwendungsbeispiele*, Diss. St. Gallen 1992.

Hars, A.: *Referenzdatenmodelle: Grundlagen effizienter Datenmodellierung*, Wiesbaden 1994.

Keller, *Informationsmanagement in objektorientierten Organisationsstrukturen*, Wiesbaden 1993.

Keller, G.; Nüttgens, M.; Scheer, A.-W.: *Semantische Prozeßmodellierung auf der Grundlage "Ereignisgesteuerter Prozeßketten (EPK)"*, in: Scheer, A.-W. (eds.): Veröffentlichungen des Instituts für Wirtschaftsinformatik, Heft 89, Saarbrücken 1992.

Kruse, C.; Scheer, A.-W.: Dezentrale Prozeßkoordination in Planungsinseln, in: *Information Management*, 9(1994)3, S.6-11.

Neis, B.: Vernetzte administrative Inseln, Prozeßorientierte Organisationsformen zur Restrukturierung der Ablauf- und Aufbauorganisation, Diss. Saarbrücken 1994.

Remme, M.; Allweyer, T.; Scheer, A.-W.: Implementing Organizational Structures in Process Industry Supported by Tool-Based Reference Models, in: Boucher, Elsayed, Jafari (eds.): *Computer Integrated Manufacturing in Process Industries*, Proceedings of the CIMPRO #94 Conference, East Brunswick NJ, 25.-26. April 1994, S. 233-247.

Scheer, A.-W.; Hars, A.: The Leitstand - A New Tool for Decentral Production Control, in: Fandel, G.; Zäpfel, G. (eds.): *Modern Production Concepts, Theory and Application*, Proceedings of an International Conference, Fernuniversität Hagen, August 20-24, 1990, Berlin et al. 1991, pp.370-385.

Scheer, A.-W.: *Business Process Engineering, Reference Models for Industrial Enterprises*, 2nd ed., Berlin et al. 1994.

Scheer, A.-W.: *Architecture of Integrated Information Systems, Foundations of Enterprise-Modelling*, Berlin et al. 1992.

7 BIOGRAPHY

Dipl.-Kfm. Markus Remme has finished his studies of business administration in 1992 and is now working as a scientific assistant at the Institut für Wirtschaftsinformatik at the University of the Saarland, Germany.

Prof. Dr. August-Wilhelm Scheer is director of the Institut für Wirtschaftsinformatik at the University of the Saarland, Germany, and has established the CIM Technology Transfer Centre in Saarbrücken. He is honorary professor at the University of Vienna. Prof. Scheer is the editor of various serial books, journals and the author of more than 300 essays and 20 books.

2

Re-engineering through fractal structures

Dr.-Ing. Wilfried Sihn
Fraunhofer Institute for Manufacturing Engineering and Automation (IPA)
Silberburgstr. 119a, 70176 Stuttgart, Germany
Tel: ++49 711 970-1964
Fax: ++49 711 970-1002
e-mail: whs@ipa.fhg.de

Abstract

Many branches of European industry have had to recognise that their lead in the world market has been caught up with, particularly through Asian competition. In many cases a deficit of up to 30% in costs and productivity already exists. The reasons are rigid, Tayloristic company structures. The companies are not in a position to react flexibly to constantly changing environmental conditions.

This article illustrates the methods of the "fractal company" which are necessary to solve the structure crisis. The fractal company distinguishes itself through its dynamics and its vitality, as well as its independent reaction to the changing circumstances.

The developed methods, procedures and framework conditions such as company structuring, human networking, hierarchy formation and models for remuneration and working time are explained. They are based on practical examples from IPA's work with the automobile industry, their suppliers and the engineering industry.

Keywords

fractal company, organisational structures, hierarchy formation, re-engineering

1 THE SHIFT OF GOALS

There has been a clear shift concerning the objectives of industrial production. See Figure 1. Goals of production that have been held up for decades are giving way in favour of speed objectives. This is mainly due to a shift from purchaser markets to seller markets. The cause for this shift is the saturation, or even the over-saturation, of markets. According to surveys, profit losses during commercialisation are highest if there is any delay in placing the product on the market. Other factors are less important. There are fewer products whose production is

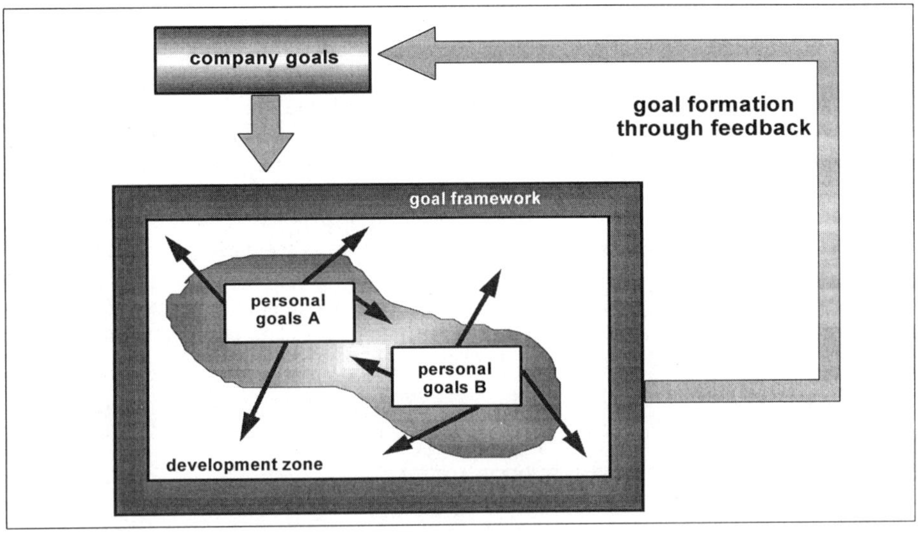

Figure 1 Goal formation process in the fractal factory.

complicated in the conventional sense. On the one hand, fully matured components whose assembly results in final products contribute to this. On the other hand, manufacturing technologies, becoming increasingly reliable, ensure consistent quality and allow the production of complicated parts by normally trained staff. This means there will be more suppliers serving the market with similar products. The result is an over-saturation of the market. See Figure 2.

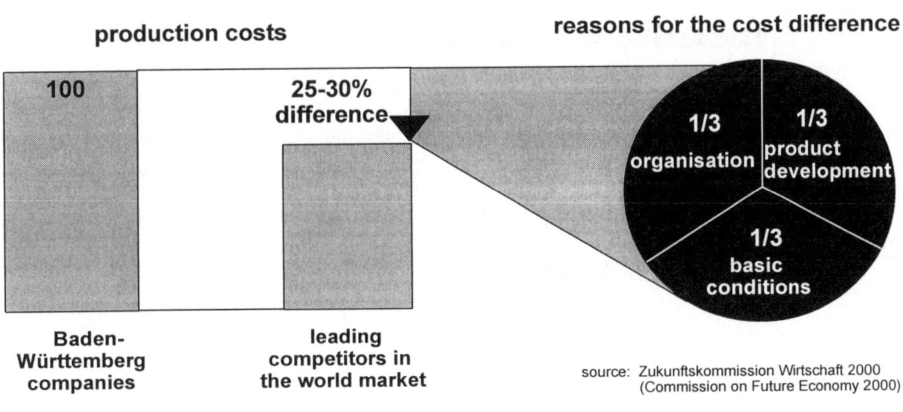

Figure 2 Cost disadvantages of companies in Baden-Württemberg.

2 HISTORICAL CHANGE

Taking a historical look at production development, the number of employed factors and their availability has been steadily increasing. This is remarkable, especially concerning the availability of mechanical energy and information which is crucial for industrial production, allowing the effective and timely usage of resources which results in reduction of the stock of resources. Because of this reduction it is not always possible to have a stable situation, that is, to always run production according to events calculated in advance. Rather, the dynamics of the events will increase. One will have to abandon certain processing stages without prior planning and leave them to their own devices, which means entrusting them to higher control circuits. As a consequence, it is of utmost importance to incorporate into the company and manufacturing strategies the greatest closeness to customers and an extreme ability to react promptly. The speed objective must be pursued vigorously, even regarding the development of operating technologies.

3 PRODUCTION FACTOR INFORMATION

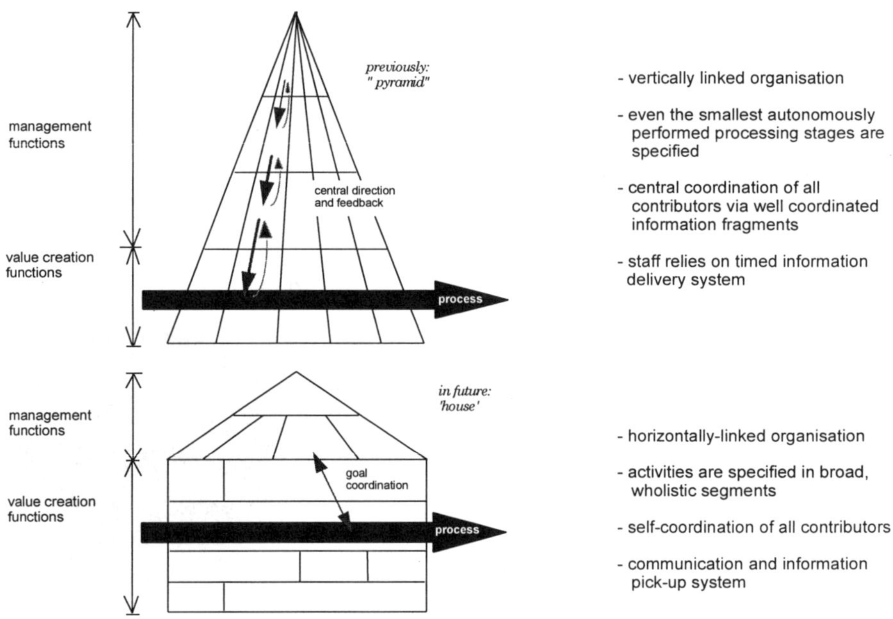

Figure 3 Interplay of organisation, information and value creation.

The decline in prices for electronic components, improvements in efficiency of data processing units and progress concerning communication technologies make any desired quantity of information available at any place at any time. Despite this fact, the processes in our industrial companies are still organised according to the old model in which the division of labour has also been used for minimising the effort for drawing up information. Although each worker could be included in the fetch-principle, as far as the required information is concerned, our organisational structures work according to the bring-principle for information by staff functions and through hierarchy levels. See Figure 3. Overcoming this discrepancy is the task of the coming years for each industrial company. These developments could enable the organisational structure to become flatter, and will result in more scope for decisions and more responsibility for each individual in the organisation. In the past, one believed the secret of effective performance specification consisted in Taylorism, applying appropriate coordination. Conditions have now radically changed. The summing up of contents of labour is required. Tasks must be interlinked and turned into processes. Also, flow design is called for. The organisational structures will increasingly turn into project organisations. See Figures 4 and 5.

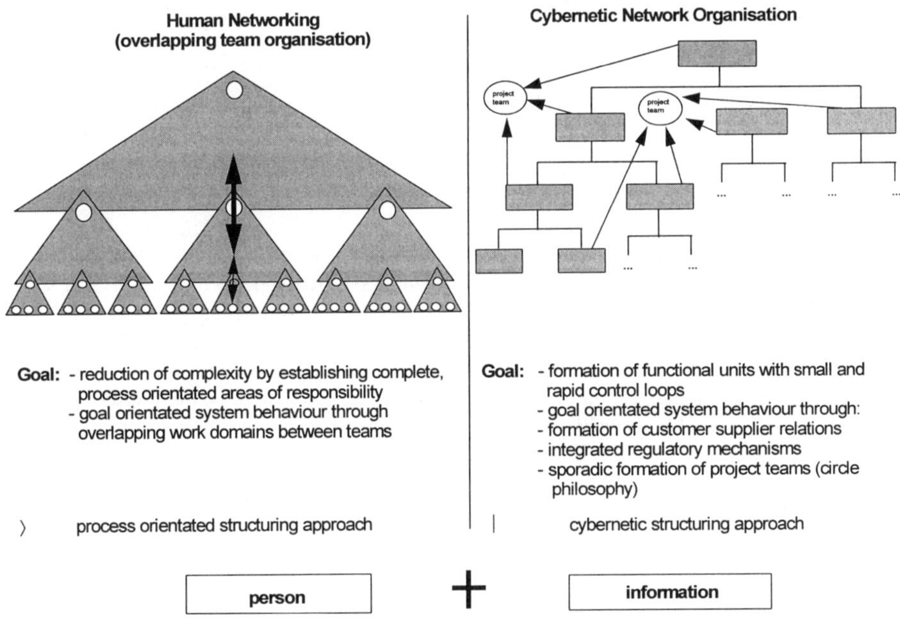

Figure 4 Organisation structures in the fractal factory.

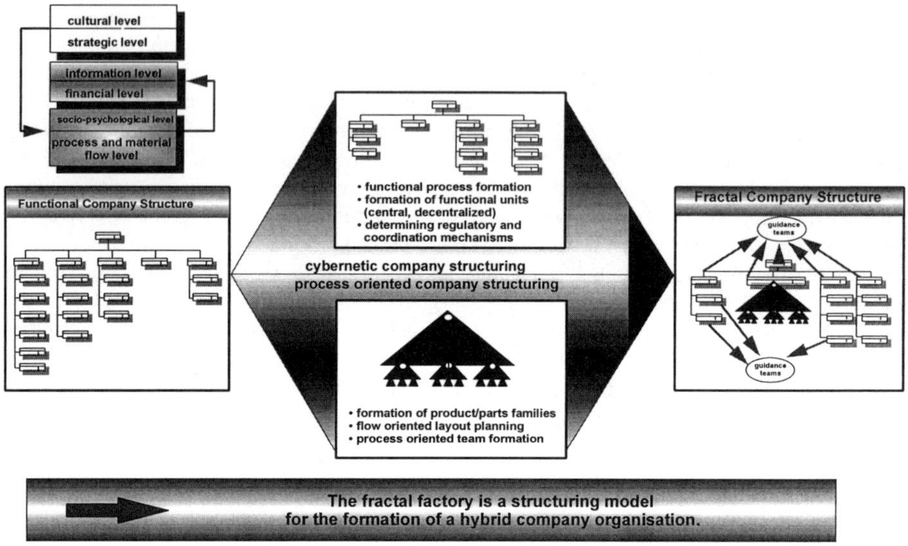

Figure 5 Company structuring in the fractal factory.

4 LIMITATIONS OF CURRENT SYSTEM CONSIDERATIONS

The faster moving markets, technologies and information systems available everywhere lead to higher pressure on the companies. There is an increased demand for innovation and speed concerning processes and adaptation. As a reaction to the environmental dynamics and fast-moving events, strong self-dynamics and the ability to evolve are required. However, our current view of a manufacturing company is static. We assume there are definite, clear-cut tasks and that a need exists for authorities who issue instructions on a rigid structure scale. A company develops in a linear fashion, and the economic and arithmetic models are static.

5 CHARACTERISTICS OF DYNAMIC CHANGE

With such above-described changes concerning important parameters for a manufacturing company, conventionally structured companies are no longer able to fulfil the required self-dynamics with the necessary speed of evolution. The results of this rigidity become increasingly perceptible to everyone in the company. Phenomena such as rising pressure concerning costs, a rapidly worsening profit situation, a waning ability to adapt to market situations and an increasing rush regarding the sequence of events should be mentioned. In such cases additional tasks, overtime and reorganisation cannot put things right. They represent a specific consideration of single phenomena like operating times, wages or productivity which can only have a limited and temporary effect. A paradigm change is required — a new conceptional way of thinking which is the basis of the manufacturing company. It must consider that these changes take place with enormous dynamics. In order to survive, a company must learn to secure market shares and profitability in a turbulent environment by developing adaptability. In such an environment it does not develop linearly but with leaps in development and with transformations according to the laws of probability which, although they can be controlled, cannot be accurately predetermined. It is therefore a matter of establishing dynamic systems whose development can become complicated. The dynamics which a company has to develop under these conditions is called *evolution*. Only a continuous development secures an appropriate adaptation to the changing environment. As a key statement: a company is an open, complex, dynamic system whose logic — based on chaotic system thinking — is one of fractals, and whose dynamics are shaped according to laws of evolution.

6 DESIGN OF INDUSTRIAL PERFORMANCE PLANNING

In order to master the future, ways of thinking and points of view must be changed radically. The considerations must move away from static systems toward dynamic systems, away from mechanistic toward organic models of explanation and away from monocausal toward multidimensional explanations. We must learn to understand a manufacturing company as an integrated system with its own processes and structures, a system which does not develop in a linear way, which is not accurately predictable and whose interior and exterior limitations are fuzzy and permeable. The model of the *fractal company* is suggested.

The fractal company is an open system which consists of independently acting self-similar units — the fractals — and is a vital organism due to its dynamic organisational structure. This approach does not describe the world anew, but emphasises the dynamic and multicausal relations of the real world. The question is, how can you get units into the factory which, regarding their objectives, are self-similar, self-organising, and able to act independently? This is a structuring task of an integrated kind. Structures must be created which support and develop the above-mentioned abilities.

7 THE BASIS FOR THE FORMATION OF FRACTALS: THE LEVEL CONCEPT

The methodology must on the one hand contain a sequence of steps, and on the other, enable integrated views. A continuous development process must result. In order to reduce the complexity in a company, and with the goal of creating a "fractal company," only horizontal divisions into levels are imaginable. The intermediate results of all projects currently dealt with at IPA show that the basic pattern is a division into six levels which can be treated as a whole, but are separate: 1. Cultural, 2. Strategic, 3. Socio-psychological, 4. Financial, 5. Informational (Information Flow) and 6. Technological. See Figure 6.

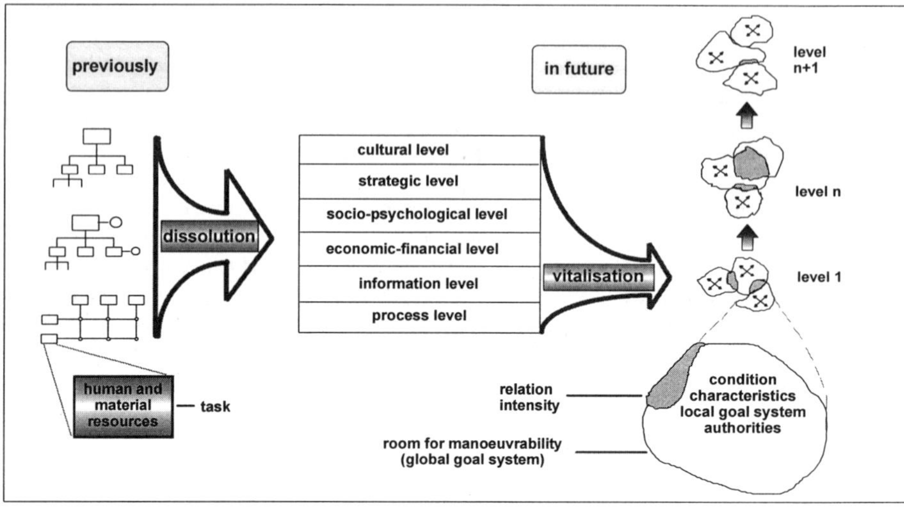

Figure 6 Dynamic organisation structures.

Organisation culture is the generic term which includes the company culture. Values are an essential part of organisation cultures. They express desires and contain the whole field of human preferences. Behaviour is value oriented. It becomes clear that a goal oriented company cannot get along without values or cultural elements. Ideals are developed, as are common views of values and principles concerning internal and external harmony.

Important *strategies* can be strategies of innovation, "me too" strategies, specialisation and diversification strategies, etc. However, the system of objectives must first be defined. A goal oriented approach is the basis for successful structuring which requires a minimum of time and cost. One must generate a system of objectives based on the philosophy, culture and strategic orientation of the company. According to the goal structures which are evaluated in this way, concrete planning can be established in order to compare the alternative principle solutions which have been worked out. See Figure 7.

The *social-informal level* includes all kinds of psychic, social and psychological factors which determine and influence the structure of relations of the entire staff. Organisational structure, communication and ability for teamwork can be identified as central variables on this level. The corresponding methods which must be adapted to the context of the fractal company would be: development of organisation, formation of teams and working groups, guidance of the teams, information and communication management and coaching.

The *financial level* of the fractal company deals with methods concerning account settlement of performance. Business management data must be judged with regard to their economic and efficiency related viabilities. This means that sales/cost considerations must be connected with the process design.

The *informational level* deals with the design of technical information flow. The central term is therefore process organisation. The main problem exists in maintaining continuity and

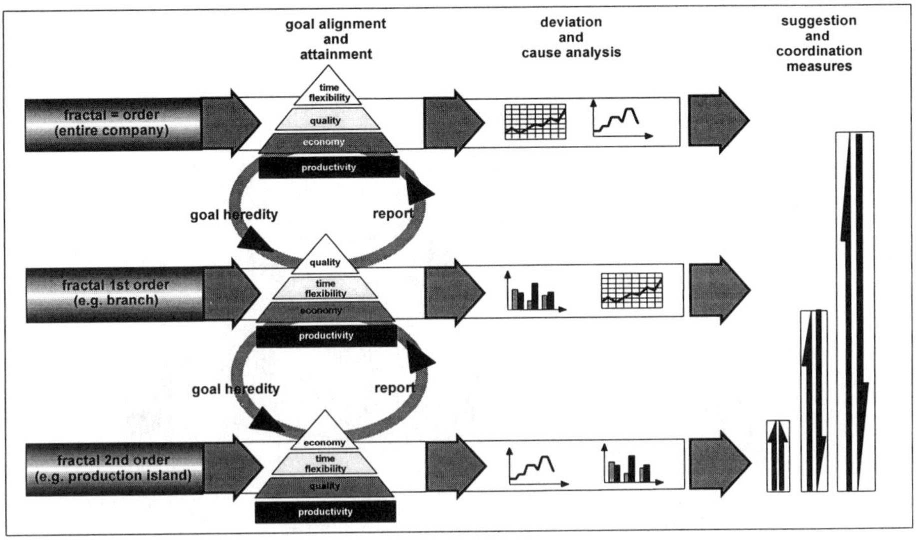

Figure 7 Interconnected cybernetic structures.

integrating information systems without hindering the dynamics of the structures. There are many possibilities for a realisation apart from computer-integrated manufacturing. The use of information must be adapted to the processes and not vice-versa.

The *technological level* of the fractal company is responsible for the technical design of material flow equipment. The whole complex of logistics and material management, including all kinds of parts of components, belongs to this.

Aimed variables, such as the increase of productivity and flexibility and maintaining schedules concern the entire order processing as well as the decrease of throughput time. They are becoming increasingly important and are the focus of attention. Pilot projects at IPA have already brought desired results such as transparent manufacturing processes, reduction of throughput times and reduction of stock value with simultaneous tripling of output. See Figures 8 and 9.

8 LEVEL MODEL AND VITALITY

The word *vitality* was coined as a superordinate term to measure the viability and efficiency of a fractal. The variables to be taken into account should have the ability to develop a dynamic system behaviour. Therefore, vitality must record and evaluate essentially those variables which are included in the individual characteristics and which can be used as a measure for the change or changeability of the individual characteristics of the levels. See Figure 10.

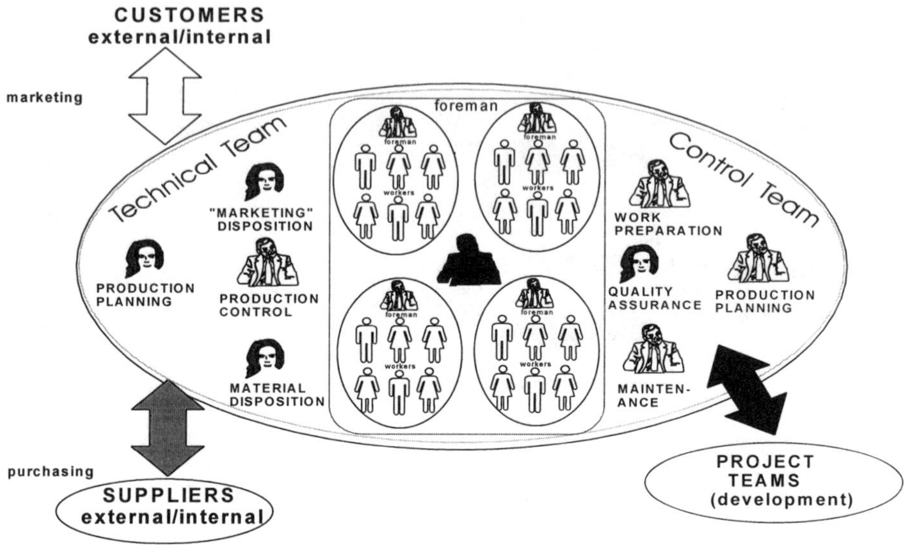

Figure 8 The production team at the point of manufacturing.

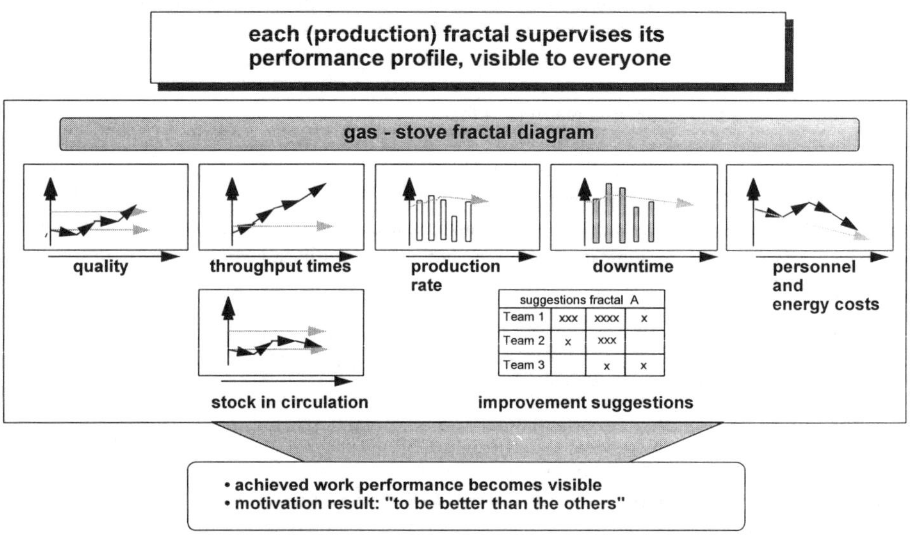

Figure 9 Optical Control

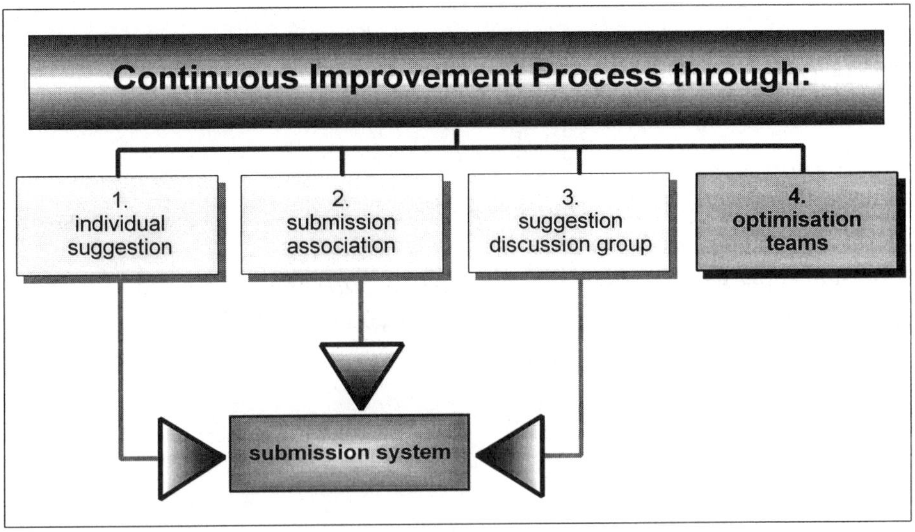

Figure 10 Continuous improvement process system.

9 SUMMARY

The concept of the fractal company, based on western structures, mentality and strengths, is a European answer to Japanese "lean production". With reference to natural organisms, vital and dynamic structures reacting flexibly to environmental conditions were created.

In addition to the fundamental elements and structures of a fractal company, the practical relevance gained from various companies was elucidated.

10 BIOGRAPHY

Dr Wilfried Sihn was born in 1955 in Pforzheim, Germany and studied commercial industrial engineering at Karlsruhe Technical College. He has been with the Fraunhofer Institute for Manufacturing Engineering and Automation (IPA) in Stuttgart since 1982 and is the director of the Corporate Management Department. Dr. Sihn is the author of numerous publications and is a member of various committees. He has worked on over 200 research and industrial projects. During the last 3 years he has been principally involved with the concept of the "fractal factory" which has been successfully implemented in over 50 companies.

3

Software benchmark design and use

Maya Daneva
Institute of Business Information Systems
at the University of Saarland
Altenkesseler str. 17, Geb.B2, D - 66115 Saarbrücken, GERMANY
phone:++49 (0681) 97 62 224, fax: ++49 (0681) 7 75 16
E-mail: maya@iwise1.iwi.uni-sb.de

Abstract

This paper presents a holistic view of software benchmarking. We elucidate the concepts of the problem area by studying relevant benchmarking aspects like why software products are benchmarked, what benchmark is useful, how the benchmarks can be designed and used, and how the benchmarking process can be modelled and supported by means of software tools.

Keywords

Software benchmark, benchmark design, computer-aided software benchmarking

1 INTRODUCTION

The software community characterizes the performance of the marketed products in a variety of confusing metrics. The confusion is due to a lack of quantifiable performance information (Ghezzi et al., 1991), accurate models of software workload profile (McDonell, 1995), comprehensible software benchmark design methodology (Maneva et all., 1994) and credible interpretation rules (Gray, 1993).

In this paper we try to overcome these difficulties by proposing an unifying approach to software benchmark design and use. To ensure the feasibility of this approach, we develop the notion of computer-aided software benchmarking.

2 INTEREST IN SOFTWARE BENCHMARKING

To develop an approach to software benchmark design we selected as a starting point the analysis of goals the organizations try to achieve by running benchmarks. This is used to characterize the useful software product benchmarks and to classify them.

Software products are benchmarked by the following classes of organizations each of which follows a different set of goals:

- Software developers: To uncover software bottlenecks and optimize software processes.
- Software distributors: To highlight strengths of the marketed products.
- Certification offices: To check a compliance between the required and the actual product.
- Software testing laboratories: To evaluate user satisfaction with the available software.
- Marketing agencies: To determine the market power of software products.
- Commercial computer magazines: To promote or to position a certain software product.
- Software researchers: To control software development through benchmarking.
- Academic institutions: To select the most appropriate software for a given project.
- Independent benchmarking companies: To learn and announce the truth to the user.

3 CRITERIA TO SOFTWARE PRODUCT BENCHMARKS

To be useful a software product benchmark should meet six important criteria:

- Pertinence: It must be closely related to customer's business problem.
- Functional coverage: It must be chosen so as to involve the software functionalities the user expects to use most in the system.
- Understandability: If it is complex to be understood, it will lack credibility (Gray, 1993).
- Ease of use: The benchmarking procedure should be relatively simple so as the costs including design time, hardware resources, and lost opportunity are optimized.
- Interpretability: The software product benchmark must provide decision rules for identifying the best software product among the competitors.
- Reproducibility: Each time the benchmark is run the results should remain unchanged.

4 CLASSIFICATION OF SOFTWARE PRODUCT BENCHMARKS

We suggest to group the software product benchmarks in two classes - generic and optional. By analogy to hardware system ratings, we consider the performance and the price/performance ratio as generic benchmarks. We follow the software engineering practice to equate performance with efficiency (Ghezzi et al., 1991). A software system is efficient, if it uses the computing resources economically.

To involve other product characteristics in benchmarking, we propose to use optional software benchmarks which fall into two classes - application-neutral and application-specific. The neutral benchmarks are constructed on the base of software domain-independent characteristics, for example: reliability, quality of documentation, ease of use, etc. The application-specific benchmarks are a response to the variety of software product uses, and thus, they are closely related to the domain the software is created for. For example, Set Query Benchmark is used for decision support systems, Cattel Benchmark - for CASE and CAD tools (Gray, 1993), and BECAUSE benchmarks - for compilers (Porscher, 1995).

Since a software characteristic can be defined by a set of metrics, it is reasonable to differentiate between elementary and complex benchmarks. A benchmark is elementary if a predefined value exists in a form of a software characteristic. A benchmark is complex, if it is determined from subordinate benchmarks. For example, data base resource utilization for a query (Set Query Benchmark) is defined by three elementary benchmarks: elapsed time, CPU time and Input/Output use (Gray, 1993).

Finally, the benchmarks can be differentiated on the base of its granularity, i.e. in dependence on what software aspect they refer to. For example, the benchmarks for a data base software might concern the whole data base, the data block, the record or the field.

5 PROCEDURE FOR SOFTWARE BENCHMARK DESIGN AND USE

After analyzing some benchmarking practices conducted by the worldwide known computer associations Standard Performance Evaluation Corporation, Transaction Processing Performance Council and Business Application Performance Consortium (Gray, 1993), we suggest the following procedure for software benchmark design and use (Figure 1):

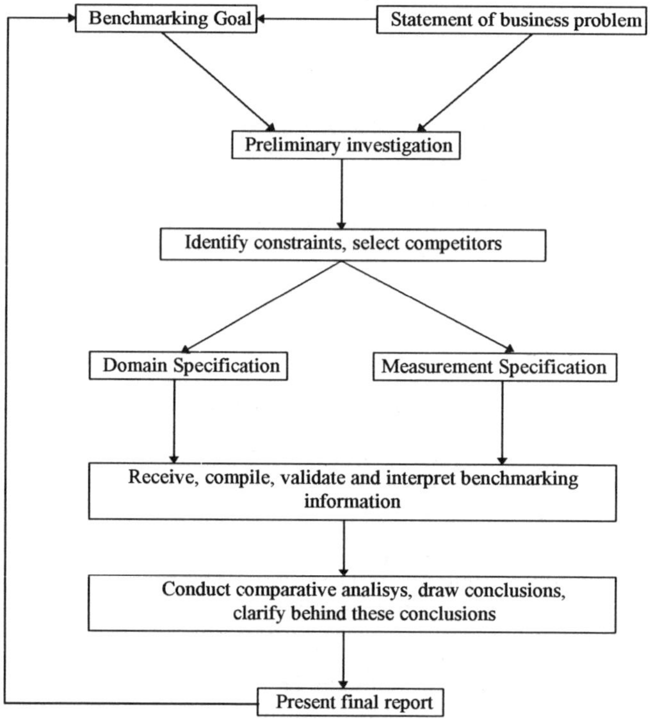

Figure 1 Designing and using software benchmarks.

The procedure involves eight steps:

1.State the goal. Formulate the business problem and identify decisions which would be made on the base of software benchmarking results.

2.Identify constraints. Establish the software-hardware environment where the benchmark should be run. Determine deadlines and allocate resources. Identify benchmarking information sources, types of reportable information, report forms.

3.Select competitors. Choose products of a certain software class according to the goal and the constraints identified.

4.Specify the software benchmark. Define two major benchmark components - the domain specification and the measurement specification. The first one describes completely the software product aspects (granularity) the benchmark should deal with and the second specification establishes the relevant software product characteristics to be measured. It also describes all expected value ranges of all characteristic metrics.

5.Develop a measurement plan. Implement the measurement specification by constructing a set of measured functions. Carry out the measurement, e.g., assign actual values to characteristics involved in benchmarking.

5.Interpret results. Establish rules for using the benchmarking results to solve the business problem formulated.

6.Validate the benchmark. Check the benchmark's pertinence, functional coverage, understandability, ease of use, interpretability and reproducibility.

7.Compare the software products. Conduct analysis and assessment on the base of the benchmarking information. The analysis implies the comparison of target products against a preliminary established model, which serves as benchmark, and the assessment means a study of a set of products so as to rank them with respect to their capability of supporting a predefined objective (Maneva et al., 1994).

8.Develop documentation. Report the results of each stage of the procedure.

6 COMPUTER-AIDED SOFTWARE BENCHMARKING

The productivity of the software benchmark investigator can be enhanced by means of a software tools carrying out much of the drudge work associated with benchmark design and use. Thus, computer-aided software benchmarking is the term we introduce to refer to the support of the benchmark investigator using software tools. These tools should involve the following three capabilities: benchmarking situation modelling, integrated information handling, and activity support (reflecting the active aspects in the procedure for benchmark design and use). The aim of situation modelling facilities is to represent the benchmarking universe of discourse in the terms of a certain information model. Integrated information handling is needed to provide the data and knowledge flow throughout the procedure. Finally, the activity support refers to the analysis and the assessment in software benchmarking. It also provides the benchmark investigator with methods for software product comparing and ranking.

7 SOFTWARE BENCHMARK DESIGN AND USE THROUGH MIC

To put the notion of computer-aided software benchmarking into the practice we have developed the prototype system MIC (Daneva, 1994). It has been designed to compare software objects (products, processes, resources, software engineering and marketing items) by means of multicriterial methods. The system involves a combination of logical reasoning procedures and analytic/algorithmic procedures together with a modularized rule base and a data base. The rule base, the data base and the reasoning procedures were implemented by means of the software package GURU, a tool suitable for intelligent systems development, and the algorithmic procedures were implemented in C++.

Logically, the system comprises two main modules, that provide the construction of benchmarking models and activity support, respectively. Next we discuss the methodology used in each of these components.

7.1. Methodology

Applying Modelling Technique to Software Benchmarking

For conceptualizing the universe of discourse an object-oriented information model has been proposed (Daneva, 1994). We have extended the concept of semantic data modelling (Farmer et al., 1985) by introducing the notion of treating both data and knowledge items in a uniform way. The suggested model is anchored on six key concepts: class, role, simple and composite attribute, function and heuristic. To define them, a set of primitive terms (Table 1) is used.

Table 1. Primitive terms used for concept definitions

Notation	*Meaning*
{}	Set
Name	String
Domain, Range	Type or Role
Restriction	Element of the set {Single, Multi, Var, Const}
Object	Abstraction unite
Body	Algebraic expression
Implication	Rule in the form: Algebraic expression → Algebraic expression
Constraint	Function or Heuristic

Next, the modelling concepts are specified:

- A Class is an ordered 5-tuple: (Name, {Object}, {Role}, {Simple Attribute}, {Function}).
- A Role is an ordered 6-tuple:
  (Name, Function, {Object}, {Simple Attribute}, {Role}, {Function}).
- A Simple Attribute is an ordered 5-tuple:
  (Name, Domain, Range, {Restriction}, {Constraint}).
  The simple attributes can be multi-valued (Multi) and single-valued (Single), also their values can be changeable (Var) or constant (Const).
- A Composite Attribute is an ordered pair: (Name, {SAttribute}).
- A Function is an ordered 4-tuple: (Name, Domain, Range, Body).
  Functions are used for constructing roles (IS-A and PART-OF hierarchies), defining nested functions and computing values. The Body of a function is an algebraic expression built by using operations (+,-,*,/, etc.) and the predefined functions CARDI (returning the cardinal number of a set), SUMA (computing the sum of the element of a set), EXE (executing a function), APPLY (returning values of a selected attribute), MAKEVAL (assigning value to an attribute), X (denoting an object-argument for EXE operator), and INFER (inferring a heuristic).
- A Heuristic is an ordered 4-tuple: (Name, {Domain}, {Range}, {Implication}).
  The Heuristics are aimed at defining nested heuristics and inferring attribute values.

A simple example of how to specify a benchmarking situation in the terms of the model is given in Figure 2. Let us benchmark Documentation Retrieval Service Software by means of Full Text Document Benchmark defined by DEC experts as follows (Gray, 1993):

Price/Performance=Five Year Cost of Ownership / Throughput
Throughput = Search Transactions Completed per Minute Database Size in Partition*

We consider four software products attributes: Brand, Overall Quality, Workload Search Expression Size and Transaction Mix. Next, the class Benchmark is characterized by the attributes: Name, Value, Min, Max and Subordinate.

```
CLASS: Doc_Retrieval_Service_Software
INSTANCES: [list of references to software products of the studied class]
S_ATTRIBUTE: (Brand, string, {Single, Const}, {} )
             (Quality, string, {Single, Var},{H})
             (Workload_Search_Expression_Size, integer, {Single, Const}, {})
             (Transaction_Mix, integer, {Single, Const}, {}).
ROLE: ().
FUNKTION: ().
HEURISTIC: ().
END.

CLASS: Benchmark
INSTANCES: [list of references to benchmarks]
S_ATTRIBUTE: (Name, string, {Single, Const}, {})
             (Value, real, {Single, Const}, {H1})
             (Min, real, {Single, Const}, {})
             (Max, real, {Single, Const}, {})
             (Subordinate, Benchmark, {Multi, Const}, {}).
ROLE: (Elementary, F1, {}, {}, {}).
FUNKTION: (F1, bool, CARDI(APPLY(Subordinate,X))=0).
HEURISTIC: (H, Doc_Retrieval_Service_Software,
      APPLY(Name,X)="Performance" and APPLY(Value,X)>0.9 → MAKEVAL(Quality,X,"high"),
      APPLY(Name,X)="Performance" and APPLY(Value,X)<0.2 → MAKEVAL(Quality,X,"low")})
             (H1, Benchmark, {APPLY (Name, X)="Performance" →
                   MAKEVAL(Value,X,EXE (*, APPLY (Value, APPLY (Subordinate, X)))),
                         APPLY (Name, X)!="Performance" →
                   MAKEVAL(Value,X,EXE (/, APPLY (Value, APPLY (Subordinate, X))))}).
END.
```

Figure 2 A specification of benchmarking situation for Documentation Retrieval Software.

The function F1 defines which benchmark can be elementary. The heuristic H captures rules for interpreting the performance benchmark. It specifies the proportionality between the performance measure and the overall software quality. Finally, the heuristic H1 checks the benchmark's name and prescribes the way of computing the benchmark value. The last can be calculated either by multiplying the subordinate benchmarks, or by dividing them.

Software Product Comparing

We suggested to use two multicriterial approaches for software product comparing. Anderson's algorithm (Anderson, 1989) was selected for software product assessment. It fits excellently with this problem, but is possesses some peculiarities which impede the support of benchmarking analysis (Daneva, 1994). Thus, to resolve the situation we proposed the QR approach. To present it we introduce the following definitions: let n be the number of the examined products and m - the number of the characteristics considered. Let $E_{(nXm)}$ represent a product-characteristics matrix, each element e_{ij} of which is the evaluation of the ith product in regard to the jth characteristic. Let $w_{(m)}$ be a vector corresponding to the importance weights assigned to the characteristics.

The algorithm QR involves the following steps:

Step.1.Define an anchored model (standard, benchmark). This is an abstract representation of a software product with artificially constructed attribute values which must be achieved or avoided (Maneva et al., 1994). Let the elements of the vector $t_{(m)}$ describe the anchored model on the characteristics considered.

Step.2.Determine the numbers *tplus* and *tminus* given as follows:

$tplus=max(e_{ik}-t_k)$, where $e_{ik}>t_k$

$tminus=max(t_k-e_{ik})$, where $e_{ik}<t_k$, $i=1,...,n$, $k=1,...,m$.

There are four cases to be considered:
- values are assigned to both the numbers *tminus* and *tplus*;
- a value is assigned to the number *tminus*, but not to the number *tplus*;
- a value is assigned to the number *tplus*, but not to the number *tminus*;
- values can not be assigned to both the numbers *tminus* and *tplus*.

In the last case the algorithm stops. In the other cases it continues.

Step3. For each row i of the matrix $E_{(nXm)}$ we build the set S and calculate the elements sup_i of the vector $sup_{(n)}$ and inf_i of the vector $inf_{(n)}$, as follows:

$$S=\{k \mid e_{ik}>=t_k, k=1,...,m, i=1,...n\}$$

$$sup_i=\sum_{k\in S} w_k*(e_{ik}-t_k) / \sum_{i=1} w_k*tplus$$

$$inf_i=1-\sum_{k\notin S} w_k*(t_k-e_{ik}) / \sum_{i=1} w_k*tminus, \text{ where } i=1,...,n.$$

The elements of the set S are the characteristics for which the evaluation received by the ith product is equal to or greater than the evaluation of the anchored model. Next, the element sup_i shows the extent to which the ith product exceeds the anchored model. In contrast, the element inf_i is a measure which indicates the extent to which the ith product is inferior to the model. The greater the elements sup_i and inf_i, the more likely that the ith product exceeds the anchored model.

Step4. Define the sets R_{1i} and R_{2i} for each software product:

$$R_{1i}=\{j \mid sup_i>=sup_j, j=1,...,n\}$$

$$R_{2i}=\{j \mid inf_i>=inf_j, j=1,...,n\}, \text{ where } i=1,...,n.$$

Determine the cardinal numbers r_{1i} and r_{2i} of the sets R_{1i} and R_{2i}, respectively. The numbers show how many products exceed the ith software package. The comparison is with respect to the elements of the vectors $sup_{(n)}$ and $inf_{(n)}$.

Step.5.Calculate the bellow defined cumulative measure *com* for each product in order to capture the differences across the products - on the one side, and those between the products and the anchored model, on the other side:

$com_i = alfa * r_{1i}/n + (1-alfa) * r_{2i}/n$, where $i = 1, \ldots, n$, $0 <= alfa <= 1$.

The element com_i of the vector $\mathbf{com}_{(n)}$ represents the extent to which the *i*th product exceeds both the anchored model and the other products examined.

Step.6.Determine the overall quality rating (*oqr*) for each software product on the base of elements of the vectors $\mathbf{sup}_{(n)}$, $\mathbf{inf}_{(n)}$, and $\mathbf{com}_{(n)}$. The *oqr* is considered in dependence on predefined weights corresponding to the contribution of each measure to the overall rating. Let aw_1, aw_2 and aw_3 represent the weights chosen for the vectors **sup**, **inf** and **com**, respectively. The numbers aw_1, aw_2 and aw_3 should be in the interval [0,1], and its sum has to equal to 1. The *oqr* for the *i*th product is given by:

$oqr_i = aw_1 * sup_i + aw_2 * inf_i + aw_3 * com_i$, where $i = 1, \ldots, n$.

Step.7.Rank the products according to their *oqr*. The rank is a number between 1 and *n*, where 1 means excellent and *n* means poor. The product with the greatest *oqr* is the best.

7.2. Application

Next, we give examples of how the system MIC could be used. We consider two software benchmarking problems described in the terms of objectives, competitive software products, comparative characteristics and importance weights.

CASE STUDY I:

Objective: To determine the appropriateness of CASE tools to build a Quality Management System according to the requirements of the standard ISO 9000.
Software Class: CASE tools.
Software Characteristics: The analyzed features are given in Table 1(Herzwurm, 1994).
Software Products: CASE tools available on German market (Herzwurm, 1994).

Table 1 Characteristics and importance weights for CASE tools

Characteristics:	*Weights:*	*Characteristics:*	*Weights:*
Team work	1	Documentation Development	9
Openness	4	Quality Assurance	9
Analysis Support	3	Project Management	5
Design Support	3	Configuration Management	7

Two cases have been studied - with equal characteristic weights and with weights, defined by researchers from the University of Köln (Herzwurm, 1994). QR approach was used. The anchored model was constructed by using the maximal software evaluations on the characteristics considered. The results are presented in Table 2.

Table 2 Ranking of CASE tools

CASE tools:	*Ranking I:*	*Ranking II:*
Maestro II	1	1
Teamwork	3	2
I-Case Yordon	2	3
Innovator	4	4
SDW	6	5
I-Case OMT	7	6
ADW	5	7
Systems Engineering	8	8
Excelerator II	10	9
DDB CASE	11	10
IEF	9	11
case/4/0	12	12

CASE STUDY II

Objective: To rank software vendors.
Software Class: Data Base Management Systems (DBMS).
Software Characteristics: The analyzed features are given in Table 3 (Datapro, 1993).
Software Products: Ingres, Oracle, Sybase, Focus, Informix (Datapro, 1993).

Table 3 Characteristics and importance weights for data base software

Characteristics:	*Weights:*	*Characteristics:*	*Weights:*
Price/Performance	8	Quality of Output	6
Reliability	7	Understandability	8
Ease of Learn	10	Userfriendness	9
Flexibility	4	OS Compatibility	3

Two cases have been studied - with equal characteristic weights and with weights defined by means of the system MIC. The software packages have been ranked by using QR approach. The anchored model was constructed by using the average values on the characteristics analyzed. The results are presented in Table 4.

Table 4 Ranking of data base software

DBMS:	*RankI:*	*RankII:*
Informix	1	1
Sybase	2	3
Ingres	3	2
Focus	4	5
Oracle	5	4

8 CONCLUSIONS AND FUTURE DIRECTIONS

Software practitioners, marketers and academicians have all emphasized different software performance issues and have used different approaches to solve particular software benchmarking problems.

In this paper, we have developed a holistic view of software benchmarking which has not been previously reported. The key research contribution of our work is the presentation and the implementation of an unifying approach to software benchmark design and use. The focus of the approach is on what steps a software benchmarking process consists of and how they are to be performed. We have proposed a model for representing benchmarking situations, an algorithm for software product ranking and a prototype intelligent software system MIC. Moreover, by applying the approach to CASE-tools and data base assessment, the power of its features to capture subtle benchmarking issues are fully illustrated.

Our investigation suggests a variety of topics for further research referring to the gap between current product-centered software benchmarking and the process dimension of software engineering. The future work will be concentrated on:

- testing the adequacy of the approach to benchmark software processes;
- relating software benchmarking and software process re-engineering;
- developing practical algorithms so as to contribute towards a set of tools to aid in the systematic software process benchmarking.

9 REFERENCES

Anderson, E. (1989) Heuristics for software evaluation and selection. *Software Practice and Experience*, **8**, 707-17.

Ghezzi, C., Yazayeri, M., Mandriorli, D. (1991) Fundamentals of software engineering. Prentice-Hall, New Jersey.

Gray, J. (1993) Benchmarking handbook for database and transaction processing systems. Morgan Kaufman Publ., California.

Daneva, M. (1994) Knowledge-based approach to software marketing modelling and support. *PhD Dissertation*, Institute of Mathematics with Computing Center, Sofia.

Datapro Research Corporation, Report on Software/Computer System, Series Software.

Farmer, D., King, R., Mayers, D. (1985) The semantic database constructor. *IEEE Transactions on Software Engineering*, **11**, 1985, 583-90.

Herzwurm, G., Hierholzer, A., Kung, M. (1994) The appropriateness of the conventional and object-oriented CASE-tools to construct a Quality Management Systems according ISO 9000. *Information Management*, **3**, 72-6.

Maneva, N., Daneva, M., Petrova, V. (1994) Benchmarking in Software Development, in *Benchmarking - Theory and Practice*, IFIP Working Group 5.7, Trondheim.

McDonell, K. (1995) Benchmarking frameworks and tools for modelling the workload profile. *Performance Evaluation*, **1**, 23-42.

Porscher, T. (1995) Benchmarking the POM PC compiler on the Connection Machine CM26. *Future Generation Computer Systems*, **11**, 19-26.

I thank Ralf Heib for comments and suggestions that led to the improved presentation of this paper. This research is funded by German Academic Exchange Service under grant A/94/10247 and Ministry of Science and Education under the contract I-406.

Maya Daneva received her M.Sc. degree in Computer Science from the University of Sofia, 1991. PhD Dissertation, Institute of Mathematics, BAS, Sofia, 1994. Currently with the Institute of Business Information Systems, University of Saarland.
Her scientific interests include: Software Engineering, Software Benchmarking, Software Marketing Modelling, Knowledge-Based Marketing Systems.

4

Requirements, Methods and Research Issues for Modeling the Product Realization Process

K. W. Lyons
Manufacturing Systems Engineering Division
National Institute of Standards and Technology
Gaithersburg, MD, USA
301-975-6550, FAX 301-258-9749, klyons@cme.nist.gov

M. R. Duffey
Department of Engineering Management
George Washington University
Washington, D.C., USA
202-994-7173, FAX 202-994-4606, duffey@seas.gwu.edu

Abstract

This paper presents a bottom up introduction to process modeling requirements for product realization which should drive new methods and computer tools. Though relatively high-level and organizational in scope, these models are attracting interest among design engineers, system engineers and technical managers whose decisions affect the complex downstream task interactions that cross organizational boundaries. A recent study of industrial usage, industrial requirements, and research issues is presented.

Keywords

Process Modeling, PERT, Product Realization Process, Enterprise Modeling, Business Process Re-engineering, BPR, IDEF, Petri Net, Project Management, PM

1 OBJECTIVE

Manufacturing firms in the U.S. and worldwide use various types of process models to study the business and engineering processes they use to "realize" their products, often in an effort to "Re-engineer the Enterprise". These process models are documented in a variety of forms and levels of detail yet the underlying objective is establish a strategy for assuring that the essential business and technical considerations related to a product's development are considered, evaluated, and understood by the total enterprise[1]. These models which typically represent the current ("as is") process can, in some cases, be used to explore and predict the impact of proposed ("to be") process changes. This paper explores how information and process models are being created and utilized by industry and attempts to identify key industry needs to assist in setting directions for future process modeling functionality extensions and standards efforts.

2 STATEMENT OF PROBLEM

Most process models in use by industry are relatively high-level and organizational in scope, yet their importance is just now being recognized by enterprises. Unfortunately, many companies fail to emphasize process modeling activities until the company is in a crisis thus looking for immediate solutions to problems that developed over extended periods. Enterprises most often use these methods when they require major restructuring to respond to current market place requirements yet they are also used by companies to support continuous quality improvement. An example of the latter might be in developing an approach to achieving improved concurrent product and process development which emphasizes process models for better coordination of design and manufacturing thus faster product development[2].

Although a significant effort is required to model processes, enterprises acknowledge its necessity. Identification and re-engineering of problem processes require as complete and unambiguous an understanding as possible of current product realization processes. For example, realization of electro-mechanical products involves complex task interactions across functional groups and departments which profoundly impact cost, quality, reliability and time-to-market. However, these interactions are extremely hard to visualize during early design stages. This raises a question: how might new computer-based tools enhance process modeling?

This report presents some key industry requirements and research issues for process modeling in design and manufacturing and identifies important functionality. Observations on the current state of process modeling and its limited acceptance by industry include the following:

■ Process models are very tedious to develop and difficult to maintain currency. Most companies that have committed a major effort in this area are still using non-computer interpretable models (static) that are used to help "guide" new process development. This representation further compounds maintainability concerns.

■Process models are reasonably useful for visualizing process flow at multiple levels of abstraction. But beyond simply documenting activity precedence in an acyclic directed graph, process models have only very limited capabilities to characterize design iteration, allow simulation for schedule and cost, and represent time-dependent information flow in the enterprise.

■ Process model accuracy and precision are inherently limited by subjective descriptions (in contrast with process planning paradigms in manufacturing) of human tasks and task interactions. There are also significant trade-off between clarity of the model (by limiting details of complex interactions) and the increased modeling effort required for precise output metrics.

■ Process modeling techniques (current and emerging) require access to actual business and technical data from diverse functional groups within commercial manufacturing organizations. For many industrial firms, both the methodology and data in their process models have been understandably proprietary because of its competitive significance. Hence, there has been relatively little dissemination which inhibits meaningful research. Also, new mechanisms to validate advanced model concepts are required.

■ Process models for electro-mechanical systems are inherently more difficult than for VLSI design, and probably even software design. Product complexity breeds process complexity, and the mechanical component interactions of geometry, heat, vibration, diverse fabrication constraints, etc. can make it difficult to predict process steps and sequences beforehand.

Despite these caveats, there is a resurgence of research efforts in developing new process modelers because of their potential pay-off for many different applications: documenting existing best practices, identifying bottlenecks (e.g., resource constraints) and task redundancy; "what if" analyses of design alternatives; risk assessment for schedule and cost; archiving processes; training; and many others.

3 SOLUTION METHODOLOGY

It is well known and quoted often that to "re-engineer a process" requires a thorough understanding of the existing process. To support this effort it is essential that adequate tools and methods are used to provide structure and a means to capture and document the process knowledge.

3.1 Definition of Process Realization Process (PRP) modeling

Although process modeling methods have been applied to many types of development efforts (e.g., software engineering, VLSI) our sole focus in this report is realization of discrete electro-mechanical products. The term PRP model (Product Realization Process model) will be used in this report to help reduce the ambiguity of the more generic term "process model", particularly as it relates to strictly physical processes in discrete and continuous flow manufacturing. Definitions of "process model" which have relevance include [3-5]. For purposes of discussion, the following working definition is presented:

> *A PRP model is a computer-interpretable description of the human and machine activities and their interactions required to realize a mechanical or electro-mechanical product. This may include early concept and configuration design activities, detailed design, prototyping, testing, tooling, fabrication, assembly and the many other activities within the scope of the realization process.*

What it is ...

A PRP model should at least be a procedural model which documents precedence relationships between activities in a directed graph, and serve as a visual aid. A more robust PRP model is parametric, and its activity representations contain attribute/value pairs for assigned resources, duration times, cost rates, etc. By using stochastic values in a parametric PRP model, simulation techniques can generate estimates of total completion time, total cost, resource utilization, and other aggregate metrics for the entire process [6]. However, there are serious obstacles to valid parametric models given the complexity, partial definition, and uncertainty of real-world product realization efforts, which are discussed later in this report.

What it is not ...

■ It is not an organization model. A process model's scope typically crosses departmental boundaries and in fact may point to mismatches between departmental responsibility and task requirements.

■ It is not an information model. Its primary representation of process sequence and related time-dependencies distinguishes a process model from an information model.

■ It is not just a flowchart of the generic sequence of design reviews and go/no-go decisions mandated within a particular corporate environment, without reference to a specific design and manufacturing effort.

■ It is not a discrete event simulator for machine-executable production processes.

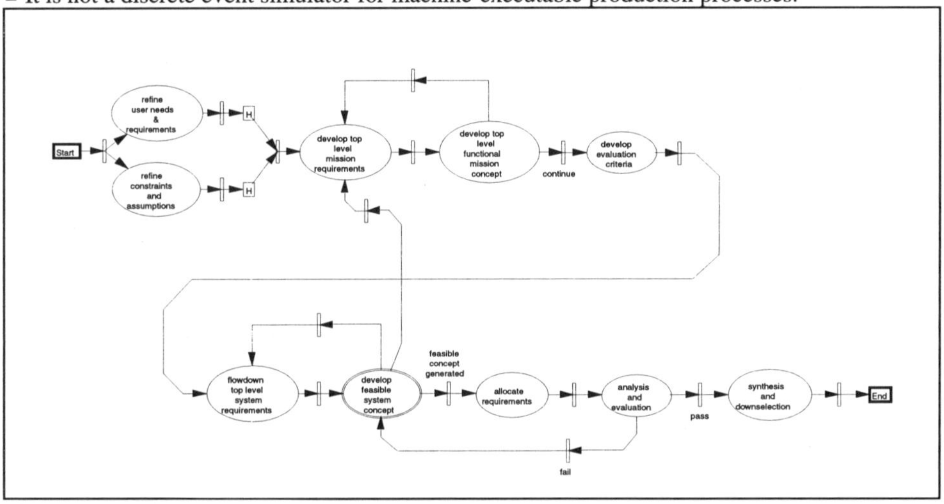

Figure 1: NASA satellite Pre-Conceptual Design PRP model using Modified Petri Net methodology (Part of an on-going collaborative effort with NASA GSFC)

3.2 Modeling concepts and methodologies employed for study

To assist in this study a tool that was recently developed and is currently being employed in several government research projects was used (Figure 1). The tool uses a Modified Petri Net (MPN) methodology and notation and has some advanced (yet largely untested) modeling features, such as representation of iteration cycles, conditional paths, shared resources and interdependencies between parallel or concurrent activities [7]. AN MPN graph offers a hierarchical decomposition similar to IDEF representations: each activity within an MPN can be made a "parent tasknet" which points to a subnet of sub-activities at a lower level of abstraction. Associated with each activity and event is a "frame" which can include information such as activity duration, resource requirements, knowledge-based rules, and conditional or probabilistic expressions for controlling activity parameters or process flow during simulation. A resource hierarchy is represented separately and can be linked to appropriate frames in the MPN. Various resource classes can be defined such as people, machines, and tools, along with multiple subclasses and, ultimately the resource instances that perform process activities. Once constructed, this MPN model can be executed via a built-in discrete event simulation routine that can be visualized as a process flow animation. Various activity states during simulation are monitored so that things such as resource utilization patterns, conflicts, and process bottlenecks can be identified.

4 MAJOR RESULTS

4.1 Advanced PRP computer functionality and capabilities

Companies that have made major commitments supporting continuous quality improvement usually have like efforts aimed at understanding their business and engineering processes. Such efforts, unfortunately, are still using non-computer-interpretable models to help "guide" new process development. These process development guidelines are very tedious and time consuming to develop and their static representation further compounds maintainability concerns (e.g., forces extensions and enhancements of the models to conform to scheduled update procedures). Companies are also looking at ways that the models can go beyond documentation of current ("as is") operations to supporting exploration of proposed ("to be") process changes and predict the result of these changes. To do this will require the development of new PRP applications and the extension of current PRP applications with advanced functionality.

What are the decision making needs of engineers and managers which should drive development of advanced PRP modeling tools? These tools should help evaluate the downstream implications of complex design process interactions that span traditional engineering departmental responsibilities. Some areas that are key to industrial acceptance of the tools are:

Open Systems

A major deficiency in currently available systems is that they are not "open." Very important product and process information remains locked up in the application and unavailable for use or access by others. This is very evident today as companies are increasingly concerned with capturing and retaining corporate knowledge to maintain their edge in a competitive marketplace. It is this corporate knowledge that is key to a company's success and typically not the applications and tools that the company uses. With the investment that companies are making to develop these models and the key information that they contain there needs to be some common representation or access (i.e., open architecture) so that companies have the capability to migrate or share information with other process modelers and engineering/manufacturing applications.

One approach to the sharing of information is conversion algorithms. Industry demand for conversion algorithms is expected to rise just as Computer Aided Design (CAD) file conversion routines have become a persistent (and still poorly met) demand of industry CAD users in the last fifteen years. Vendors of IDEF-based software have begun to provide output capabilities for both popular project management software and simulation software [7]. From the other end, several simulation software vendors are developing routines to input legacy IDEF models for their proprietary activity network representations. In these specific instances information or established links between information items can still be lost since most process models consist of both the formal model and supporting editorial and technical materials [8]. Beyond these vendor-specific efforts, there has evidently been little research in this area. Other information exchange mechanisms such as neutral files and open-architecture based applications can also assist industry in addressing this major acceptance and implementation barrier.

Advanced Representations and Accuracy Issues

Iterations: With all the uncertainties on designing products and the processes required to support the product, iterations remain as one of the most difficult topics to concisely represent in a model. Iterations are difficult events to predict and thus model, yet they are important components of engineering and manufacturing processes. While the general perception of iterative activities is negative there are times when it is desired. What is really required is a way to control iteration. To control iterations there needs to be recognition of when it occurs; both productive occurrences such as when it aids in the creative design process [9], and non-productive occurrences such as major redesign of products due to the late identification of assembly problems.

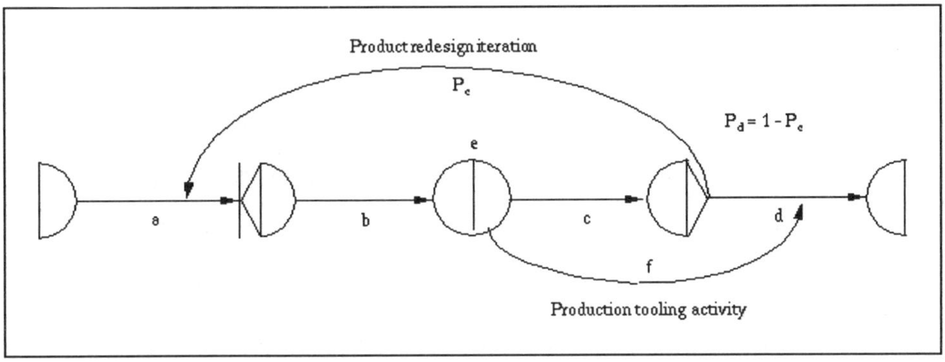

Figure 2: Example of a design iteration

Design iteration is a loosely defined term which describes the cycling of subgroups of activities typical in most PRP processes. Two types of activity interrelationships that are difficult or impossible to model with existing PRP methods are: i) iteration which occurs after concept reviews, prototype testing and other early phases of engineering design, and ii) changes to or cancellation of other concurrent manufacturing activities when redesign iterations occur. The use of iterative looping in a stochastic activity network is an imperfect yet perhaps potentially useful representation of the uncertainty associated with activity overlapping as described by [10] and identified by many companies as concurrent or simultaneous engineering of product and process. Consider modeling process flow in a simplified network for concurrent design and tooling in Figure 2 (an activity-on-arc representation described by Elmahraby [11]). When the redesign iteration occurs, the production tooling activity *f* must terminate at the same point in system time when the iteration activity *e* is realized (i.e., the duration for *f* is not statistically independent). When activities *b*, *c*, and *f* then occur for the second time, the parameters for the distribution functions of the duration times of *b*, *c*, and *f* have changed. Similarly, the branching probability p_e is typically reduced after each successive redesign loop.

<u>Complexity:</u> As product complexity increases a like increase in modeling complexity occurs. In addition, the complexity of the product to be designed affects the company's approach to defining the engineering and manufacturing processes to realize it. If product complexity and related issues can be modeled earlier in the design cycle at an adequate accuracy level it could significantly impact a company's success in the marketplace. One example might entail the use of a model to make key business decisions such as whether to make a specific product sub-system in-house or purchase from another company that has specific knowledge and core competency related to the sub-system. These make-buy decisions often are made with limited consideration of system level knowledge taking into account only component or sub-assembly attributes. Explicit representation of iterations based on historical patterns experienced by companies can capture complexity related systems problems.

<u>Fidelity:</u> A satisfactory model should contain the appropriate detail that is required to arrive at the correct decision. Too detailed a model requires substantial resource and time commitments that may not be reflected in savings to the company (ROI, Return on Investment). Knowing the right level of detail in creating the model depends on the experience of the person modeling the process.

<u>Uncertainty:</u> Using simulation to model uncertainty in a PRP has some unique problems. While manufacturing simulations typically model steady-state workflow conditions in a sequence of production operations, simulation of a product realization process must model a single unit, the design itself, as it increases in complexity of informational and physical detail by undergoing tasks within multiple organizational departments. Unlike in a manufacturing process, the description of the unit itself is not predefined, and is actually being transformed during its development. Because of this, there is uncertainty inherent in task duration and iteration that is not present in manufacturing simulation. Time variability in manufacturing simulations is typically aggregated from delay times for queuing, machine down-time, etc. and largely deterministic machine times in a production sequence. However, in a PRP simulation, virtually all task times have variability and iterative loops must be characterized. Nonetheless, PRP simulation merits research since traditional computer-based project management metrics for time, cost, resources, etc. are often inadequate.

<u>Resources:</u> To model a process the resources associated with the activities and events is key information. Many times comparison of "as is" and "to be" conditions pertain to the elimination, addition, or modification of certain resources (e.g., draft documents that are refined into final form). The resource can be a person, computer tool, documentation, or equipment. PRP systems need to be able to dynamically model resources usage and dependencies and minimize the need for apriori definition. In addition, the model should link the associated statistics that define the resource's availability and anticipated usage during the activity [12].

Work Flow Management

When a PRP model of sufficient detail and accuracy has been developed, and simulations run and validated, it can serve as a basis for exploring work flow management (WFM). This area needs further study to determine how specific issues unique to enactment of activities (e.g., generation of, and receipt of messages) and model complexity differ between simulation and enactment models [13]. Since simulation of engineering process models is based on a simplification/approximation of real processes there needs to be excellent correlation to what actually occurs under all conditions prior to its use in WFM as a controller of these engineering processes.

Initially it is expected that only portions of the PRP model will be used in the WFM environment where timing and coordination of activities are well defined. An example of WFM is the calling of design applications and automatic loading of specific models such as CAD 3-D models. Current application of WFM technologies have been most successfully applied in areas that have good structure and are deterministic (e.g., activity duration that are sequential with no iteration). Until accurate representations of processes are possible, WFM applications tend to be restrictive to users, and to address this condition most systems that are in production use today typically support a limited set of activities.

4.2 Other issues relevant to PRP computer functionality and capabilities

Subjectivity

Because PRP models represent sequences of human-executable as well as machine executable activities, the structure of a PRP model is inherently subjective in its capture of information flow. Modeling implications of subjectivity can be categorized into two areas: 1) the configuration of the model (e.g., defining activities and their precedence relationships in a directed graph), and 2) the content of the model (e.g., values of activity attributes which allow parametric characterization of activity duration, branching probabilities, and quantitative evaluation of time, cost, etc.). Beyond simply documenting activity precedence, how effectively can PRP models provide aggregate, quantitative metrics such as total cost and project completion time? This raises issues about the subjectivity of content in a process model such as estimation of activity duration time and assignment of branching probability values (for example, the pitfalls of assuming beta distributions for modeling time uncertainty in probabilistic PERT models).

Information flow

Unlike the clear sequence of fabrication activities in physical process planning, PRP models must often characterize a very ambiguous information flow. The natural tendency for those constructing the model is to define activities in terms of tangible inputs and outputs such as written specifications, analysis results, and material transformations. However, much information flow in a PRP is not easily specified, such as the informal communication network built up in an interdepartmental concurrent engineering team. Documenting these transactions is somewhat analogous to problems with the knowledge acquisition process acknowledged for expert systems development.

Organizational issues

The nature of organizational behavior makes a rigorous mathematical definition of its process semantics difficult. In addition, PRP models are very dependent on the imposed organizational constraints which will determine the optimal PRP considering company or industry specific constraints.

Cultural and institutional values: A company's or organization's culture, established principles and history often determines the approach and resulting "look" of the process models. Some organizations use specific groups skilled in enterprise modeling whereas other companies spread the task throughout the different organizations that comprise the enterprise. Company culture and institutional values can also determine the company's "inertia" which can prevent change from occurring, or at a minimum, slow it down.

Organization structure: Organizational structure can have significant effects on the development of a PRP model. Inherent within organizational structures are imposed constraints and limits that direct how process models will be developed. For organizations that are undergoing a major restructuring or re-engineering of the enterprise, the organizational aspects are many times less significant when it has been recognized that the current organizational structure is not effective. With smaller scoped re-engineering efforts this is usually not the case and models will tend to follow the along guidelines established to conform to organizational infra-structure. In many cases this organizational infra-structure is what provides the competitive edge for companies in responding to customer requests.

In some cases, a company's competitive environment can impair its ability to recognize or respond to change and make it vulnerable to new competitive attacks. An example of this is when a company competes against a select group of competitors and becomes complacent in this environment. The company knows what to expect and usually has sufficient knowledge to know when they are competitive. Their competitors often have very similar organizational structures and this structure in a static marketplace enables them to compete effectively as long as the marketplace can support them. Yet, when a new competitor enters the marketplace this company realizes that to be competitive they must approach the product development process differently. This change in the way "the game is played" can have severe consequences on the established companies if they fail to recognize this change and take actions to predict its impact on them. This example demonstrates where imposed organization structure can actually impede the recognition of new methods and approaches in developing the product and can have major repercussions.

Model acceptance: Modeling engineering processes is a very tedious and difficult activity and most often is done by individuals skilled primarily in process modeling, not typically in the engineering or manufacturing process being modeled. This approach has been traditionally viewed as the best practice since modeling required very specific skills to successfully interview engineering personnel and obtain the process knowledge and information in a form conducive to the modeling methodology. This was effective in that this skilled individual could rapidly construct a model that, from one view point, captured and accurately represented the process. Unfortunately, this approach can lead to models that fail to adequately represent the process in a way that is of value to the enterprise. Three reasons for these model inadequacies are:

- Lack of domain knowledge by the expert process modeler which prevents the modeler from being able to realize the inaccuracies of the model,
- Lack of commitment of engineering personnel ("buy-in") to validate the models and apathy toward the real application or need for the models, and
- Limitations of the modeling methods and tools which prevent adequate representation of key process attributes.

To initiate a modeling effort it can be useful to start with existing process information for assisting in model creation. This can help achieve acceptance by jump starting the modeling effect yet can also have some unanticipated and potentially significant effects. Many times the most useful documents are PERT charts used for project management (PM). Although these materials help reduce the time in generating a model there is some concern that it can bias a PRP model by filtering the intended enterprise view with a project management perspective. It is difficult to surmise the impact of biasing PRP models with PM information. Depending on how the PRP models are to be used and what results are required will dictate what information should be used for model creation. Awareness of the potential for biasing the model is the first step in minimizing this effect.

4.3 Future Directions

Many of these have been touched on in previous sections. Others include:

- Standards-related research
- Methods to assign different levels of confidence, different representational constructs, with different levels of abstraction
- Identifying feedback loops and iteration
- Multi-media process models
- Multiple viewpoints of a process an aid to DFX evaluations.
- Relationship of process model to both product representation (CAD models) and external data of the firm (e.g., financial data in accounting software)
- Explicit methodology for fine-tuning cost and time estimates as the project progresses.
- Application of advanced statistical techniques to eliminate distribution dependencies and improve predictions of complex series of activities.

5 SUMMARY

Product realization for electro-mechanical assemblies is a very complex, interdisciplinary process. It requires decision-making not only about attributes of the design itself, but also about resource allocation and scheduling for many product and manufacturing engineering activities, as well as extensive coordination with concomitant activities in marketing, finance, purchasing, and other functional groups inside and outside a company structure. Typically, complex interdependencies exist among these disparate activities, and it is difficult to predict how decisions will affect overall organizational objectives of low cost, high quality and short time-to-market.

Unfortunately, progress towards a comprehensive model of the product realization process falls in a gap between established research disciplines and industry practice. The value of modeling processes is still not realized, in that most companies still under utilize the information and knowledge contained in the models. Causes of this are found in the enterprise's approach to process modeling as well as the inefficiencies of currently available tools.

To gain better insight into these problems, case studies are critical in that they are based on actual industry experiences with supporting justification on the benefits that can result. Engineering designers have developed some very good models that guide decisions about product attributes as they affect cost and/or quality for an individual manufacturing activity. However, these models generally neglect interdependencies between activities and organizational constraints. At the other extreme, PRP managers have generally modeled the product development process as a comprehensive but highly abstract network of activities (such as PERT networks). These network models ideally assist decision-making about activity scheduling and resource allocation by providing very rough estimates of cost and time-to-market. However, the lack of detail and other limitations in such models obscure many important issues, such as the effects of activity interdependencies and sensitivity to changes in product attributes.

6 REFERENCES

[1] National Research Council, *Improving Engineering Design: Designing for Competitive Advantage*, National Academy Press, Washington, D. C., 1991.

[2] Dertouzos, M., Lester, R., and Solow, R., *Made in America: Regaining the Productive Edge*, The MIT Press, Cambridge, Massachusetts, 1989.

[3] Busby, J.S. and Williams, G.M. "The value and limitations of using process models to describe the manufacturing organization." *International Journal of Production Research*, vol. 31, p.39, 1993.

[4] Curtis, Bill, Kellner, Marc, and Over, Kim. "Process Modeling", *Communications of the ACM*, vol.35, no.9, September 1992.

[5] Malone, Thomas W., Crowston, K., Lee, J. and Pentland, B. "Tools for Inventing Organizations: Toward a Handbook of Organizational Processes, "Proceedings of the 2nd IEEE Workshop on Enabling Technologies Infrastructure for Collaborative Enterprises, Morgantown, WV, April 20-22, 1993.

[6] Duffey, Michael R., and Dixon, J.R. "Managing the Product Realization Process: A Model for Aggregate Cost and Time-to-Market Evaluation, "*Journal of Concurrent Engineering: Research and Applications*, no.1, vol.1, 1993.

[7] Madni, Azad M. "Computer-Aided Concurrent Engineering/Process Modeler: User's Manual,"Perceptronics, Inc., Woodland Hills, CA., 1994.

[8] Design/IDEF/ServiceModel: The relationships between Workflow Analysis Models and ServiceModel/ProModel. Meta Software Corporation, Cambridge, MA, February 1995.

[9] Schenck, D., and Wilson, P., *Information Modeling: The EXPRESS Way*, Oxford University Press, Oxford, U.K., 1994.

[10] Clausing, D., *Total Quality Development: A Step-By-Step Guide to World-Class Concurrent Engineering*, ASME Press, New York, NY, 1994.

[11] Elmaghraby, S.E., *Activity Networks: Project Planning and Control by Network Models*. John Wiley & Sons, New York, 1977.

[12] Clark, Kim, and Fujimoto, Takahiro. "Overlapping Problem-Solving in Product Development," *Managing International Manufacturing*. Elsevier Science Publishers, New York, 1989.

[13] Bonney, M., Barson, R., Head, M., and Huang, N., UNISON - A Tool for Enterprise Integration, in Enterprise Integration Modeling: Proceedings of the First International Conference (ed. C.J. Petri), MIT Press, Cambridge, Massachusetts, 1992.

7 BIOGRAPHY

Kevin W. Lyons - Mr. Lyons' current assignment involves design methodology, modeling methodologies for product realization processes, and design system integration in the domain of mechanical and electro-mechanical components, assemblies, and systems. He is also working with the ARPA MADE (Manufacturing Automation and Design Engineering) Program which involves coordination activities and applied research in design knowledge capture.

Mr. Lyons came to NIST with 15 years of industry experience with IBM. In IBM's Lexington, KY manufacturing facility. He had assignments in Product Assurance, Quality Engineering (Electrical and Mechanical), Development Engineering, and Computer-Aided-Design and Analysis Support, with key focus on assemblies.

Michael R. Duffey - Dr. Duffey is an Assistant Professor at George Washington University (GWU), Department of Engineering Management. Dr. Duffey joined the faculty at GWU in September 1992 after completing his Ph.D. in Mechanical Engineering at the University of Massachusetts (Amherst).

Dr. Duffey's current research interests include modeling methodologies for the product realization process, engineering economics, and curriculum development for adult technical training. He works closely with the National Institute of Standards and Technology (NIST) Engineering Design Technologies Group where he does collaborative research as a faculty researcher.

5

Application of Information Technology in the Redesign and Implementation of Business Processes

Dr. G. Lyons
Director, I.T. Centre, U.C.G., Ireland.
Tel.: 353-91-750382 (Direct)
Fax.: 353-91-750501
E-mail: Gerard.Lyons@UCG.IE

Abstract

From the first appearance of the term "business process reengineering (BPR)" in the literature, information technology (IT) has been viewed as the central feature of the reconfigured enterprise. As the practice of process reengineering matures, a host of software tools and enabling technologies has besieged the lucrative BPR marketplace. This paper attempts to demystify the role of IT in enterprise reengineering, and offers a taxonomy of relevant applications, as well as some lessons from practice.

In broad terms, IT has two discrete functions in process reengineering:

1. it can support the mapping, analysis and modelling aspects of BPR projects; and
2. it can provide essential communications and systems infrastructure to facilitate the integration and automation of redesigned work processes.

This paper will explore both of these application areas and provide an assessment of their contribution to BPR methodology and process realisation in practice. The paper will finally comment on the difficulties of migration to a process based IT infrastructure and the constraints imposed by legacy systems and data.

Keywords

Business process redesign, information technology, process modelling, IT architecture, legacy systems.

1 INTRODUCTION

Business processes are the collections of logically linked activities which coalesce to create tangible value for a customer. As such, they represent the central nervous system of an enterprise and cut across defined organisational boundaries with a single minded focus on creating customer satisfaction in the delivery of goods or services. For most enterprises today, business processes remain an abstraction of "the way things work around here", and have no physical manifestation in terms of organisation structure, IT architecture or building design. (The central production processes of manufacturing companies are a notable exception in this respect). But, with the lure of short cycle time delivery, dramatic reductions in overhead costs, and exceptional increases in product and service quality, "business process reengineering (BPR)" (Davenport, 1990; Hammer, 1990; Hammer and Champy, 1993) has created a religious zeal among business managers world wide in search of the optimised "core processes" (Kaplan & Murdock, 1991) that drive enterprise competitiveness.

2 BUSINESS PROCESSES AND IT

Business process design, or indeed redesign, represents a fundamental departure from conventional wisdom on the way the essential work of an enterprise is viewed, examined and organised. While functional bureaucracy had its place in controlling and coordinating a complex array of discrete skill sets, its rigid structure and internal preoccupation have become barriers to performance improvement in a climate of "time based competition" and "customer value". The process based enterprise should not suffer from such maladies. But, the transition from a functional organisation will create its own difficulties, as the cornerstones of structure, strategy, skill sets and information systems are undermined in favour of a new enterprise order. For some companies in the financial services sector in the US and UK, this challenge has been too great, and a simpler solution was found: they've started afresh on green field sites with totally new processes, people, structures and systems.

From its conception, BPR has been synonymous with the rational application of information and communication technologies. For its part, the IT industry in the early 1990's desperately needed new demand drivers, especially ones that were visibly linked to business performance. BPR has already spawned a host of software tools to aid in redesign efforts and has promoted a lucrative market for enabling information technologies. The function of IT in business process redesign is thus two fold:

1. it can support the mapping, analysis and modelling aspects of BPR projects; and
2. it can provide essential communications and systems infrastructure to facilitate the integration and automation of redesigned work processes.

But, as the existing systems infrastructure is a durable component of a functionally designed business, it too must be reconfigured or replaced. The reengineering of IT architectures must, then, be an essential part of BPR initiatives.

3 BUSINESS PROCESS ANALYSIS AND DESIGN

3.1 Redesign methodology

During the past four years, various attempts at defining "BPR methodology" have appeared in the technical literature and at public business seminars. These methodologies typically emanate from one of the following two sources:

1. large consulting groups who have restructured their "performance improvement" products as cook book guides to BPR projects; or
2. academic systems analysts, software engineers and industrial engineering researchers, exploring new applications for their cherished modelling techniques.

Regardless of parentage, both of the resulting approaches have technical merit. But in application terms, the consulting methodologies come closer to the spectrum of needs of BPR practitioners.

The BPR project is primarily an holistic and radical performance improvement intervention. It carries responsibility for assessing and redesigning the full array of enterprise components (Waterman, et al, 1980), although centred on the **work process** and **IT** dimensions of the organisational diamond (Scott Morton, 1991). The organisational development (OD) method, applied by some practitioners, supported by the modelling techniques of IT and industrial engineering, collectively provide the breadth of understanding and rigorous analytical capability needed for effective process redesign.

In broad terms, the BPR project traverses four discrete phases:

Phase 1 Process identification, mapping and diagnosis;
Phase 2 Process baseline measurement;
Phase 3 Process benchmarking; and
Phase 4 New process design.

Alternative approaches are favoured by some practitioners, most notably the radical, "clean sheet" method, followed by early disciples of Hammer (1990), which completely ignores existing process configurations, and proceeds directly to the creation of a "no constraints" vision of the ideal process, as conducted in Phase 4 above. The difficulties encountered in such approaches are reflected in a report published last year by Hammer's own company, CSC Index (1994), which reports on a survey of over 600 US and European corporations. The report highlights poor success rates for BPR projects (with over 67% providing zero or marginal results) conducted by these corporations, and lists the most commonly found difficulties with reengineering, the top two of which were:

1. Getting the information systems and technology infrastructure in place; and
2. Dealing with fear and anxiety throughout the organisation.

The phased approach suggested above is based on the author's own practical experience of large scale BPR in a range of industries, and has parallels in other systematic methods found in consulting practice (Kaplan & Murdock, 1991; Johansson et al., 1993). This approach comes close to the OD model of change management (Cummings and Huse,

1989) and pragmatically addresses the difficulties inherent in implementing radical process improvements which are not grounded in the organisation's own perception and experience of "reality". Thus, the real structural and behavioural impediments to process improvement are uncovered, and personal (as well as organisational) fear and resistance are confronted openly. The rôle of the BPR practitioner in this case is to:

- develop a detailed understanding of the current business process, its performance levels and limitations;
- create a tangible and realisable model of the ideal, simplified business process, together with proof of its performance capabilities; and
- stimulate the climate for organisational change, within which transition to the new process design is likely to succeed.

3.2 BPR project support tools

A host of IT support tools are now available to aid the BPR project team. For the most part, these are based on established techniques, which have been applied successfully in the analysis and design of information systems in both commercial and manufacturing settings for a number of years. Table 1 presents a summary of the tools relevant to each phase of the BPR project, following the methodological framework outlined.

Discussion of each of these techniques is beyond the scope of this short paper. However, it is clear that a variety of modelling and analytical approaches are available to support process analysis and redesign. Within the past four years, a host of automated tools have been developed, or "repackaged", to enable the application of these techniques in BPR studies. These range from the simplest of graphics applications, which support process mapping (usually IDEF or process block diagramming), to integrated tool sets capable of: mapping, spreadsheet based evaluation, simulation, and even code generation.

A study of over 40 software tools which offer BPR features, conducted during the past year, suggests that most of the available tools are based on enhancements to software developed earlier for other, albeit related, purposes, including: software engineering, workflow modelling, manufacturing simulation, and general graphics. Consequently, they have not been designed for a particular methodology, nor do they support all phases of BPR projects. It is difficult to generalise about current BPR tools, as they represent such a wide range of functionality and cost. But, while most provide process mapping features, the more basic packages are limited to single level analysis and do not allow multi level "explosion" (or levelling) of the processes being analysed. Only a few are capable of dynamic simulation and evaluation, and code generation is restricted to the top end of the cost/functionality spectrum. Tool integration is also a weakness, and only the more sophisticated proprietary packages offer full integration of mapping, simulation and analysis functions. However, a number of the simpler graphical packages are now supporting DDE linkages to common spreadsheet software, to enable activity based analysis and process measurement.

Table 1. IT as a process reengineering **project support tool**

Project Phase	IT Support Rôle	Tools
1. Mapping and diagnosis	Process capture and appraisal	• Data flow diagrams • IDEF models • Role Activity diagrams • Work Study process flow diagrams • IS flowcharting • Functional entity diags. • Soft systems methods
2. Baseline measurement	Process evaluation	• Activity based costing • Value added flow analysis • Shop floor data collxn. • Spreadsheets
3. Process benchmarking	Information retrieval	• Benchmarking databases • Library info. services
4. Process design	Process description and development	• Mapping tools as in 1. • Simulation • Workflow modelling • Code and schema generation
BPR Project management	Coordination and control	• GANTT, PERT, CPM • PM software

More broadly, the available tools reflect a poor understanding of the practical needs of BPR teams, whose members are drawn primarily from Business functions and have little experience of IT terminology, techniques and concepts, and consequently find sophisticated tools inaccessible and highly redundant. Like the early CASE tools, upon which many BPR products are now based, these early packages:

- offer poor coverage of the full BPR project cycle;
- lack transparent tool to tool integration (although several support DDE linkages);
- provide single user support and don't allow BPR teams to work together;
- offer poor GUI interfaces;
- are difficult to master quickly for the novice process designer, and entail a lengthy learning curve.

While the development of fully automated BPR support environments remains technically challenging, there is a danger that, once again, IT specialists will become embraced by technical elegance rather than application value. The limited success to date in the application of advanced CASE technology, to the improvement of the software development process, should serve as a timely reminder that process redesign is primarily a group learning activity. While fully automated solutions provide discipline and precision, and go some way to identifying optimal solutions, they rely too heavily on the individual designer's breadth of experience and business understanding. In practice, this has proven to be a poor substitute for less automated, but more "involving" approaches to process redesign, software development (such as Joint Application Development - JAD), and indeed hardware product design.

In short, automated analysis and modelling tools perform useful point solutions in the appraisal and redesign of business processes. The tools are capable of removing much of the drudgery from process mapping and evaluation, and support more sophisticated analyses and higher standard presentation. However, these are only valuable when used in support of a systematic and challenging consulting methodology, where innovation and willingness to explore new "ways of working", at high levels of abstraction, are more important than model precision.

3.3 Analysis and design of process based systems

Apart from exploring the processes themselves, the specification and development of a software infrastructure needed to implement the new process designs is an essential product of the BPR project. Viewed from this perspective, the automated modelling tools discussed above perform a more central function -- that of translating high level process abstractions into tangible specifications of realisable software modules.

Information Systems (IS) practitioners argue that conventional approaches to systems analysis and design are built on the modelling of enterprise processes, and that BPR does not challenge the underlying principles of software engineering. This view is misguided in a number of respects. Conventional software engineering is based on the "stepwise refinement" of high level abstractions of an enterprise's procedures and information needs, and is achieved through:

1. the functional decomposition of high level statements of procedure, until detailed definitions of functionality can be specified as software modules; and
2. data modelling and analysis, of a high level interpretation of an enterprise's subject areas, or data "entities", until relational data structures can be specified.

However, the "processes" examined by the software engineers are the defined activities performed **within the boundaries** of conventional organisational functions, for which IS applications are being developed.

Established software engineering approaches to enterprise level modelling and IS planning take existing organisational structures and functions as a "given" (Martin, 1982 & 1989; Finkelstein, 1989) The Information Engineer is thus not empowered to question existing organisational structures and procedures, as is required in the analysis of horizontal business process information flows. Clearly then, the term "process" has a different meaning in Information and Software Engineering than it has for BPR practitioners, where business processes ignore the existence of today's functional and business areas.

Thus, while Software Engineering gave birth to the modelling techniques found today in BPR, application of these within the context of its traditional IT methodologies will not yield business process based solutions and systems. Yet, software engineering tools and concepts, such as Data Flow Diagrams, IDEF modelling, Levelling, etc., can be used effectively in process redesign, as discussed earlier. What is called for, then, is a realignment of the "process" dimension of Software Engineering and the process models yielded by BPR. This realignment should be a prerequisite for the identification of systems' boundaries, early in the application development work, and should produce high level conceptual models of the enterprise (both process and data models) which reflect the process based enterprise orientation.

4 BUSINESS PROCESS ENACTMENT

4.1 Enabling information technology

Information technologies have been deployed in the service of business for over three decades. In recent years, developments in both telecommunications and computer technologies have seen an unprecedented rate of progress, as well as a convergence to yield new demand drivers and application potential. Yet, despite the enormous investment in information and telecommunications technologies throughout the 1970's and 1980's, there has been little tangible result in terms of observable economic productivity (Drucker, 1991), at least not before the emergence of BPR.

That IT can fundamentally alter the way business works, indeed the way entire industries are organised, is well argued in an authoritative text on this topic by Peter Keen (1991), and was studied at length through the MIT Management in the 1990's research programme (Scott Morton, 1991). The current BPR phenomenon owes much of its early momentum to this recognition, and to the related need to derive visible return from investment in IT, following a prolonged period of doubtful benefits and growing disenchantment with the technology.

Perhaps fortuitously, the business need to re-think core processes in response to competitive factors has coincided with the commercial maturity of a powerful array of advanced information and communication technologies. There is every likelihood that the BPR momentum will, thus, be maintained and organisational transformation will give rise to significant improvements in business performance through the application of enabling IT.

Table 2 presents a summary of the information technologies which are currently available as infrastructure components for the process based enterprise, and illustrates the principal business process requirements which these address.

Table 2. IT as a **process enactment technology**

Process Requirement	IT Enabling Technologies
Process integration & communication	• LAN, WAN, Telecomms., EDI, EFTPOS, etc.
Process co-ordination & control	• Workflow, process scheduling, etc.
Front-end data capture & validation	• Imaging, ATM, Minitel, SFDC, lap-top
Integrated work support	• Workflow, Client-Server, Imaging
Information storage & access	• Database and query tools
Documentation & document mgmt.	• WP, DT publishing, Imaging
Process work support	• Expert systems, Client-based applicns.
Process systems development	• CASE, IPSE, RAD, Prototyping, O-O

4.2 Implementing IT enabled process re-design

It is evident that the BPR practitioner has a well equipped arsenal of enabling information technologies to support the design phase of radically new business processes. However, new process designs seldom fail due to lack of imagination and technology opportunities. But, the implementation of these technologies has been the single greatest impediment to process realisation in BPR practice to date (CSC Index, 1994). And, unless the implementation difficulties highlighted here can be overcome, IT based process redesign is in danger of becoming yet another "technology swallow hole."

Invariably, BPR projects create a huge and characteristically different set of demands on, already over-burdened, IT functions. Wishing to respond enthusiastically, these functions find themselves with an insurmountable number of additional "system change" requests, major development plans, platform migration strategies, and data integrity problems -- all on top of the daily requirement to keep the current IT platform running smoothly. With their ability to deliver new systems on time and within budget already in doubt, the IT function often becomes the principal bottleneck to enactment of redesigned business processes. Frequently, the skills of internal IT people are more relevant to obsolete technologies; there is a culture of delivering new systems in years, rather than weeks; and the application of advanced CASE techniques to improve the development process has done little to help. Viewed as a set of business processes itself, the IT function appears to be in the greatest need of redesign.

An alternative solution is to outsource the development of new process based systems, or purchase industry proven applications which meet the needs of the new process designs. While this might appear to overcome the failings of internal IT functions, the solution is far from ideal. In many cases, the redesigned processes are so new, or unique, that external IT systems developers face the same design uncertainties as internal groups. Costs are high to begin with, and the popular press has reported several high profile court cases involving projects which have run significantly over budget and delivery date. Where "packaged" solutions are available, these typically impose their own process designs, as these have been developed as generic solutions for a wide marketplace. These purchased processes may not support all the requirements of the idealised new process design, and internal customisation may be inevitable.

In many situations, BPR endeavours will render the entire current systems infrastructure redundant or flawed. An enterprise's information systems, like its processes, grow organically as the immediate needs of the business change. After many years, these systems have acquired a thick covering of "barnacles" -- patches to the core systems to accommodate variations in procedures, product designs, etc., as the business evolves. While not meeting high standards of architectural design, these patched systems work, keep the business running, and embrace a wealth of knowledge developed by the business over many years. Apart from the enormous costs involved in replacing the entire systems infrastructure, there are risks associated with losing much of the "hidden intelligence" of the business, currently coded in legacy systems. In most cases, then, the enterprise will not be able to afford the luxury of an overnight conversion to new process based systems, and legacy systems will need to be maintained as new applications are developed and installed.

The migration to new hardware platforms, database technology and process based applications, thus, remains one of the principal challenges facing the enterprise as it attempts to enact redesigned business processes. The IT infrastructure may be the most difficult organisational component to change, whereas structure, performance appraisal systems and other seemingly intractable issues, can be addressed in a shorter time frame. Implementation planning of BPR must, then, recognise the centrality of IT architectures and migration schedules to the delivery of redesigned business processes. Solutions must be found which will not impair the achievement of greatly improved performance results while new technologies and applications are being assimilated. These temporary "transition solutions" may be paper based, or low technology patches to existing systems, and might well be part of a well planned migration strategy. The "surround approach", discussed by Heygate (1993), is a feasible alternative, which preserves the integrity of legacy systems while building new process functionality and user interfaces "around" the existing infrastructure.

5 CONCLUSIONS

Business process redesign can produce enormous improvements in operational performance of the enterprise through the **deliberate design** and creation of organisational structures, systems and core work activities, as distinct from evolving these as needs dictate. As an holistic and radical performance improvement intervention, BPR brings together the consulting and organisational skills of organisational development and the analytical techniques of IT. Software tools, based on IT and industrial engineering methods, can play an important support rôle in BPR projects. While the sophistication of these tools is developing rapidly, currently available products remain technologically immature.

A more important issue for the application of IT in BPR is the enactment of redesigned processes in new software and hardware applications. In this respect, IT has been an impediment to early realisation of BPR benefits, due to the difficulties associated with software development, inappropriate skills, business reliance on legacy systems, and platform migration. These issues must be addressed in BPR implementation planning, if the intentions and efforts of enterprise process redesign are not to be discredited before they've had a chance to deliver results.

6 REFERENCES

CSC Index. 1994. State of Reengineering Report. Pub. CSC Index.

Cummings, T.G. and Huse E.F. (1989) Organization Development and Change 4th edn. West Publ. Co., St. Paul, Mn.

Davenport, T.H. and Short, J.E. (1990) The new industrial engineering: information technology and business process re-design. Sloan Management Review. Summer 1990. Vo. 31, No. 4.

Drucker, P. (1991) The new productivity challenge. Harvard Business Review, Nov.-Dec.

Finkelstein, C. (1989) An Introduction to Information Engineering. Addison-Wesley Publ. Co., Wockingham, England.

Hammer, M. (1990) Reengineering work: don't automate, obliterate. Harvard Business Review. .

Hammer, M. and Champy, J. (1993) Reengineering the Corporation: A Manifesto for Business Revolution. Harper Collins Pubs., Inc.

Heygate, R. Accelerating the payback from information technology: the "surround" approach. In: Managing Change to Get Results. McKinsey & Co. Inc., (1993).

Johansson, H.J., McHugh, P. , Pendlebury, A.J. and Wheeler, W.A (1993) Business Process Reengineering: Breakpoint strategies for market dominance. John Wiley & Sons, New York.

Kaplan, R.B., and Murdock, L. (1991) Core process redesign.The McKinsey Quarterly, No. 2.

Keen, P. (1991) Shaping the Future - business design through information technology. Harvard Business School Press, Boston, Mass.

Martin, J. (1982) Strategic Data Planning Methodologies. Prentice Hall, N.J.

Martin, J. (1989-90) Information Engineering (3 volumes) Prentice Hall, N.J.

Scott Morton, M. (ed. 1991) The Corporation of the 1990's - Information technology and organisational transformation. Oxford Uni. Press, New York.

Waterman, R.H., Peters, T.J.and Phillips, J.R. (1980) Structure is not organization. Business Horizons.

7 BIOGRAPHY

Gerard Lyons is Director of the Information Technology Centre, at University College Galway, where he is responsible for academic programmes and applied research in Information Technology. He joined UCG in 1991, having previously worked in manufacturing, with Digital Equip. Corporn., where he managed groups in the Information Systems, IT Strategy, Logistics Process Design and R&D areas. Gerard's early career was spent with a national research institute, where he specialised in systems analysis and financial modelling. He has published widely in management and technical journals, has co-authored a text on manufacturing systems, and is a frequent speaker at international conferences.

In addition to his academic position, Gerard is an independent Management and IT Consultant, specialising in BPR, Supply Chain Management, IT and Organisation Design, and IT Strategy Planning. His clients are large companies in the Service and Manufacturing and sectors, and his work with these is at the leading edge of BPR practice in Europe.

6

Re-engineering the Sales Department

B. E. Hirsch, M. Krömker, K.-D. Thoben and A. Wickner
BIBA - Bremen Institute of Industrial Technology and Applied Work Science
Hochschulring 20 - D 28359 Bremen - E-mail: kro@biba.uni-bremen.de

Abstract

Successful bid preparation for complex products requires cooperation among specialists of different domains. To enable efficient cooperation between employees of different departments a computerized system capable of supporting distributed processes on possibly different hardware platforms is necessary. This paper presents a concept for cooperative bid preparation as well as concept and infrastructure designed to support the pre-sales phase for companies producing capital goods. Within the ESPRIT project No. 7131 "BIDPREP - An Integrated System for Simultaneous Bid Preparation" enterprises, research institutes and a software vendor cooperate to develop a generally applicable solution.

Keywords

Bid preparation, teamwork, concurrent engineering, distributed processes, object orientation.

1 INTRODUCTION

The preparation of a bid for a one-of-a-kind product tailor-made to customer demands is a complex task as the product specifications are usually incomplete, contain contradictory requirements and undergo last-minute changes. Reliable concepts are hard to produce as the customer sets a tight deadline for submitting the offer. Furthermore, the cost calculation has to be performed precisely in order to ensure a realistic chance of being awarded the contract as well as to realize a profit.

The task may become enormously complex as soon as all EU member countries are fused into one "home market" and — in a broader perspective — when international competition turns into a global encounter. Especially small- and medium-sized companies will face difficulties in their attempts to exploit the market potential. To be competitive requires both optimized organizational structures as well as computerized tools for efficient operation.

A promising concept helping companies keep their competitive edge is *concurrent engineering* [1] involving cross-functional teams from sales, marketing, product and process design, manufacturing as well as R&D. Active cooperation between in-house experts spreads the responsibility for the bid specification to all relevant enterprise activities.

Within the ESPRIT programme the Commission of the European Communities (CEC) is backing the project entitled "An Integrated System for Simultaneous Bid Preparation" (BIDPREP) which aims at developing both a methodology as well as a computerized system capable of supporting the bid preparation process by applying the concurrent engineering concept [2]. In this project, multinational companies are cooperating with research partners from Norway, Denmark, Italy and Germany.

2 THE PROBLEM SITUATION

A bid for a complex product generally consists of a technical concept as well as information about the salesprice, the delivery date, warranty, etc. Fig. 1 depicts the workflow of a typical bid preparation process:

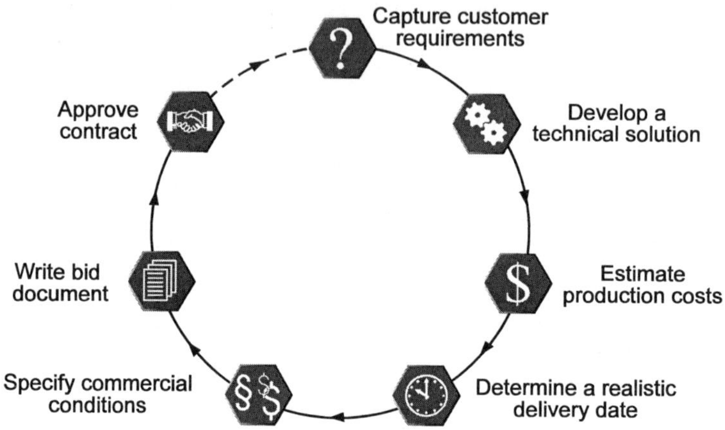

Fig. 1: *The Bid Preparation Process*

Executing the different activities requires specific knowledge of different domains. Therefore, to prepare a bid for a complex product it is necessary to involve experts from different departments of the enterprise. Fig. 2 describes the relation between activities and resources:

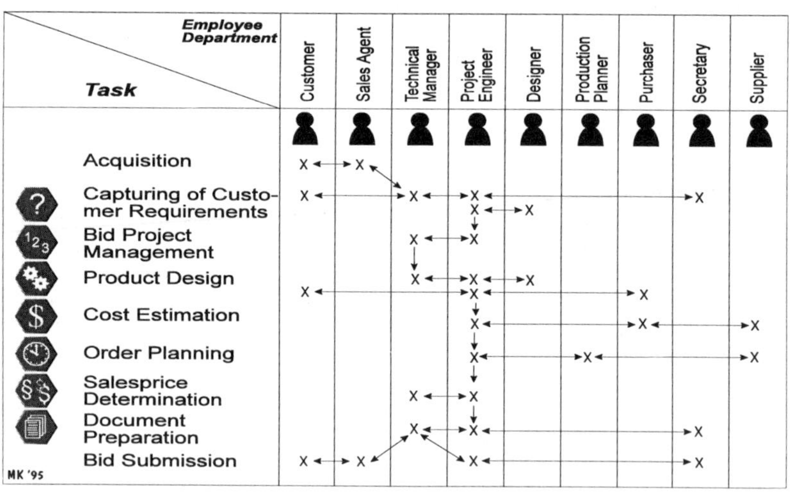

Fig. 2: *Involved personnel in bid preparation*

As depicted in Fig. 2, each bid passes a number of experts stemming from different departments which are possibly spread all over the enterprise. Conventional organizational structures have set up borders between these departments which impede communication and interaction as depicted in Fig. 3.

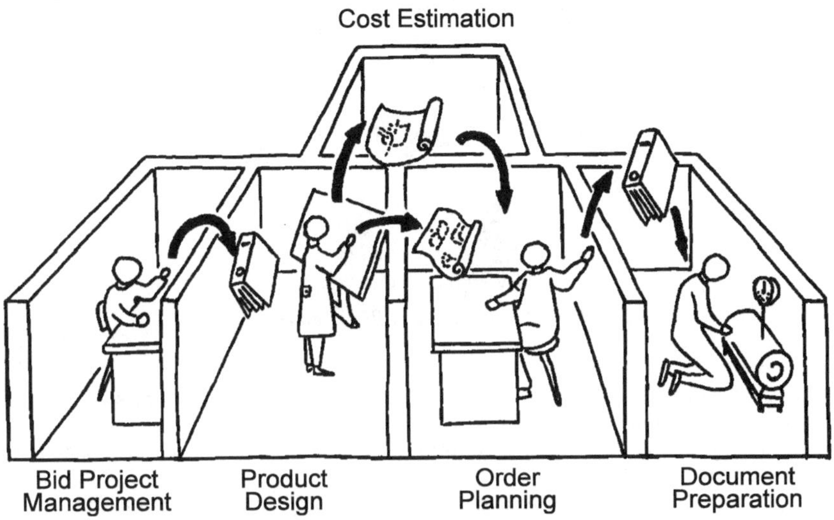

Fig. 3: *Sequential Bid Preparation*

As a result of this, the final bid often contains inconsistencies. Furthermore, due to a lack of suitable arrangements, redundancies occur in the process which increase the cost of bid preparation. Another aspect to be taken into account is the overall lead time for bid preparation. Performing all activities sequentially, it is often not possible to keep tight submission deadlines which can result in the rejection of the bid by the customer.

3 THE BIDPREP STRATEGY

In times of increasing competition and growing complexity of customer demands in the capital goods industry, "customer orientation" has become a buzz word. We have to understand "customer orientation" as a strategy for fulfilling customer wishes to the utmost extent. This not only affects the quality of the technical solution but also leads to a focus on cost minimization. Due to tight submission deadlines for the bid, time — in this case the lead time of the bid — also has become an important criterion.

Considering these aspects, measures have to be taken in order to enable efficient cooperation, interaction and communication within the bid preparation. Teamwork is a promising concept increasing both the quality and reliability of the bid as well as optimizing the price/performance relation by involving experts from different domains. By performing tasks in parallel, lead time can be reduced. "Simultaneous Bid Preparation" as depicted in Fig. 4 is a concept aiming at the formation of interdisciplinary, temporary teams comprising internal experts of different domains as well as external subsuppliers and the customer himself.

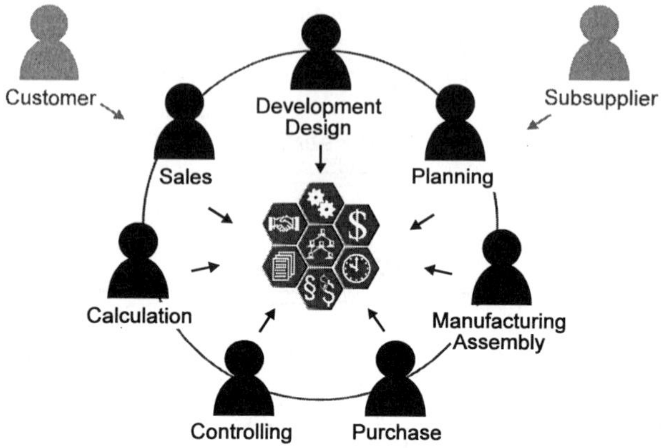

Fig. 4: *"Simultaneous Bid Preparation"*

To enable the team to perform efficiently, several constraints have to be satisfied. Firstly, the team has to be given the authority to make decisions. Secondly, an experienced employee has to lead the team. Through the cooperation within a team, individual tasks will change. For example, the designer's role as an inventor turns into that of an active team member, coordinator and project manager.

As the described bid preparation team is usually widely spread over different enterprise departments, an efficient computerized tool comes into focus. Acting as a bridge crossing the borders between individual departments, the system has to serve as a medium for communication and information exchange. It has to support access to existing solutions, former bids and available information about the customer.

4 THE CONCEPT OF AN INTEGRATED SYSTEM

The envisaged software system has to fulfill two criteria: Functionality in terms of following an overall bid preparation methodology as well as enabling cooperative work in an enterprise-wide network which might also include subsuppliers.

In order to capture the overall requirements of the enterprises regarding the bid preparation methodology functional modelling was performed by applying IDEF0, a technique based on the SADT approach (hierarchical decomposition of processes). Firstly, the "as-is" situation was captured and analyzed. Both bottlenecks and best practices were determined, following *benchmarking* [3] concepts. Finally, a generally applicable "to-be" situation was defined. As a result of the analysis, requirements in relation to the system have been identified. The system has to:

- reduce the effort of bid preparation,
- enhance cross-functional aspects of the process,
- provide access to former bids & existing solutions,
- support the precise estimation of all costs entailed,
- improve the presentation quality of the bid,
- complement existing systems like MRP, CAD, etc.

In order to support the various activities of the overall bid preparation process, the consortium has decided on a modular structure as depicted in Fig. 5:

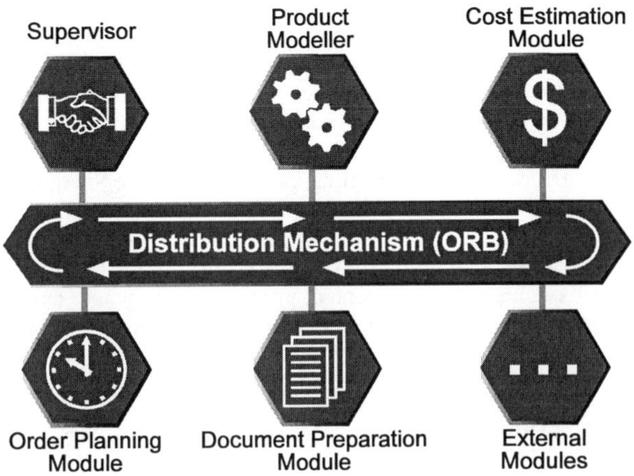

Fig. 5: *The BIDPREP System Architecture*

A *supervisor module* was specified to support the coordination as well as the controlling of all bid-related activities which are performed in several departments. Its main purpose is to serve as a tool for the organization and the management of a bid preparation project.

For the specification of the product design a *bid product modeller* was designed. This module allows the efficient generation of initial product designs based on previously successful designs. The modelling concept is based on an object-oriented "meta-product model" that can be adapted to product models already in use at specific companies.

The cost estimation is realized by applying the *cost-estimation module.* The cost-estimation module employs the successive calculation principle developed by Lichtenberg [4] which presumably improves the reliability of the estimation and accelerates the cost-estimation process. The estimation process is partly based on the design structure of the product, which is provided by the product modeller.

An *order-planning module* determines realistic delivery dates. It provides the traditional features of planning systems for creating and maintaining an order plan. Information about former projects and available resources can be accessed and "what-if" scenarios can be simulated. The order-planning module also accesses the product modeller to retrieve manufacturing information.

Finally, the *document preparation module* eases the compilation of all relevant data into a bid document for submission to the potential customer. As the user configures the various bid sections into a relevant and attractive sequence, the document-preparation module accesses the results of the other modules and transforms them into the necessary formats. The final result is handed over to a standard desktop publishing application, which can — if necessary — be used to fine tune the bid document's layout before it is printed.

External modules may have to be integrated into the BIDPREP system in order to perform certain standard tasks. For example, the document preparation module will use a standard word-processing system to compile the final document.

4.1 The Supervisor

The Supervisor supports the responsible project manager to coordinate the different concurrent bid preparation activities some of which are depicted in Figure 1. The Supervisor informs the project manager about the current status of each task, the responsible team member and the deadline. Critical tasks can easily be ascertained and necessary actions initiated.

In the beginning of the bid preparation process, based on characteristic inquiry data the supervisor informs about similiar former bids and orders. Thereby, it eases reuse in order to reduce the production effort, minimize technical risks and improve the quality.

Access to the BIDPREP modules is gained through start-up buttons. In this way, the user needs not know how to start the different modules or where to store the results. Another major task of the Supervisor is its role as a communication tool. Each team member fills in relevant information like changes implemented, problems anticipated, customer reaction, etc. This ensures that each team member is provided with up-to-date information and allows the replacement of a member in case of holiday, illness or obligations for other tasks.

4.2 The Product Modeller

The aim of the Product Modeller is to support the designer in the effective generation of reliable product structures in the bid preparation phase. Tasks to be supported by the Product Modeller include:

- Fast configuration of an initial product design by means of existing product templates containing formalized design knowledge.
- Creation and maintenance of product templates.
- Fine-tuning of initial product designs until the inquiry requirements are met.
- Structural as well as geometrical manipulation of product models.
- Checking of product designs for inconsistencies to be reported to the designer.
- Providing product model data to outside modules.

The Product Modeller meets these requirements by using a flexible, object-oriented product modelling concept and offering a choice of three different views of the product model:

- A *structure view* that extends full control of the object-oriented product model and allows complex structures to be created and maintained.
- An *assembly view,* resembling a structured bill-of-materials, that is especially suited for acquiring an overview of a product model and for initial configuration tasks.
- A *geometry view* for the manipulation of geometrical aspects of a product model.

Any number of views can be open for the same product model at the same time. Changes made in one view will be reflected in the other views immediately.

Similarities between parts and part families can be captured in *part classes* and class inheritance hierarchies. A product model with a very intricate class hierarchy and only a few fixed root parts can be used as a *product model template,* which generates concrete product models on demand. Such concrete product models are examples of *product model instances,* which are characterized by a complete part decomposition. Another important feature of product models is the use of expressions and rules. The main advantage of these features is automation: Expressions automate the recalculation of attributes while rules allow product structures to be changed automatically.

4.3 The Cost Estimation Module

The chance of winning an order and the possibility of obtaining an acceptable contribution margin depend very much on the ability to calculate a reliable cost price of the inquired product or service. The ability of calculating a reliable cost price is often hindered by the lack of exact information available in the bid phase of a product. Therefore, a major part of the calculated cost price depends on estimated values.

Different cost estimation techniques exist. Often, the price is determined by estimation (rule of thumb method). Another widely used method is the weight method i.e. the cost is calculated on the basis of the weight (or some other parameter) of the whole product or a part of it. These methods are suited for a quick assessment of the cost of a product in the early stage of a design but are limited in relation to their reliability. The accuracy of the costs calculated by applying these methods depends on the similarity of the calculated products to previous products as well as on personal judgement and the experience of the calculator.

To overcome these bottlenecks, the BIDPREP project has specified a cost estimation module which is based on the successive calculation method [4] in which the calculator can estimate costs by decomposing the cost factors stepwise until an acceptable result is obtained. The uncertainty of any cost component is quantified by means of statistical figures such as mean value and standard deviation. These figures guide the cost calculator in identifying the weak areas of a calculation task and thereby help the calculator to concentrate the efforts on the critical cost factors. The Cost Estimation Module supports its users with functions like:

- Creating hierarchical cost structures comprising a top level summary sheet and connected sub-level sheets.
- Performing a min-max estimation using minimum, most likely and maximum cost. Results are a mean value and a standard deviation allowing to calculate the uncertainty.
- Adding information concerning parts and activities.
- Easy search and retrieval of existing cost information.

4.4 The Order Planning Module

An important issue within bid preparation for tailor-made products is the planning of the activities needed to implement the product, i.e. how to execute the order project. This "order planning" includes a specification of the necessary activities, related responsibilities, the duration and interdependence of the activities, an estimation of involved engineering and manufacturing resources, the type of subsuppliers and their services, and other equipment needed. In this way, input for the cost calculation is provided. The planning is constrained by the customer requirements in relation to the delivery date.

The Order Planning Module supports its user with functions like:

- Creation and manipulation of the project/manufacturing plan by textual as well as graphical input/output, bar charts or activity networks, resource allocation diagrams, etc.
- Providing access to multi-project planning in order to take into account limited capacities by allowing resource levelling across projects.
- Facilitating access to former plans in order to search and extract information.
- Providing access to shared resource information, i.e. resource pools.
- Allow simulation, i.e. what-if scenarios.

4.5 The Document Preparation Module

The purpose of the Document Preparation Module is to facilitate the compilation of all relevant information generated in the previous steps into an attractive, comprehensive bid document to be submitted to the customer. The document is represented by a hierarchical data structure:

- Document level. Attributes describing this level are project/product reference, responsibility, status, due date, etc.
- Section level. Attributes describing this level are section type, author, status, dates, etc.
- Item level. Attributes describing this level are item typc, author, pathname, etc.

5 IMPLEMENTATION

5.1 Multi-platform implementation

The BIDPREP system has been implemented in two different environments. Firstly, to demonstrate the operability of the system in a distributed environment, a multi-platform environment has been selected. The *Common Object Request Broker Architecture* (CORBA) [5] as proposed by the *Object Management Group* (OMG) has been adopted as the base architecture for module integration. CORBA enabled the development of the individual modules in four different countries on different platforms. Fig. 6 depicts the resulting architecture.

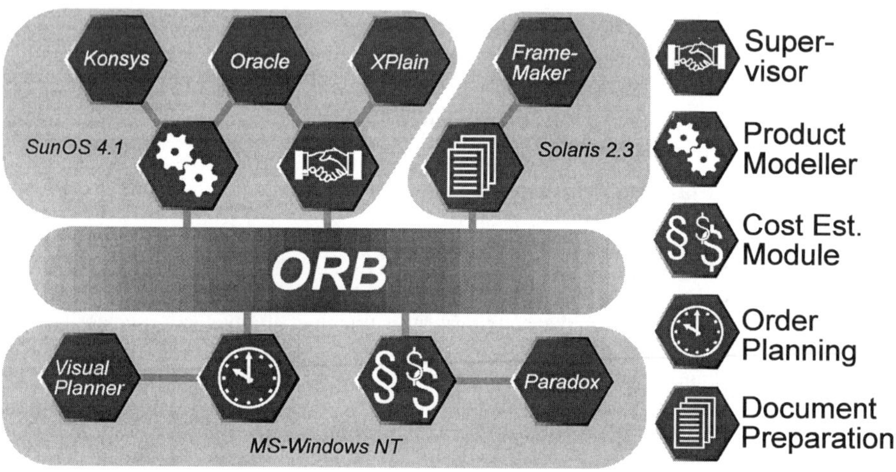

Fig. 6: *Implementation of BIDPREP in a multi-platform environment*

The Supervisor is implemented on a Sun workstation equipped with SunOS 1.4 using the event driven language XPLAIN. The Product Modeller is implemented on a Sun workstation equipped with SunOS 4.1.3 The data is stored in an Oracle database. The Product Modeller provides a complex programming interface based on the CORBA architecture allowing access and manipulation of product models from within other applications and BIDPREP modules.

Both the cost calculation as well as the order planning module operate on a WINDOWS NT platform. The implementation of the cost estimation module is based on the relational data-base Paradox, while VisualPlanner from ViSolutions Oy has been adapted in relation to the specification of the order planning module. The Document Preparation Module has been implemented on a Sun workstation using Word Perfect as a text processor.

Other applications which may already be in use at a companies site can also be integrated into the system. Since most commercial applications available today do not provide a CORBA interface, application shells will have to be developed that translate CORBA service requests into the appropriate application protocol.

5.2 MS-Windows implementation

The previously described multi-platform implementation of the BIDPREP system provides a powerful infrastructure capable of supporting bid preparation in a heterogeneous environment. The applied CORBA architecture has been evaluated as a future-oriented integration mechanism allowing interaction between various platforms with different operation systems.

However, small and medium-sized companies may not intend to invest in a UNIX environment. Since they might already be using various applications operating under MS-Windows, they would prefer a pure PC-based solution. Therefore, the BIDPREP methodology has with some restrictions also been implemented on a MS-Windows platform [6] as depicted in Fig. 7.

Fig. 7: *The MS-WINDOWS implementation of BIDPREP*

As Fig. 7 displays, for most of the activities standard software packages are available on the market. The main aim of BIDPREP system development in this environment was to add value to these software packages in terms of system adaptation (programming of procedures defined within the BIDPREP methodology) as well as system integration. CORBA as a distribution mechanism of the UNIX-based solution has been replaced by OLE 2.0 as well as by the use of Application Programming Interfaces (APIs).

6 CONCLUSION

Efficient preparation of bids for complex products requires both organizational measures like

- overcoming department boundaries
- concurrent execution of the related activities in temporary teams
- early involvement of subsuppliers and customers

as well as the introduction of innovative information technology

- based on an integrated concept
- supporting communication and interaction between all team members
- allowing quick access to existing information from former bids or orders executed

Within the BIDPREP project, a methodology addressing these aspects has been defined. A software system based on this methodology has been specified and implemented in two different ways. Within the following months it is envisaged to develop a commercial software packages based on these prototypes.

7 REFERENCES

[1] T. Kuhlmann, K.-D. Thoben: *"Concurrent Engineering in One-of-a-Kind Manufacturing"*, CIM Management 2/93, p. 10-16.

[2] ESPRIT Project 7131: *"BIDPREP – An Integrated System For Simultaneous Bid Preparation"*, project brochure, March 1993.

[3] K. Jennings, F.Westfall: *"Benchmarking for Strategic Action"*, Journal of Business Strategy 2/92, p. 22-25.

[4] S. Lichtenberg: *"The Successive Principle"*, Report from the Department of Construction Management, Tech. University of Denmark, September 1980.

[5] Object Management Group: *"The Common Object Request Broker: Architecture and Specification"*, Object Management Group, San Diego, CA, USA, 1992.

[6] M. Krömker: *"Efficient bid preparation for complex products"*, Mainz Verlag, Aachen 1995

8 BIOGRAPHY

Prof. Dr.-Ing. Bernd Hirsch, professor for production systems design, management and control at the University of Bremen and Director of BIBA, received his degree in mechanical engineering in 1965, his doctorate at Aachen Technical University in 1969 and a honorary professorship at Hannover University in 1979. His professional career includes fourteen years of corporate production management experience in the European aerospace industry. He has authored many published papers and is a member of the IFIP Working Group 5.7. His research interests focus on production and automation technology, human-centered CIM, one-of-a-kind as well as distributed manufacturing and technology management and transfer functions.

Dipl.-Ing. Mathias Krömker joined BIBA in 1991 after concluding his studies of production engineering at the University of Bremen. As a member of the VDI committee "Bid Preparation", his special interests are the early phases of the product life cycle with emphasis on one-of-a-kind products. His research activities involve various national and international projects aiming at the overall improvement of bid preparation.

Dr.-Ing. Klaus-Dieter Thoben studied mechanical engineering at the TU Braunschweig. After finishing his studies, he was employed at the University of Bremen in the Faculty of Production Engineering. He received his doctor of engineering degree in CAD applications in 1989. In the same year he joined BIBA, where he led the CAD/CAM laboratory. Since 1991 he has headed the Department of Computer-Aided Design, Planning and Manufacturing. His special interests are computer-aided techniques and applications in product development, CIM and simultaneous engineering.

Dipl.-Inform. Andreas Wickner studied computer sciences at the University of Bremen. After finishing his studies he was employed as a software developer for several years before joining BIBA as researcher in the Department of Computer-Aided Design, Planning and Manufacturing. His special interests include object-oriented software engineering, distributed computing and programming languages.

7

Re-engineering Inter Company Processes Through Partnership Networks

S. Crom, J. Julian, T. Kuhlmann, B. Hirsch
BIBA - Bremen Institute of Industrial Technology and Applied Work Science at the University of Bremen
Hochschulring 20, 28359 Bremen, Germany
Tel. +49 421 2185582; Fax: +49 421 2185510
e-mail: km@biba.uni-bremen.de

Abstract

In the future the potential of many firms for improvement lies in the inter-company processes of the supply chain. The objective must be to create partnership networks. This paper outlines their strategic benefits and explains the re-engineering process required in order to achieve these new structures.

Keywords

Business process re-engineering, inter company processes, partnership networks, supply chain

1. INTRODUCTION

In the last decade, companies have made enormous progress in becoming more globally competitive by implementing Total Quality, re-engineering processes and restructuring their organizations. Simultaneously, customers are enjoying more and more choices in their purchasing decisions, they have access to information about competitive offerings never before available, and they expect nothing less than impeccable customer service.

In short, it is now a competitive necessity to **both** increase efficiencies **and** deliver greater service to customers by providing unique value. These dual management imperatives, among others, have led to these times being described as "an era of no tradeoffs" (Ciampa, 1994). Although many companies have focused on increasing efficiency through re-engineering, there has been a management bias towards downsizing and retrenchment as opposed to seeking new opportunities to grow by providing unique value to customers (Hamel & Prahalad, 1994).

The necessity of providing superior customer value, although almost a truism today, has recently received a great deal of attention, most likely because traditional re-engineering is by nature internally focused[1]. As defined by John Guaspari, value as the ratio of what the customer "gets" to what it "costs" over the life of the product or service (Guaspari,). What the customer "got" includes the combination of product, service, and intangibles and is relative to customer expectations. The "cost" includes price, time, and other intangibles. Each company must choose the unique blend of gots and costs most suited to winning over the customers it chooses to serve. More important, each company must structure its operations - or re-engineer if necessary - to be able to deliver its value it promises most effectively.

Recently, companies have begun to find that they can not only reduce the costs in the value equation, but that they can increase customer gots through the improvement of key processes that manage and deliver the flow of products, services, and information to customers (Christopher, 1994). In fact, it is becoming increasingly clear that the next "step function" improvement for many firms lies in looking at these processes across company boundaries in the supply chain (Crom, 1994). We call this inter-company process re-engineering.

The opportunity to improve performance by simplifying, harmonizing and integrating activities between companies, inter company re-engineering, is potentially great. Bringing new products to market in half the normal time, cutting total inventories by 75%, reducing scrap costs by 50%, 100% on-time deliveries are all realistic performance targets if we work within the context of a partnership network of all the players involved in developing and delivering a product or service to a specific customer.

Because of the strategic opportunities - and the potential risks - that partnerships present, several issues must be addressed by senior management before undertaking such an effort. Specifically, how will it contribute to the firm's value proposition to customers? What are the major benefits to the firm? With whom should partnerships be developed and how many should be developed at one time? We will look at answers to these questions next.

2. RE-ENGINEERING INTER COMPANY PROCESSES

Partnerships Must Support each Firm´s Value Proposition

In their book "The Discipline of Market Leaders", Treacy and Wiersema (1994) point out that companies cannot succeed by trying to be "all things to everybody". They must find unique value that they alone can deliver to customers. The authors have identified three "value disciplines", each focusing on a dimension of value which companies can use "to stake market reputation over the long-term": operational excellence, customer intimacy, and product leadership. According to the authors, operationally excellent companies are not innovators or relationship builders, but they provide the best price and the most convenience for customers. Product leaders simply find ways for their products to perform better and be more innovative than the products of others. Customer intimate companies cultivate relationships and specialize in satisfying unique needs through close relationships with customers.

Clearly, as Treacy and Wiersema claim, an organization's core processes, culture, and management systems must support the chosen value proposition to customers. The failures of re-engineering efforts are well documented and lie in large part with management's inability to fully understand the changes needed to become more competitive and the level of commitment

---

1 Treacy and Wiersema (1994) suggest that market leaders redefine value "by raising customer expectations in the one component of value they choose to highlight".

these changes sometimes require[2]. It follows, then, that before pursuing partnerships for inter-company re-engineering, the firm must fully understand its value proposition and how inter-company re-engineering will bolster its ability to deliver this value to its customers.

Figure 1: Benefits of Partnership Networks

Strategic Benefits of Partnership Networks		
Value Discipline of the Firm	Customer Partnership Benefits	Supplier Partnership Benefits
Customer Intimacy	. Continuous learning about customer's unique requirements . Stronger cultural understanding of the specific customer's needs . Support the customer to implement new products and services	. Collaboration with external expertise to design and introduce individual customer solutions . Supplier responsiveness to the individual needs of the firm's customers
Operational Excellence	. Source for understanding demand patterns to optimize logistics and inventory costs . High-volume, low variety opportunities with specific customers are a steady source of cashflow to fund greater economies of scale	. Accurate, predictable incoming shipments that enable low inventories . Streamlined, low cost purchasing and receiving processes . Purchased product conformance quality
Product Leadership	. Source for understanding demand patterns to ensure product availability . Source for market feedback information on product performance . Source for active experimentation with new product ideas	. Responsive, flexible deliveries that support end-product availability . Collaboration with suppliers' expertise in the design and introduction of new products . Supplier responsiveness to product modifications

Major Benefits of Developing Supply Chain Partners

The figure above outlines the strategic benefits that companies following each of the three value disciplines can expect from pursuing inter-company re-engineering partnerships. We

[2] James Champy (1995) caustically describes fallen companies like GM, IBM, and PanAm: "Like arrogant navigators, their managers have to feel the surf crashing right over the bows before they change course. Until then, the stick to the methods that gave them mastery."

define strategic benefits as those that directly support a company's value discipline. Secondary benefits are those that do not directly support the firm's value discipline but nonetheless benefit the firm.

For example, the strategic benefits of supply chain relationships to a customer intimate firm lie in the close relationships it develops with customers. Secondary benefits for this firm would include productivity improvements and cost reductions. On the other hand, the operationally excellent company would find strategic value in the productivity improvements associated with supply chain partnerships (see Figure 1).

The questions of which firms to select and how many with which to partner can be answered within the context of understanding the above benefits for individual firms.

Selecting Partners for Inter-company Re-engineering

The process of selecting and working with supply chain partners is critical to its success. Thought of as an inter-company team effort, we can build on the well researched criteria for successful teams (Shunk), namely: a common goal, well defined and accepted roles, effective procedures for decision making and communication, compatible and complementary values of the individuals involved. That is, provided that the parties involved have similar management philosophies (e.g., Supply Chain Management) and values i.e., the "chemistry" between senior managers is right, well defined goals, roles and procedures for working together go along way to supporting the success of a cooperative venture. Companies can begin the process of selecting partners by proactively looking at the compatibility of their potential goals and roles within the context of a supply chain partnership. While we suggest pursuing a pilot project with an existing supplier with whom a positive relationship already exists, the entire supply base should be reviewed with an eye to developing partnerships with the best suppliers available not just the ones you are currently doing business with.

Company Goal and Role Compatibility

Each firm should clarify its value proposition to its customers and make the benefits they expect from the relationship explicit before the goals of the relationship are established. The more the benefits to both parties support each party's intended value discipline as outlined above, the more value the relationship will have for each party and the longer it will thrive.

For example, a company pursuing product leadership will value a supplier partner who can provide responsive, flexible deliveries that support end-product availability and who can offer innovative solutions to their sourcing needs. A company pursuing customer intimacy would value the opportunity to satisfy these needs through a close partnership because of its ability to provide unique solutions and grow its business with this customer.

Providing the value propositions of each firm are well understood, the goals can be developed for mutual benefit. As business results begin to move towards the established goals, the relationship will reinforce itself allowing the partners to grow their respective businesses together.

How Many Network Partners Should a Company Develop?

Deciding on the number of supply chain partners to develop depends entirely on each company's business needs. The first question is which suppliers would you consider as sole source suppliers? Of those, which are best suited as partners for inter-company re-engineering? What can be said is that, because of the time and energy involved with developing partners, the rule of thumb that "the fewer the better" probably holds true here. For example, a company with a limited product line pursuing customer intimacy may require only one or two customer partnerships. However, a large corporation with a broad product line pursuing customer intimacy may require many such partnerships in order to sustain its investment in a broad array of products.

How can a team be developed and maintained across company and national boundaries?

Having identified a concrete, measurable benefit two companies can achieve more easily by working together, the next step is to build the working teams. Before tasking the teams to further diagnose and re-design a process, they should learn more about the individual learning styles of the members of the teams. An inter-cultural team should be more aware of the stereotypes they have of each other because of national origin (Trompetaars). Once different learning styles, behaviors and norms have been identified and discussed, the group establishes its own groundrules for working together (e.g., when consensus is appropriate and when it isn't, how to manage conflict, how to manage meetings, etc.) Likewise a contract of expectations and groundrules is established between the teams and an inter-company Steering Committee which sponsors the teams, provides over-sight. The Steering Committee plays a critical role in providing the teams "air cover" as they encounter the inevitable political barriers and resistance that come with breakthrough improvement efforts.

What re-engineering methodology should the inter-company teams use?

The methodology we subscribe to is based on the following approach: first eliminate non-value added steps, second simplify what remains, third systematize it, then finally automate. As with CIM and CAD/CAM the temptation is all to great to see new technology as the panacea of all our process problems. One manager involved in a re-engineering effort even tried to "instill a sense of discipline" into people and the process by automating it, i.e., forcing people to complete various screens on the computer before passing on to the next step (transaction) in the process. Our experience is that listening to the process itself by collecting data about non-valued added steps and quality problems is the more pragmatic approach. With an "as-is" process map in hand, created by the people who do the work, one can then imagine (brain storm) the ideal process. At that point team members should be made aware of the impact new technology could have on eliminating non-value added steps (customers entering their own orders on-line) and/or creating new capabilities heretofore unimagined (configuring new products in front of the customers using a lap-top computer). In most cases, breakthrough improvements come from getting those closest to everyday problems together to solve the myriad of small problems that have accumulated over time. Because problems are often due to lack of coordination between departments it is especially important that middle managers be involved in the steps described below.

The specific steps an re-engineering team should go through are:

1. Diagnose existing processes, practices and the organization.
2. Benchmark best practices in critical processes, capabilities and organization structures.
3. Design the ideal processes, practices and organization emphasizing the integration of supplier and customer activities.
4. Identify implementation barriers that have to be overcome.

5. Specify the best processes, practices and organization and the right implementation sequence given the company's climate and circumstances.
6. Develop implementation plans with subgroups of the entire company as well as customers and suppliers.
7. Implement new processes on a pilot basis.
8. Develop and implement plans for full implementation including organizational and other changes needed (e.g. information systems) to support new material and information flows throughout the new inter-company processes.

Rather than "blowing a process up" (along with the people involved) and starting over again with "a clean sheet of paper," our experience is that given the right tools, structured approach and management support, people involved in the process today can redesign processes while improving the organization climate - call it "accelerated evolution" rather than revolution.

What training and consulting tools can be used to build and maintain an inter company community, dialogue and consensus building?

The principle to keep in mind is that partnerships are based on relationships, relationships are based on trust, and trust is build over time through common experience. To that end a workshop design is presented in Figure 2 that forms the foundation for re-engineering inter company processes. The intention is that this series of re-engineering workshops is one element, usually a first step, in an overall strategy for building a partnership network. In the workshop steps 1 through 4 outline above can be accomplished at a high level. Thereafter, both companies can appoint project teams to diagnose, benchmark, design and overcome problems at a detailed level.

There are several criteria that have to be met for the following inter company re-engineering workshops to make sense, the companies involved have to:

- have an existing relationship that is significant and strategically important to both companies,
- senior managers from both companies must share the belief that more is to be gained by working collaboratively than working in traditional adversarial ways,
- both companies must be ready to commit resources to a re-engineering effort.

If those pre-requisites are met, then the following workshops help identify the potential benefit of working collaboratively and motivates key stakeholders in both companies to participate in a joint re-engineering effort.

Objective: To identify the potential overall benefits that could come from jointly working on improvement opportunities.

Scope: The process to be focused on starts with the receipt of an order from a Company A customer through placing an order with Supplier B for product through production and into Company A's finished good inventory.

Workshop Design: Using Product XYZ as an example, the three one day workshops are devoted to identifying opportunities, quantifying them and implementing "quick win" ideas.

Day 1:

1. Conduct Workshop I - The "As Is" & Ideal Process.
   - review the history and current business needs of each company
   - map existing processes
   - discuss the principles of supply chain management
   - imagine the ideal process
   - brainstorm quick wins
   - identify further data to be gathered by small inter company teams
2. Walk the process together as inter company teams to complete process maps, verify quick wins ideas and collect additional data on opportunities.

Day 2

3. Conduct Workshop II - Redesign:
   - present existing processes and data gathered
   - expand on the ideal process
   - present "best practices" in managing similar inter company processes
   - expand the list of quick win ideas
   - prioritize the quick win list
   - select quick wins to pursue as small inter company teams
   - agree on how data should be summarized to show
   - potential improvement benefits
4. Work on quick win ideas and data summarizing as inter company teams.

Day 3

5. Conduct Workshop III - Opportunities and Quick Wins.
   - present the summaries of improvement potential and draft plan to capture those opportunities
   - report progress and results of quick win teams
   - decide whether and how to pursue other opportunities and quick win ideas
   - identify next steps

Figure 2: Inter-Company Supply Chain Management Workshop

3. SUMMARY

To succeed, inter-company re-engineering efforts are based on companies working together with clear and compatible business strategies, either operational excellence, product leadership or customer intimacy. Critically important is that the companies involved have a relationship that is significant and strategically important to both companies. Senior managers from both companies must share the belief that more is to be gained by working collaboratively than working in traditional adversarial ways. Both companies must be ready to commit resources to a re-engineering effort. If that is the case, the opportunity to improve performance by simplifying, harmonizing and integrating activities between companies, inter company re-engineering, is potentially enormous. Start with one or two core processes and relationships since the first teams will encounter inevitable political resistance that will require the time and attention of senior managers to resolve as well as a considerable dedication of manpower to inter-company working teams. To demonstrate the potential benefits and create a common understanding of inter-company re-engineering conduct a three day workshop with key managers from the companies involved. This starts the process of relationship and trust building. As working groups are established to capture the biggest opportunities, train them in how to work effectively as a team. Take the time to identify the natural tension between learning styles and nationalities that any inter-company, inter-country team will encounter. Give them the tools and support (Steering Committee) to be self-managing. Follow a pragmatic re-engineering approach and first eliminate non-value added steps, second simplify what remains, third systematize it before automating. Look for information technology to play a support role in both the process of working together and the introduction of new capabilities that help deliver the value promised to customers.

4. REFERENCES

Champy, J. (1995) Re-engineering Management: The mandate for new leadership. Harper Collins Publisher, New York

Ciampa, D. (1994) The era of no tradeoffs. *Leadership report*. Rath&Strong, Inc., Boston.

Cristopher, M. (1994) Logistics & supply chain management. Richard D. Irwin Inc, New York.

Crom, S.; Kuhlmann, T.; K. Thoben and Hirsch, B. (1994) Partnership networks: A model for future production structures, in Proceedings on Advanced Logistics, 27th ISATA, Aachen.

Hamel, G. and Prahalad, C. (1994) Competing the future. Harvard Business School Press, Boston.

Treacy and Wiersema (1994) The discipline of market leaders, Addison Wesley Publishing, New York

5. BIOGRAPHY

Steven E. Crom

Mr. Crom received his Bachelor of Arts degree in economics and management with honours from Beloit College. He studied international economics at the London School of Economics and Political Science and received his Masters in Business Administration from the University of Chicago. Mr. Crom works with Rath & Strong clients in leading and implementing Just-In-Time and Total Quality efforts in order to help them create greater customer value.

Since 1994 he is a research scientist at BIBA, where he is completing his Ph.D. in Economics.

Jerry L. Julian
Mr. Julian earned his Bachelor's degree in Industrial Engineering from the School of Engineering & Applied Science at Columbia University. He also received an M.B.A. in General Management from the Kenan-Flagler Business School at the University of North Carolina in Chapel Hill.

Mr. Julian has particular expertise in the strategic application of information technology as an enabler of organization change, improved resource productivity, and market responsiveness.

Mr. Julian is Certified in Production & Inventory Management (CPIM) by the American Production and Inventory Control Society (APICS). At the moment he is working as a consultant at Rath&Strong.

Thorsten Kuhlmann
Dr. Thorsten Kuhlmann is research scientist at the BIBA since 1989. At the BIBA, he is in charge of the department "Logistics, Distributed Production".

Dr. Kuhlmann received his M.Sc. degree in business und engineering in 1989 at the University of Hamburg and Technical University of Hamburg-Harburg and his Ph.D at the University of Bremen. His research field covers logistics, production planning & control and the design of software systems for production management. His special interests are focused on the problem of distributed, inter-organisational production.

He worked as segment manager and workpackage manager in a number of national and European research projects.

Bernd E. Hirsch
Prof. Dr.-Ing. Bernd E. Hirsch, professor for production systems design, management and control at the University of Bremen and Director of BIBA, the Bremen Institute of Industrial Technology and Applied Work Science at the University of Bremen.

He received his degree in mechanical engineering in 1965, his doctorate at Aachen Technical University in 1969 and a Honorary Professorship at Hannover University in 1979. His professional career includes fourteen years of corporate production management experience in European aerospace industry. He has authored many published papers and is member of the International Federation of Information Processing (IFIP) Working Group 5.7.

His research interests focus on production and automation technology, human-centered CIM, one-of-a-kind as well as distributed manufacturing and technology management and transfer functions.

8

How Benchmarking Supports Reengineering

Dr.-Ing. K. Mertins, Dipl.-Ing. S.Kempf, Dipl.-Ing. G.Siebert
Fraunhofer-Institute for Production Systems and Design Technology (IPK). Division of Systems Planning
Pascalstr. 8-9, 10587 Berlin, Germany,
Tel: +49 (30) 39006 234, Fax:+49 (30) 3911037,
E-Mail: kai.mertins@ipk.fhg.de

ABSTRACT

Far-sighted enterprises have always used to collect information about competitors and have always compared themselves. Nowadays, the combination of a strategic alignment with central processes, Reengineering and Benchmarking (Mertins, Kempf, Siebert, 1994-2) is an equipment which supports each enterprise effectively in the process of continually improving their performance.

An enterprise-wide, communicable illustration of processing and organizational structures in variable models (Jochem, Schallock, 1994) facilitates this process. On the basis of business processes benchmarking of processes and organizational structures across business lines is made possible.

Market changes are reflected in constantly modified objectives. A well-directed benchmarking supplies the standard values to attain, Reengineering initiates the accomplishment of the objectives by setting up new business processes. The business process indicates the improvements. At the same time it represents the initial position for the next restructuring process. With the outlined instruments the management of change will be easier.

KEYWORDS

Benchmarking, organizational benchmarking, business process reengineering, enterprise modeling.

STARTING POSITION

In a turbulent environment with accelerated innovation cycles of products and manufacturing processes, with increasing customer demands in regard to quality at favorable terms and an outstanding after-sales-service and with regulative demands for an ecological production prices alone do not determine the chances of an enterprise on the market. In order to be successful in the long term enterprises have to be able to react at short notice to changes occurring on the market.

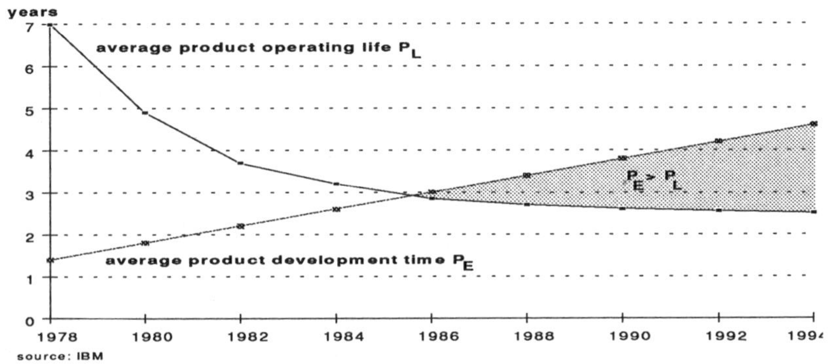

Figure 1 Change of product life and developing times.

While the development time of a product is becoming longer and the operating life is simultaneously becoming shorter (see Figure 1) enterprises today are prepared for the production of tomorrow's products, even though neither the structure nor the functionality of these products is yet known in detail. As past development processes have painfully made clear this is a difficult process of searching and orientating oneself. Branches of industry such as consumer electronics and photo technology have disappeared from Germany either completely or at least in part. And this was not only due to high wages and high nonwage labor costs of the location Germany, but also due to a much higher innovative power and higher customer orientation of the international competitors.

Therefore, the market shares constantly shift, not only between the triad United States, Japan and Europe, but increasingly to East European countries and Asian threshold countries .

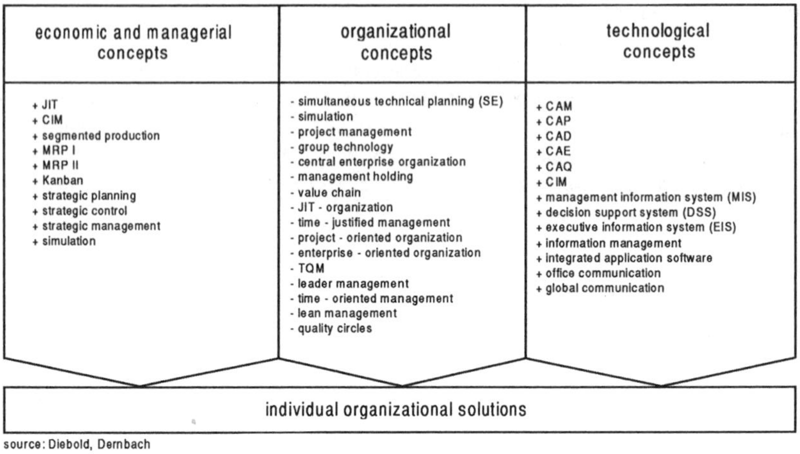

Figure 2 Available organization concepts.

Traditional enterprises have extensive central functions which work on Taylor's principle. It relies on a complex system of controls, gives away though the potentialities of well-educated and motivated employees. Superfluous hierarchies prolong the throughput and exhaust resources in double work which does not contribute to the adding of net value.

Often, an enterprise gets an impulse to adapt to the changing environment only if it loses market shares, if its profits dwindle, or if the customers are dissatisfied. The enterprises react with various methods of restructuring and slimdowns up to anorexia (see Figure 2). In this context the most discussed slogan is Business Process Reengineering, a term which should now be livened up.

REENGINEERING

All listed individual measures are basically aimed at improving a fundamental and decisive factor for the success of an enterprise: The competence to produce innovative high-quality products at marketable prices in due time. For that, enterprises should reflect upon their central competences, should formulate their strategic development objectives, and should then convert these objectives into solid concepts. Reengineering - another term for the necessary restructuring process - understood as the radical reorganization of all company-wide business processes however, implies a fundamentally new beginning without permanently looking at the way work processes used to proceed.

Hammer and Champy define enterprise processes as bundles of activities which require one or more different inputs and which produce a valuable result for a customer (Hammer, Champy, 1994). For the characterization of business processes they suggest quoting, besides the name of the business process, the beginning and end of the process in order to emphasize the nature of a process. They mention five frequently occurring business processes:

- Production: acquisition up to delivery.

- Product development: rough design up to prototype.
- Sale: prospect up to order.
- Order processing: order up to payment.
- Customer service: Inquiry up to solution of the problem.

For the design of business processes Hammer and Champy coined the term Business Reengineering. They define Business Reengineering as the fundamental reconsideration and radical Redesign of enterprises or essential enterprise process. The results are improvements in decisive, nowadays important and measurable performance figures in areas of costs, quality, service and time (Hammer, Champy, 1994).

Fundamental reconsideration is understood as the non-presumptive scrutiny of the course of operations. The objective of this activity is determining and examining those unspoken rules and assumptions which are the basis for the operations. Radical Redesign is understood as the entire reorganization of the enterprise disregarding all existing structures and procedures. An improvement, expansion or modification is rejected explicitly. Improvement is understood as an improvement of the indicators of the performance figures by more than 10%.

Improvements in this order have been made (see Figure 3). In addition, a high proportion of enterprises attained their objectives.

Naturally, the success of Reengineering is tempting, the risk of this radical approach however, is seen by enterprises likewise. One has to look for ways to minimize the risks and to utilize the potentials of Business Reengineering fully. The potential for success is to be increased by proceeding systematically and continually in all Reengineering projects.

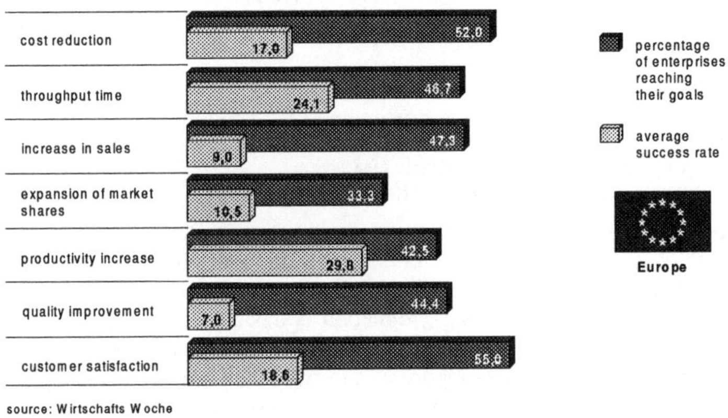

Figure 3 Improvement of results through Reengineering in Europe.

Among others, Reengineering is supported effectively by

- concentrating on the central processes,
- enterprise modeling and
- benchmarking.

CONCENTRATING ON THE ESSENTIALS

Regardless of the cause for the restructuring, a fundamental Reengineering of an enterprise should proceed from the product itself and the processes to efficiently develop and manufacture the product. Business processes serve to add value to products in the enterprise. The direct value adding takes place in production and assembly. Accordingly, the manufacturing process must the main reference point of all activities undertaken to strengthen the enterprise's competence and ultimately the profits (Mertins, Edeler, 1993).

Traditionally-matured production structures are cumbersome and thus complicate the realignment with the efficient manufacturing of innovative products. Often, they also undermine a flexible reaction on market demands. The available high level of technical knowledge of engineers and specialists is to be employed in strategically developable and innovative fields instead of in perfectly attuned operative processes. Their potential should be tapped in flexibly structured units of the factory.

The planning of flexible factories for complex products requires optimizing the strategic share of in-house manufacturing hand in hand with the design of production structures. It is necessary to find the optimal answer to the questions:

- What is the central business? and
- How do we accomplish it?

The decision which products or structural components remain in the enterprise can not occur any longer operationally, e.g. according to costs. It must be considered as a strategic policy decision which eventually determines the alignment of the enterprise with its specific central processes.

Strategic products and their central structural components include:

- bearers of constructive and manufacturing engineering know-how,
- development-determining for the innovative potential of the enterprise,
- decisively for the market position / image and
- designed optimally for the advantage of the customer.

Structural components not characterized by these criteria should be passed to external contractors in order to influence the communication and coordination costs positively. Competence deficits of the enterprise may not be accepted, though.

Concentrating the development and production competence - and with that also limited capital appropriations - on the predetermined central structural components is suitable, if sufficient products and structural components are manufactured in the enterprise, to realize significant production cost advantages towards competitors.

To produce central structural components the processes - or business processes - of products, resources and orders in the production and in operations scheduling are described model-oriented in the sequential steps of Reengineering. One can derive the central process of the respective enterprise from the survey of the corresponding processes (see Figure 5).

After a corresponding connection with spatial and organizational aspects structural units, in the sense of largely self-regulating organizational units, are formed along the value adding chain.

BENCHMARKING "TRIGGERS" REENGINEERING

Central elements of each benchmarking are:

- aiming, i.e. defining an objective for the development of the enterprise,
- measuring, i.e. permanently examining the process codes,
- comparing, i.e. examining whether the goals were accomplished or measuring against the best, and
- designing, i.e. converting the results.

In this sense benchmarking (Camp, 1989) can be a permanent inspiration, a constant incentive to continually improve the own business processes (see Figure 4).

The objective of internal and external benchmarking is to broaden one's mind, disclose deficits through the comparison with competitors and to solidify innovative processes stably (Mertins, Kempf, Siebert, 1994-1).

Usually, the comparison takes place between enterprises of the same branch of industry. An additional use may be made possible by comparing business processes from different branches of industry.

A "classifying benchmarking" across business lines is only possible if business processes can be portrayed comparably. The Kanban principle as a supply control system in the automobile industry was discovered by chance in a supermarket. It was considered abstractly as a business process and put into action. When applying benchmarking across business lines such potentials may be developed systematically.

Benchmarking, as a way to deal with predefined comparable figures, must not be limited to business processes. For many years it has been employed in EDP technology where response times are given for standardized arithmetic operations. In the other extreme there are comparable figures for huge structural units like national economies and large corporations. Statements about the efficiency of company-related structural units, e.g. central process units, are usually not carried out. Such organizational units however represent small, clearly visible and thus easily susceptible substructures, which eventually, added up with all other substructures of the respective enterprise, yield the total efficiency of the enterprise.

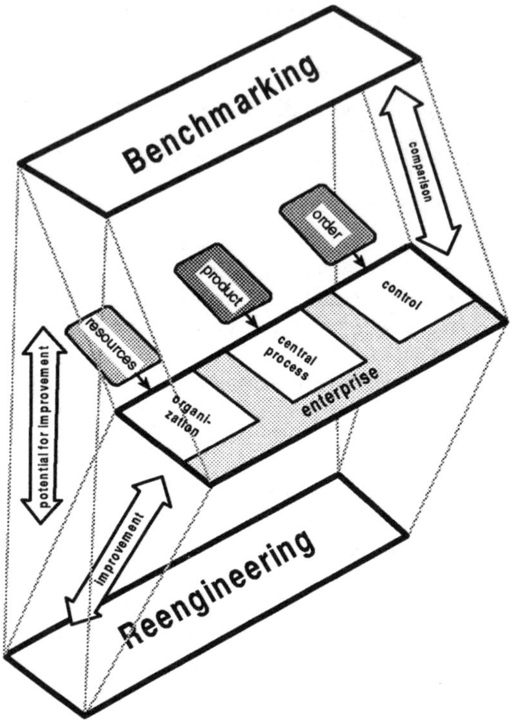

Figure 4 Comparison and improvement of the enterprise processes.

ORGANIZATIONAL BENCHMARKING

Measuring the performance of organizational units, e.g. of manufacturing segments or of islands of automation, used to be done subsequently, as a measurement of effects (Wildemann, 1988, Klingenberger, 1987). Of the effects one was able to determine partial benchmark values such as turnover rates, net value added per capita, and throughput times. There are different code systems (Troßmann, 1992) which measure the degree of efficiency. Most of these however, refer to figures which are only calculable for the whole corporation. In most cases, the determination of the efficiency of smaller organizational units, which should have been developed by Reengineering, were not related to the length of time and the cost in the indirect areas. Nowadays, the calculation of process costs improves this approach.

Comparative values do not include a description of the features which led to these effects. An organizational unit develops its efficiency through the interplay between people, machines and other resources. Staff members develop their optimal performance only if the work tasks are demanding, if social interaction is enabled and if opportunities for further education are opened up. To a certain degree this has been taken into consideration in the approach of the assessment of economical efficiency developed 15 years ago. The work system value (ASW) (Metzger, 1977) originating in the utility value analysis (Zippe, 1979) states, if one is somewhat experienced in working with codes, quite a bit about quality of a work system. These means allow an assessment of flexibility and other values which are difficult to quantify, particularly in personnel-intensive areas, such as assembly (Stiefel, 1989).

On a long-term basis it is necessary however, to develop a standard which enables production engineers to assess a production by the benchmark figures. This requires a generally accepted definition of

- the supporting functions belonging to the central process,
- the dimension figures and their reference values and
- the procedure of determining the comparative figures.

It is suggested to adopt the dimension figures and the reference values described in figure 5. In order to determine the values an enterprise should be able to gather most data from controlling. If it is unable to do so the information processing of this enterprise should be questioned. Information about the attractiveness of work is to be determined by interviews and charts of activity valuation procedures.

This organizational comparison should not be limited to the organizational unit, but should include all functions contributing to the value adding processes. For example, all programming expenditures should be included regardless of whether the actual programming takes place in the workshop or in a central service department (see Figure 6).

The need for such an approach is underlined by the MIT study of the automobile industry. For reasons of simplification only direct manufacturing processes, e.g. carcasses, were examined. The determined figures on processing times and errors were criticized, among other reasons, because the study had left out the organizational dimension, which includes high fixed costs. The continuation of this study by the International Motor Vehicle Program (IMVP) will reinforce this direction.

It is thus possible to compare the expenditure and efficiency of enterprises which are organized rather centrally with such enterprises which a more decentralized structure.

At the moment, we do not consider the departments of finance services, marketing, sales, services or product development. Relevant comparative standards, as they exist for manufacturing organizational units, would have to be developed accordingly.

evaluation criteria for an organizational design

(simplitied description)

Criteria	weight (1-10)	degree of fulfillment Alt. A (1-10)	degree of fulfillment Alt. B (1-10)	result (weight x degree of fulfillment) Alt. A	result (weight x degree of fulfillment) Alt. B
times					
- preparing an offer/planning [days] →	1	3 (20 days)	7 (7 days)	3	7
- throughput time [days]					
- delay [%; days]					
- decision duration [h]					
costs					
- valve added per staff member [DM]					
- depreciation to value adding					
- inventory and circulating stock to value adding					
- overhead to value adding					
- unit costs at selling price					
- machine utilization [%]					
- contingent expenses (e.g. project management costs) →	0,7	5 (200000 DM)	2 (500000 DM)	3,5	1,4
attractive work/work places					
- integral tasks [scale value]					
- processing volume of work systems [% percentage of all operations]					
- individual organization of time [scale value]					
- responsibility [scale value]					
- communication requirements [scale value]					
- individual flexibility [number of work places] →	0,5	5 (3 places of work)	7 (5 places of work)	2,5	3,5
- flexibility of the work system [ad. scale value]					
- absence [%]					
- staff turnover [%]					
- nealth protection and ergonomic data					
- converted suggestions for improvement [number, value]					
quality					
- reworking to value adding [h]					
- rejects to value adding [DM]					
- complaints [number, gravity]					
innovative potential/responsiveness					
- number of available production techniques					
- success rate for offers [%]					
- introduction time for new products and manufacturing processes [months] →	0,3	4 (7 months)	8 (4 months)	1,2	2,4

Figure 5 Measuring figures of organizational benchmarking.

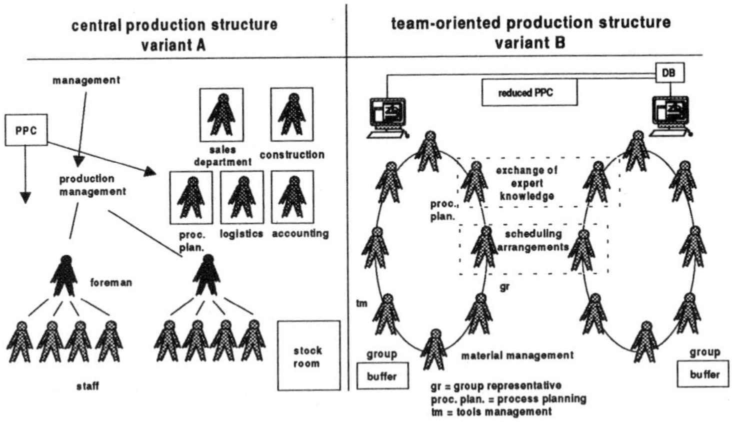

Figure 6 Delimitation of examination units through business processes.

REFERENCES

Camp, R. (1989). Benchmarking: The search for Industry Best Practices that lead to Superior Performance. ASQC Quality Press.

Hammer, M. and Champy, J. (1994). Business Reengineering. Frankfurt. New York.

Jochem, R. and Schallock, B. (1994). Partizipative Gestaltung gesamtbetrieblicher Strukturen mit PadeS und IUM. In: Mertins, K.(Hrsg.): Gestaltung teamorientierter Produktion. (Tagungsband) Berlin.

Klingenberger, H. and Kränzle, H.-P. (1987). Humanisierung bringt Gewinn; Modelle aus der Praxis; Band 2, Fertigung und Steuerung, Eschborn.

Mertins, K. and Edeler, H. (1993). Erfolg durch Konzentration auf Kernprozesse durch durchgängige Auftragssteuerung. ZwF 88, no. 10, p. 489-491.

Mertins, K., Kempf, S. and Siebert, G. (1994-1). Benchmarking Techniques. IFIP Conference, Trondheim, Norway.

Mertins, K., Kempf, S. and Siebert G. (1994-2). Benchmarking ein Managementwerkzeug. ZwF 89, no. 7-8, p. 359-361.

Mertins, K., Süssenguth, W. and Jochem, R.(1994). Modellierungsmethoden für rechnerintegrierte Produktionsprozesse. München, Wien.

Metzger, H. (1977). Planung und Bewertung von Arbeitssystemen in der Montage. Mainz.

Spur,G., Mertins, K. and Jochem, R. (1993). Integrierte Unternehmensmodellierung. Berlin, Wien, Zürich.

Stiefel, K.-P. (1989). SIGMA - Ein Instrument zur Bewertung aufbauorganisatorischer Gestaltungsmaßnahmen. IAO Sonderdruck, Stuttgart.

Troßmann, E. (1992). Kennzahlen als Instrument des Produktionscontrolling. In: Handbuch der Produktionsorganisation, Hanser.

Wildemann, H. (1988). Die modulare Fabrik - Kundennahe Produktion durch Fertigungssegmentierung. München.

Zippe, B.-H. (1979). Die betriebswirtschaftliche Beurteilung neuer Arbeitsformen. Mainz.

CURRICULUM VITAE

Dr.-Ing. Kai Mertins

Dr.-Ing. Kai Mertins was born 1947 in the Federal Republic of Germany. After studying Control theory in Hamburg and Economy together with production technology at the Technical University of Berlin, he became member of the scientific staff of the University Institute for Machine Tool and manufacturing Technology (IWF), Berlin/FRG. Since 1983 he had been head of the department "Production Control and Manufacturing Systems" at the Fraunhofer-Institute for Production Systems and Design Technology (IPK), Berlin/FRG, where he has been Director for Planning Technology (President: Prof. Dr. h.c. mult. Dr.-Ing. G. Spur) since 1988. He has more than 15 years experience in design, planning, simulation and control of flexible manufacturing systems (FMS), manufacturing control systems (MCS), shop floor control systems, and computer integrated manufacturing (CIM). He was General Project Manager in several international industrial projects and gave lectures and seminars at the Technical University Berlin, Polytechnic Nottingham/UK, Czech Republic, Indonesia and China. Special field of interest: Manufacturing strategy development, planning for production systems, shop floor control and simulation.

Dipl.-Ing. Stefan Kempf

Born 1963.
Dipl.-Ing. in Mechanical Engineering at the Technical University of Berlin, Germany.
Mr. Kempf is working for the Fraunhofer-Institute for Production Systems and Design Technology (IPK) at the department Planning Technology since 1990. Since 1994 he is Head of the Group Benchmarking and Manager of the German Information Center Benchmarking.
His working areas are the planning of production systems, management consulting, application of grouptechnology, management of teams working on large software projects and developing software and benchmarking.

Dipl.-Ing. Gunnar Siebert

was born 1963 and after studying Industrial Engineering at the Technical University of Berlin he is working for the Fraunhofer-Institute for Production Systems and Design Technology (IPK), Berlin/FRG at the department Planning Technology. Since 1992 he is the Manager of the CIM Technology Transfer Center in Berlin/Brandenburg. Since 1994 he is Manager of the German Information Center Benchmarking.
His working areas are management consulting, establishing logistics systems, introduction and establishing of CIM-Systems and benchmarking.

9

THE LOGIC OF ENTERPRISE MODELLING

M. Gruninger and M.S. Fox

Department of Industrial Engineering, University of Toronto

4 Taddle Creek Road, Toronto, Ontario M5S 1A4 CANADA

tel:+1-416-978-6823 fax:+1-416-971-2479

email:{msf, gruninger}@ie.utoronto.ca

Abstract

We present a logical framework for representing activities, states, and time in an enterprise model. We define an ontology for these concepts in first-order logic and consider the problems of temporal projection and reasoning about the occurrence of actions.

1.0 INTRODUCTION

An Enterprise Model is a computational representation of the structure, processes, information, resources, goals and constraints of a business, government activity, other organisational system. It can be both definitional and descriptive - spanning what should be and what is. The role of an enterprise model is to achieve model-driven enterprise design, analysis and operation.

A number of issues exist concerning the design of Enterprise Models. *Reusability* is concerned with the large cost of building enterprise-wide data models. Is there such a thing as a generic, reusable enterprise model whose use will significantly reduce the cost of information system building? A second issue is the *Consistent Usage* of the model: Given the set of possible applications of the model, can the model's contents be precisely and rigorously defined so that its use is consistent across the, enterprise? A third issue is model *Accessibility*. Given the need for people and other agents to access information relevant to their role, can the model be defined so that it supports query processing, both surface and shallow[1]. Lastly, there is the *Selection* issue: How do I know which is the right Enterprise Model for my application?

---

1. By surface, level processing we mean the direct retrieval of data that is represented explicitly in the model. By shallow level processing we mean retrieval that requires a small number of deductions, i.e., 1-100, in order to answer the query.

A number of enterprise models have been proposed over the last fifteen years, including those by:

- **CAMI:** A US-based non-profit group of industrial organizations for creating manufacturing software and modelling standards.
- **CIM-OSA:** A reference model developed by the European Esprit project: AMICE.
- **ICAM:** A project run by the Materials Lab. of the US Air Force [Davis et al. 83] [Martin et al. 83] [Martin & Smith 83] [Smith et al. 83].
- **IWI:** A reference model developed at the Institut fur Wirtschaftsinformatk Universitat des Saarlandes, Germany [Scheer 89].

The primary focus of these models has been on reusability; they all provide a data dictionary of object classes which can be reused, mostly for manufacturing organisations, Consistency of usage depends upon the clarity of the written documentation. Accessibility is limited to surface access provided by typical database systems. Lastly, the appropriateness of the model, i.e., selection, is left up to the user.

It is our belief that the issues of reusability, consistent usage, accessibility and selection can best be addressed by taking a more formal approach to enterprise modelling. By formal, we are not referring to analytical models as found in Operations Research, but to logical models as found in Computer Science. Towards this end, the T'OVE ontology [Fox et al. 93] has been developed. An ontology [Gruber 93] is a formal description of entities and their properties; it forms a shared terminology for the objects of interest in the domain, along with definitions for the meaning of each of the terms.

The goal of the TOVE (TOronto Virtual Enterprise) project is to create an enterprise ontology that has the following characteristics: 1) provides a shared terminology for the enterprise that every application can jointly understand and use, 2) defines the meaning (semantics) Of each term in a precise and as unambiguous manner as possible using First Order Logic, 3) implements the semantics in a set of Prolog axioms that enable TOVE to automatically deduce the answer to many "common sense" questions about the enterprise, and 4) defines a symbology for depicting a term or the concept constructed thereof in a graphical context.

The TOVE ontology currently spans knowledge of activity [Gruninger & Fox 94], time, and causality, resources [Fadel 94] [Fadel et al. 94], and more enterprise oriented knowledge such as cost [Tham et al. 94], quality [Kim & Fox 94] [Kim et al. 95] and organization Structure [Fox et al. 95] and agility [Atefi 95]. The TOVE Testbed provides an environment for analyzing enterprise ontologies; it provides a model of an enterprise and tools for browsing, visualization, simulation, and deductive queries.

In this paper we present a logical framework for the TOVE Enterprise Model. We first review our process for engineering an ontology. We then describe our logical framework which is based on Reiter's solution of the frame problem [Reiter 91] and Pinto's formalization of occurrence and the incorporation of time within the situation calculus [Pinto & Reiter 93].

2.0 ONTOLOGY ENGINEERING

For any given ontology, the goal is to agree upon a shared terminology and set of constraints on the objects in the ontology. We must agree on the purpose and ultimate use of our ontologies, We must therefore provide a mechanism guiding the design of ontologies, as well as providing a framework for evaluating the adequacy of these ontologies. Such a framework allows a more precise evaluation of different proposals for an ontology, by demonstrating the competency of each proposal with respect to the set of questions that arise from the applications. These justify the existence and properties of the objects with the ontology.

The process of engineering an ontology that we have developed, starts from scenarios describing the applications which the model is to support\. Next, we define a set of informal competency questions [Gruninger & Fox 94] that represent the types of the information the model is to provide to each application. Next, an initial terminology composed of objects, attributes and relations is designed. The terminology and questions are then iteratively refined until a precise and unambiguous set of terms are created

Motivating Scenarios: The development of ontologies is motivated by scenarios that arise in the applications. In particular, such scenarios may be presented by industrial partners as problems which they encounter in their enterprises. The motivating scenarios often have the form of story problems or examples which are not adequately addressed by existing ontologies. A motivating scenario also provides a set of intuitively possible solutions to the scenario problems. These solutions provide a first idea of the informal intended semantics for the objects and relations that will later be included in the ontology.

- **Informal Competency Questions:** Given the motivating scenario, a set of queries will arise which place demands on an underlying ontology. We can consider these queries to be requirements that are in the form of questions that an ontology must be able to answer. These are the informal competency questions, since they are not yet expressed in the formal language of the ontology.

  Ideally, the competency questions should be defined in a stratified manner, with higher level questions requiring the solution of lower level questions. It is not a well-designed ontology if all competency questions have the form of simple lookup queries; there should be questions that use the solutions to such simple queries.

  These competency questions do not generate ontological commitments; rather, they are used to evaluate the ontological commitments that have been made. They evaluate the expressiveness of the ontology that is required to represent the comptency questions and to characterize their solutions.

- **Specification in First-order Logic -- Terminology:** Once informal competency questions have been posed for the proposed new or extended ontology, the terminology of the ontology must then be specified using first-order logic.

  Recall that an ontology is a formal description of objects, properties of objects, and relations among objects. This provides the language that will be used to express the definitions and constraints in the axioms. This language must provide the necessary terminology to restate the informal competency questions.

The first step in specifying the terminology of the ontology is to identify the objects in the domain of discourse. These will be represented by constants and variables in the language. Attributes of objects are then defined by unary predicates: relations among objects defined using n-ary predicates.

- **Specification in First-Order Logic -- Axioms:** The axioms in the ontology specify the definitions of terms in the ontology and constraints on their interpretation; they are defined as first-order sentences using the predicates of the ontology. It is important to understand the significance of using axioms to define the terms and constraints for objects in the ontology. Simply proposing a set of objects alone, or proposing a set of ground terms in first-order logic, does not constitute an ontology. Axioms must be provided to define the semantics, or meaning, of these terms.

  It is also important to realize that this is not the implementation of the ontology; it is the specification of the ontology. However, the implementation of the ontology should be translatable into KIF.

  The process of defining axioms is perhaps the most difficult aspect of defining ontologies. However, this process is guided by the formal competency questions. As with the informal competency questions, the axioms in the ontology must be necessary and sufficient to express the competency questions and to characterize their solutions; without the axioms we cannot express the question or its solution, and with the axioms we can express the question and its solutions. Further, any solution to a comptency question must be entailed by or consistent with the axioms in the ontology alone. If the proposed axioms are insufficient to represent the formal competency questions and characterize the solutions to the questions, then additional objects or axioms must be added to the ontology until it is sufficient. This development of axioms for the ontology with respect to the competency questions is therefore an iterative process.

  There may be many different ways to axiomatize an ontology, but the formal competency questions are not generating these axioms. Rather, we use them to evaluate the completeness of the sets of axioms in any particular axiomatization. In this sense, we can compare the expressiveness of different sets of axioms using the competency questions. If there is a competency question that one set of axioms can represent and another cannot, then the first set is more expressive. If two different axiomatizations can represent a competency question and characterize its solutions, then they are equivalent with respect to the question, and any comparison must use other criteria.

- **Completeness Theorems:** Once the competency questions have been formally stated, we must define the conditions under which the solutions to the questions are complete. This forms the basis for completeness theorems for the ontology.

The TOVE Ontology Engineering process addresses the issues raised earlier. Reusability is addressed by defining a terminology that is generic across many domains. Consistent Usage is addressed by providing semantics for the terminology in the form of axioms in first-order logic. Accessibility is addressed by implementing the axioms in Prolog thereby providing a deductive query processing facility. Lastly, the Selection issue is addressed by providing a way of characterizing the span of an ontology by what we call competency questions. The ontology must contain a necessary and sufficient set of axioms to represent and solve these questions. It is in this sense that we can claim to have, an adequate ontology appropriate for a given task, and it is this rigor that is lacking in previous approaches to enterprise modelling.

3.0 LOGIC FRAMEWORK: SITUATION CALCULUS

At the core of the TOVE ontology lies a representation of action composed of activity, state, time and causality. The competency of this ontology is focused primarily on projecting what will be true in time. That is, given a set of actions that occur at different points in the future, what are the properties of resources and activities at other points in time. This is also called Temporal Projection and induces the following set of requirements on the ontologies:

- Temporal projection requires the evaluation of the truth value of a proposition at some point in time in the future. We therefore need to define axioms that express how the truth of a proposition changes over time. In particular, we need to address the frame problem and express the properties and relations that change or do not change as the result of an activity.
- We must define the notion of a state of the world, that is, define what is true of the world before and after performing different activities. This is necessary to express the causal relationship between the preconditions and effects of an activity.
- The time period over which the state has a certain status is bounded by the times at which the appropriate actions that change status occur. This period defines the duration of a state if the status is enabled. This is essential for the construction of schedules.
- We want a uniform hierarchical representation for activities (aggregation). Plans and processes are constructed by combining activities. We must precisely define how activities are combined to form new ones. The representation of these combined activities should be the same as the representation of the subactivities. Thus aggregate activities (sets of activities or processes) should themselves be represented as activities.
- The causal and temporal structure of states and subactivities of an activity should be explicit in the representation of the activity.

Within the TOVE project, we have adopted the situation calculus to provide a semantics to our ontology of activity and state. The intuition behind the situation calculus is that there is an initial situation, and that the world changes from one situation to another when actions are performed. There is a predicate $Poss(a,\sigma)$ that is true whenever an action a can be performed in a situation σ.

The structure of situations is that of a tree; two different sequences of actions lead to different situations. Thus, each branch that starts in the initial situation can be understood as a hypothetical future. The tree structure of the situation calculus shows all possible ways in which the events in the world can unfold. Therefore, any arbitrary sequence of actions identifies a branch in the tree of situations.

Further, we impose a structure over situations that is isomorphic to the natural numbers by introducing the notion of successor situation [Reiter 91]. The function $do(a,\sigma)$ is the name of situation that results from performing action a in situation σ. We also define an initial situation denoted by the constant σ_0.

To define the evaluation of the truth value of a sentence in a situation, we will use the predicate $holds(f,\sigma)$ to represent the fact that some ground literal f is true in situation σ. A fluent is a predicate or function whose value may change between situations. Using this predicate we can define state constraints, which are sentences that must be satisfied in all situations.

One important property that must be represented is the notion of causality, that is, the specification of what holds in the world after performing some action. As part of the logical specification of the activity ontology, we define successor axioms that specify how actions change the value of a fluent. These axioms provide a complete characterization of the value of a fluent after performing any action, so that we can use the solution to the frame problem in [Reiter 91]. Thus if we are given a set of action occurrences, we can solve the temporal projection problem (determining the value of a fluent at any time point) by first finding the situation containing that time point, and then using the successor axioms to evaluate the status of the state in that situation.

Another important notion is to represent the occurrence of actions. The work of [Pinto & Reiter 93] extends the situation calculus by selecting one branch of the situation tree to describe the evolution of the world as it actually unfolds. This is done using the predicate *actual* defined by the following axioms:

The initial situation is always actual:

$$actual(\sigma_0) \qquad \text{(EQ 1)}$$

If a situation is actual, then its immediate predecessor must also be actual:

$$(\forall\, a,\sigma)\; actual(do(a,\sigma)) \supset actual(\sigma) \wedge Poss(a,\sigma) \qquad \text{(EQ 2)}$$

An actual situation has at most one actual successor situation.

$$(\forall\, a_1,a_2,\sigma)\; actual(do(a_1,\sigma)) \wedge actual(do(a_2,\sigma)) \supset a_1 = a_2 \qquad \text{(EQ 3)}$$

Thus, actual defines a line within the situation tree.

To represent occurrences, we then introduce the predicate $occurs(a,\sigma)$ defined as actions performed along the actual line:

$$occurs(a,\sigma) \equiv actual(do(a,\sigma)) \qquad \text{(EQ 4)}$$

The notion of the actual line and action occurrences plays a crucial role in the representation of enterprises. We need to express the following class of constraints: suppose that a plan exists that violates some constraint, but we do not want to allow plans that violate the constraint. How can we distinguish between this constraint and those that must always be satisfied in order for a plan to exist? Using the notion of actual line, we can reason about hypothetical branches where we allow such constraints to be violated, but enforce these constraints on the actual line, so that branches that violate the constraints cannot be actual. For example, suppose we want to represent the constraint that no spoiled food is allowed. We cannot represent this as a state constraint which must be satisfied in all situations e.g.

$$(\forall\, r,\sigma)\; holds(spoiled(r),\sigma) \qquad \text{(EQ 5)}$$

since there can exist situations where spoilage occurs; however, we do not want spoilage to occur. Using the notion of actual line, we can represent this as

$$(\forall\ r,\sigma)\ actual(\sigma) \supset holds(spoiled(r),\ \sigma) \qquad \text{(EQ 6)}$$

We can thus represent maintenance and prevention as actual line constraints with universal quantifiers (we will call these universal actual line constraints), and represent goals and deadlines as actual line constraints with existential quantifiers (which we will call existential actual line constraints). For example, we can represent the goal of producing 50 bolts as the following actual line constraint:

$$(\exists\ r,q,\sigma)\ actual(\sigma) \wedge holds(rp(bolt,50),\sigma) \qquad \text{(EQ 7)}$$

This allows us to reason about hypothetical non-actual situations where the goal is not achieved by the deadline. This also allows us to formally characterize the conditions which must hold in the world and actions that must necessarily occur in order to achieve a goal.

4.0 TIME AND ACTION

We represent time as a continuous line; on this line we define time points and time periods (intervals) as the domain of discourse. We define a relation < over time points with the intended interpretation that $t < t'$ iff t is earlier than t'. One important property that must be represented is the intuition that some action a occurs and then some action b occurs, and that there is no intervening event between a and b. Furthermore, we want to define what holds in the world after performing some action in order to capture the notion of causality. How do we express these notions if we have a continuous time line? Since situations have duration, they can be defined as a set of distinguished intervals on the time line; they will be denoted by the letters σ. The following axioms establish the properties of the relation < over situations:

$$(\forall\ a,\ \sigma_1,\sigma_2)\ \sigma_1 < do(a,\sigma_2) \equiv \sigma_1 \leq \sigma_2 \qquad \text{(EQ 8)}$$

$$(\forall\ a_1,a_2,\ \sigma_1)\ do(a_1,\sigma_1) = do(a_2,\sigma_1) \supset a_1 = a_2 \qquad \text{(EQ 9)}$$

$$(\forall\ \sigma_1,\sigma_2)\ \sigma_1 < \sigma_2 \supset \neg\ \sigma_2 < \sigma_1 \qquad \text{(EQ 10)}$$

$$(\forall\ \varphi)[\varphi(\sigma_0) \wedge (\forall\ \sigma,\text{a})\ (\varphi(\sigma) \supset \varphi(do(a,\ \sigma))] \supset (\forall\ \sigma)\ \varphi(\sigma) \qquad \text{(EQ 11)}$$

This enables us to define the intuition of no intervening events, that is, there is no situation between a situation and its successor, which is a consequence of the axioms:

$$(\forall\ \alpha,\sigma,\sigma')\ \neg\ (\sigma < \sigma' < do(a,\sigma)) \qquad \text{(EQ 12)}$$

Situations are assigned different durations by defining the predicate *start(s,t)* [Pinto & Reiter 93]. Each situation has a unique start time; these times begin at 0 in σ_0 and increase monotonically away from the initial situation.

$$(\forall\ \sigma)\ (\exists\ t)\ start(\sigma,t) \qquad \text{(EQ 13)}$$

$$start(\sigma_0,0) \qquad \text{(EQ 14)}$$

$$(\forall\ \sigma,\ t,t')\ start(\sigma,t) \wedge start(\sigma,t') \supset t = t' \qquad \text{(EQ 15)}$$

$$(\forall\ a.\ \sigma,\ t,t')\ start(\sigma,t) \wedge start(do(a,\sigma),t') \supset t < t' \qquad \text{(EQ 16)}$$

To define the evaluation of the truth value of a sentence at some point in time, we will use the predicate *holds*(f,σ) to represent the fact that some ground literal f is true in situation σ. Using the assignment of time to situations, we define the predicate $holds_T(f, t)$ to represent the fact that some ground literal f is true at time t. A fluent is a predicate or function whose value may change with time.

Another important notion is that actions occur at points in time. To represent this we introduce two predicates, $occurs(a,\sigma)$ and $occurs_T(a,t)$, defined as follows:

$$occurs(a,\sigma) \equiv \sigma_0 < do(a,\sigma) \qquad \text{(EQ 17)}$$

$$occurs_T(a,t) \equiv occurs(a,\sigma) \wedge start(do(a, \sigma), t) \qquad \text{(EQ 18)}$$

We will now apply this formalism to the representation of activities in an enterprise.

5.0 ACTIVITIES AND STATES

At the heart of the TOVE Enterprise Model lies the representation of an *activity* and its corresponding enabling and caused *states* ([Sathi et al. 85], [Fox et al. 93]). In this section we examine the notion of states and define how properties of activities are defined in terms of these states. An activity is the basic transformational action primitive with which processes and operations can be represented; it specifies how the world is changed. An enabling state defines what has to be true of the world in order for the activity to be performed. A caused state defines what is true of the world once the activity has been completed.

An activity, along with its enabling and caused states, is called an *activity cluster*. The state tree linked by an *enables* relation to an activity specifies what has to be true in order for the activity to be performed. The state tree linked to an activity by a *causes* relation defines what is true of the world once the activity has been completed. Intermediate states of an activity can be defined by elaborating the aggregate activity into an activity network.

There are two types of states: *terminal* and *non-terminal*. In Figure 1, *es_fabricate_plug_on_wire* is the nonterminal enabling state for the activity *fabricate_plug_on_wire* and *pro_fabricate_plug_on_wire* is the caused state for the activity. The terminal conjunct substates of *es_fabricate_plug_on_wire* are *consume_wire, consume_plug*, and *use_inject_mold* since all three resources must be present for the activity to occur; the terminal states of *pro_fabricate_plug_on_wire* are *produce_plug_on_wire* and *release_inject_mold*. The activity *assemble2 wire_switch* is enabled by the consumption of plug_on_wire (*consume plug_on_wire*) and the use of an assembly area (*use asmbly_area*); this is represented by the nonterminal state *es2_assemble_wire_switch*. This activity causes the production of wire_switch (*produce wire_switch*) and the release of the used resource (*release asmbly_area*); this is represented by the nonterminal state *pro2_assemble_wire_switch*.

In TOVE there are four terminal states represented by the following predicates:*use(s,a), consume(s,a), release(s,a), produce(s,a)*. These predicates relate the state with the resource required by the activity. Intuitively, a resource is used and released by an activity if none of the properties of a resource are changed when the activity is successfully terminated and the resource is released. A resource is consumed or produced if some property of the resource is changed after termination of the activity; this includes the existence and quantity of the resource, or some arbitrary property such as color. Thus *consume(s,a)* signifies that a resource is to be used up by the activity and will not exist once the activity is completed, and *produce(s,a)* signifies that a resource, that did not exist prior to the performance of the activity, has been created by the activity. We define use and consume states to be enabling states since the preconditions for activities refer to the properties of these states, while we define release and produce states to be caused states, since their properties are the result of the activity.

Terminal states are also used to represent the amount of a resource that is required for a state to be enabled. For this purpose, the predicate *quantity(s,r,q)* is introduced, where *s* is a state, *r* is the associated resource, and *q* is the amount of resource r that is required. Thus if *s* is a consume state, then *q* is the amount of resource consumed by the activity, if *s* is a use state, then *q* is the amount of resource used by the activity, and if *s* is a produce state, then *q* is the amount of resource produced.

In this section, we formalize the relationship between states and activities. First we examine the notion that an activity specifies a transformation on the world; this requires that we introduce fluents for states and activities, and the actions that change these fluents. The axioms presented adequate for solving the temporal projection problem for these properties of states and activities.

To formalize the notions of nonterminal states and aggregate activities, we introduce occurrence axioms for a set of actions.

5.1 Successor Axioms for Status of Terminal States

The primary fluents we will consider are the values assigned to states to capture the notion of the status of a state. We define a new sort for the domain of the status with the following set of constants: {*possible, committed, enabled, completed, disenabled, reenabled*}. The status of a state is changed by one of the following actions: *commit(s,a), enable(s,a), complete(s,a), disenable(s,a), reenable(s,a)*. Note that these actions are parametrized by the state and the associated activity.

The next step is to define the successor axioms that specify how the above actions change the status of a state. These axioms provide a complete characterization of the value of a fluent after performing any action, so that we can use the solution to the frame problem in [Reiter 91]. Thus if we are given a set of action occurrences, we can solve the temporal projection problem (determining the value of a fluent at any point in time) by first finding the situation containing that time point, and then using the successor axioms to evaluate the status of the state in that situation.

The status of a state is committed in a situation iff either a commit action occurred in the preceding situation, or the state was already committed and an enable action did not occur.

$$(\forall\, s,a,e,\sigma)\; holds(status(s,a,\, committed),\, do(e,\sigma)) \equiv (e = commit(s,a) \wedge holds(status(s,a,possible),\sigma)) \vee \neg(e = enable(s,a)) \wedge holds(status(s,a,\, committed),\sigma) \quad \text{(EQ 19)}$$

The status of a state is enabled in a situation iff either an enable action occurred in the preceding situation, or the state was already committed and a complete action or disenable action did not occur.

$$(\forall\ s,a,e,\ \sigma)\ holds(status(s,a,\ enabled),\ do(e,\ \sigma)) \equiv (e{=}\ enable(s,a) \wedge holds(status(s,a,committed),\ \sigma)) \vee \neg[(e{=}complete(s,a) \vee e{=}disenable(s,a)) \wedge holds(status(s,a,\ enabled),\ \sigma)] \quad \text{(EQ 20)}$$

The status of a state is completed in a situation iff either a complete action occurred in the preceding situation, or the state was already completed.

$$(\forall\ s,a,e,\ \sigma)\ holds(status(s,a,\ completed),\ do(e,\ \sigma)) \equiv [e{=}\ complete(s,a) \wedge (holds(status(s,a,enabled),\ \sigma) \vee holds(status\ (s,a,reenabled),\sigma))] \vee holds(status(s,a,\ completed),\ \sigma) \quad \text{(EQ 21)}$$

The status of a state is disenabled in a situation iff either a disenable action occurred in the preceding situation, or the state was already disenabled and a reenable action did not occur.

$$(\forall\ s,a,e,\ \sigma)\ holds(status(s,a,\ disenabled),\ do(e,\ \sigma)) \equiv [e{=}\ disenable(s,a) \wedge (holds(status(s,a,enabled),\ \sigma) \vee holds(status\ (s,a,reenabled),\sigma))] \vee \neg\ (e{=}reenable(s,a)) \wedge holds(status(s,a,\ disenabled),\ \sigma) \quad \text{(EQ 22)}$$

The status of a state is reenabled in a situation iff either a reenable action occurred in the preceding situation, or the state was already reenabled and a complete action or disenable action did not occur.

$$(\forall\ s,a,e,\ \sigma)\ holds(status(s,a,\ reeenabled),\ do(e,\ \sigma)) \equiv (e{=}\ reeenable(s,a) \wedge holds(status(s,a,disenabled),\ \sigma)) \vee \neg(e{=}complete(s,a) \vee e{=}disenable(s,a)) \wedge holds(status(s,a,\ reenabled),\ \sigma) \quad \text{(EQ 23)}$$

Note that in each of these axioms we also specify the precondition for the action. These preconditions can equivalently be expressed as the following occurrence axioms:

$$(\forall\ s,a,\ \sigma)\ occurs(commit(s,a),\ \sigma) \supset holds(status(s,a,possible),\ \sigma) \quad \text{(EQ 24)}$$

$$(\forall\ s,a,\ \sigma)\ occurs(enable(s,a),\ \sigma) \supset holds(status(s,a,committed),\ \sigma) \quad \text{(EQ 25)}$$

$$(\forall\ s,a,\ \sigma)\ occurs(complete(s,a),\ \sigma) \supset (holds(status(s,a,enabled),\ \sigma) \vee holds(status(s,a,reenabled),\ \sigma)) \quad \text{(EQ 26)}$$

$$(\forall\ s,a,\ \sigma)\ occurs(disenable(s,a),\ \sigma) \supset (holds(status(s,a,enabled),\ \sigma) \vee holds(status(s,a,reenabled),\ \sigma)) \quad \text{(EQ 27)}$$

$$(\forall\ s,a,\ \sigma)\ occurs(reenable(s,a),\ \sigma) \supset holds(status(s,a,disenabled),\ \sigma) \quad \text{(EQ 28)}$$

How are these incorporated into the activity-state clusters, which only represent the causal relationships among states and activities? The occurrence of a commit action is not explicitly given in the specification of an activity. However, since the status fluents can only be changed by the above set of actions, the following sentences can be derived from the axioms:

$$(\forall\ s,a,\ \sigma)\ occurs(enable(s,a),\ \sigma) \supset (\exists\sigma')\ occurs(commit(s,a),\ \sigma') \quad \text{(EQ 29)}$$

$$(\forall\ s,a,\ \sigma)\ occurs(complete(s,a),\ \sigma) \supset (\exists\sigma')\ (occurs(enable(s,a),\ \sigma') \vee occurs(reenable(s,a),\ \sigma')) \quad \text{(EQ 30)}$$

$$(\forall\ s,a,\ \sigma)\ occurs(disenable(s,a),\ \sigma) \supset (\exists\sigma')\ (occurs(enable(s,a),\ \sigma') \vee occurs(reenable(s,a),\ \sigma')) \quad \text{(EQ 31)}$$

$$(\forall s,a,\sigma)\ occurs(reenable(s,a),\sigma) \supset (\exists \sigma')\ occurs(disenable(s,a),\sigma') \qquad \text{(EQ 32)}$$

Similarly, the precondition for the *commit* action is that the state be *possible*. In [Fadel 94] it is shown how the *possible* status is defined in terms of the availability of a resource for the activity. This includes the configuration or setup of a resource as well as capacity constraints for the concurrent execution of activities with a shared resource. Axioms similar to those above would be used to express the occurrence of the appropriate setup activities for some activity. This is necessary for formalizing time-based competition, where the occurrence of setup activities is minimized.

5.2 Status of Non-Terminal States

In TOVE, non-terminal states enable the boolean combination of states. We will consider four non-terminal states: *conjunctive, disjunctive, exclusive, not.* What precisely does it mean for a non-terminal state to be a boolean combination of states? For example, how do we define the status of a non-terminal state given the status of each substate? To define this notion, we must refer to the occurrence of the actions that change the status of the states.

Disjunctive states are used to formalize the intuition of a resource pool. We may have a set of resources, such as machines or operators, that can possibly be used by an activity. The activity only requires one of these resources, so the activity only needs to nondeterministically choose one of the alternative resources in the pool. Thus, the status of the disjunctive state changes if one of the resources has been selected and its status has been changed. For example, we have

$$(\forall s,s_1,\ldots,s_n,a,\sigma)\ disjunctive(s,a) \wedge substate(s_1,s) \wedge \ldots \wedge substate(s_n,s) \supset occurs(enable(s,a),\sigma) \equiv occurs(enable(s_1,a),\sigma) \vee \ldots \vee occurs(enable(s_n,a),\sigma) \qquad \text{(EQ 33)}$$

The successor axioms for the other values of status are defined in the same way. In other words, the occurrence of an action for a disjunctive state is equivalent to a disjunctive sentence of occurrence literals for each disjunct substate.

Similarly, we have the following constraints on conjunctive states:

$$(\forall s,s_1,\ldots,s_n,a,\sigma)\ conjunctive(s,a) \wedge substate(s_1,s) \wedge \ldots \wedge substate(s_n,s) \supset occurs(enable(s,a),\sigma) \equiv occurs(enable(s_1,a),\sigma) \wedge \ldots \wedge occurs(enable(s_n,a),\sigma) \qquad \text{(EQ 34)}$$

The occurrence of an action for a conjunctive state is equivalent to a conjunctive sentence of occurrence literals for each conjunct substate. Note that we make the assumption that all conjunct substates change their status at the same time.

For exclusive states we have constraints of the form

$$(\forall s,s_1,\ldots,s_n,a,\sigma)\ exclusive(s,a) \wedge substate(s_1,s) \wedge \ldots \wedge substate(s_n,s) \supset occurs(enable(s,a),\sigma) \equiv occurs(enable(s_1,a),\sigma) \vee \ldots \vee occurs(enable(s_n,a),\sigma) \qquad \text{(EQ 35)}$$

$$(\forall s,s_i,s_j,a,\sigma)\ exclusive(s,a) \wedge substate(s_i,s) \wedge substate(s_j,s) \wedge occurs(enable(s,a),\sigma) \supset (occurs(enable(s_i,a),\sigma) \equiv \neg occurs(enable(s_j,a),\sigma) \qquad \text{(EQ 36)}$$

so that the occurrence of an action for an exclusive state is equivalent to the occurrence of the action for exactly one of the substates.

For not states we have the constraint that the action for the substate does not occur when the action for the nonterminal state occurs:

$$(\forall\ s,s_1,a,\sigma)\ not(s,a) \wedge substate(s_1,s) \supset occurs(enable(s,a),\sigma) \equiv \neg\ occurs(enable(s_1,a),\sigma) \quad \text{(EQ 37)}$$

In this way we can define arbitrary nonterminal states as occurrence axioms.

5.3 Status of Activities

Just as status was defined for states, we can define the status of an activity. We define a new sort for the domain of the status of an activity with the following set of constants:{ *dormant, executing, suspended, reExecuting, terminated* }. The status of an activity is determined by the status of its enabling and caused states.

An activity is dormant iff its enabling state is committed. In this case, the resources associated with the state are committed but not yet enabled.

$$(\forall\ a,s,\sigma)\ enabling(s,a) \supset holds(status(a, dormant),\sigma) \equiv holds(status(s,a,committed),\sigma) \quad \text{(EQ 38)}$$

An activity is executing iff either its enabling or caused state is enabled.

$$(\forall\ a,\sigma)\ holds(status(a, executing),\sigma) \equiv (\exists\ s)\ (enabling(s,a) \vee caused(s,a)) \wedge holds(status(s,a, enabled),\sigma) \quad \text{(EQ 39)}$$

An activity is suspended iff its enabling and caused states are disenabled.

$$(\forall\ a,s,\sigma)\ (enabling(s,a) \vee caused(s,a)) \supset holds(status(a, suspended),\sigma) \equiv holds(status(s,a, disenabled),\sigma) \quad \text{(EQ 40)}$$

An activity is reexecuting iff its enabling and caused states are reenabled.

$$(\forall\ a,s,\sigma)\ (enabling(s,a) \vee caused(s,a)) \supset holds(status(a, reExecuting),\sigma) \equiv holds(status(s,a, reenabled),\sigma) \quad \text{(EQ 41)}$$

An activity is terminated iff its enabling and caused states are completed.

$$(\forall\ a,s,\sigma)\ (enabling(s,a) \vee caused(s,a)) \supset holds(status(a, terminated),\sigma) \equiv holds(status(s,a,completed),\sigma) \quad \text{(EQ 42)}$$

5.4 Duration

By combining the ontology of time with the ontology of states of activities, we arrive at the notion of duration, which is essential for scheduling and the analysis of activities in time-based competition. The duration of a state is defined as the time period beginning at the time that the state is enabled and ending at the time that the state is completed. Similarly, the duration of an activity is defined as the time period beginning at the time that activity begins the status of executing and ending at the time that the activity begins the status of terminated. The duration of a state is represented by the predicate *state_duration(s,d)*, while the duration of an activity is represented by the predicate *activity_duration(a,d)*. In the representation of an activity, the duration of a state satisfies the occurrence axiom

$$(\forall\ a,s,t,t',d)\ state\_duration(s,d) \equiv occurs_T(enable(s,a),\ t) \wedge occurs_T(complete(s,a),\ t') \wedge d = t - t' \quad \text{(EQ 43)}$$

We can also define intervals for the remaining status values, such as committed:

$$(\forall\ a,s,t,t',d)\ committed\_duration(s,d) \equiv occurs_T(commit(s,a),\ t) \wedge occurs_T(complete(s,a),\ t') \wedge d = t - t' \quad \text{(EQ 44)}$$

In [Fadel 94], the committed duration is necessary to schedule the availability of a resource for a set of activities over some time interval. The resource must have sufficient capacity to support each activity at every time point in the interval and the resource may not be available to one activity if it is committed to other activities .

5.5 Aggregation of Activities

An important requirement for an ontology for activities is the ability to aggregate a set of activities to form a new activity. Activity clusters may be also aggregated to form multiple levels of abstraction. An activity is elaborated to an aggregate activity (an activity network), which then has activities [Sathi, et al. 85]. These activities are subactivities of the aggregate activity. We introduce the predicate *subactivity(a,a')* to denote that activity *a'* is a subactivity of activity *a*. For example, consider the activity clusters in Figure 1; the activities *fabricate plug_on_wire* and *assemble2 wire_switch* are sub-activities of *assemble_ws aggregation.* The states *es_fabricate plug_on_wire* and *es2_assemble wire_switch* are substates of *enable_ws aggregation*, and the states *pro_fabricate plug_on_wire* and *pro_assemble wire_switch* are substates of *pro_ws_aggregation.*

To completely specify an aggregate activity, we must define the temporal relations over its subactivities and the states of the subactivities. Indeed, the definition of status for activities allows to represent the temporal structure of an aggregate activity in terms of the enabling and caused states of each subactivity; we do this using occurrence axioms.

Essentially, the representation of an activity consists of a sequence of actions that commit, enable, and complete states. These actions may be partially ordered; once the actions in an activity have been totally ordered, we then assign times to the situations in which the actions occur. Thus activities will be represented by an occurrence axiom of the form:

$$(\forall\ a,\ s,s_1,\ldots,s_n,\sigma)\ enabling(s,a) \wedge substate(s_1,s) \wedge \ldots \wedge substate(s_n,s) \supset [occurs(enable(s,a),\sigma) \supset (\exists\ \sigma_1,\ldots,\sigma_n,t_1,\ldots t_n)\ occurs(enable(s_1,a),\ \sigma_1) \wedge \ldots \wedge occurs(enable(s_k,a),\ \sigma_i) \wedge occurs(complete(s_1,a),\ \sigma_{i+1}) \wedge \ldots \wedge occurs(complete(s_k,a),\ \sigma_n) \wedge occurs_T(enable(s_1,a),\ t_1) \wedge \ldots \wedge occurs_T(enable(s_k,a),\ t_i) \wedge occurs_T(complete(s_1,a),\ t_{i+1}) \wedge \ldots \wedge occurs_T(complete(s_n,a),\ t_n)] \quad \text{(EQ 45)}$$

Note that this specification of the activity does not place any constraints on the ordering over the situations and times in which the actions occur. The complete specification of the aggregation of a set of arbitrary activities would impose a total ordering over the occurrences. In general, this is a scheduling problem which remains for future work.

FIGURE 1. Activity Cluster

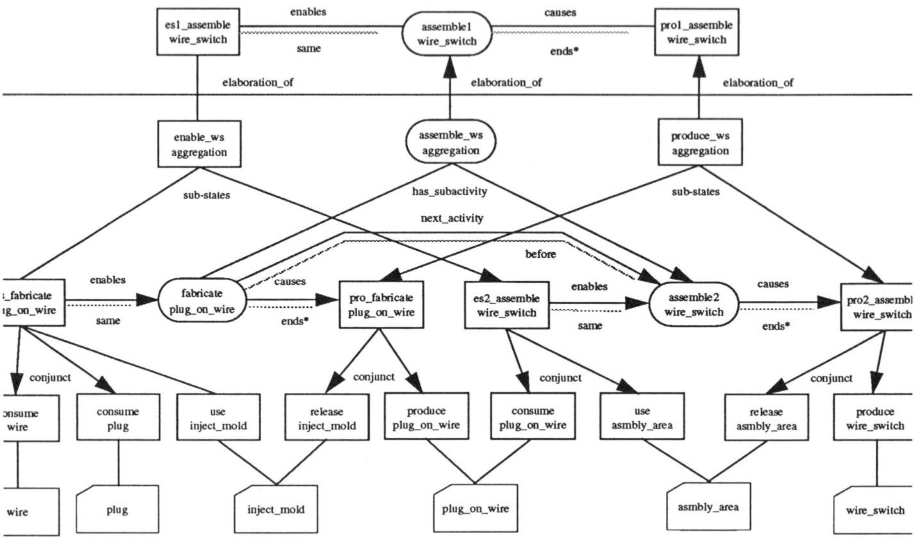

ends* - All time relations in this diagram illustrates how a state's time interval is related to that of an activity's time interval. Therefore, <activity> does not end <state>, <state> ends <activity>.

temporal relation — activity-state relation

6.0 SUMMARY

In this paper, we presented a logical formalization of the TOVE ontology of activity and time that has been designed to specify the tasks that arise in integrated supply chain management and enterprise engineering. To this end, we have defined the TOVE ontologies for activities, states, and time within first-order logic. This formalization allows deduction of properties of activities and states at different points in time by formalizing how these properties do or do not change as the result of an activity (temporal projection). The representation of aggregate activities, and the role of temporal structure in this aggregation, is accomplished through axioms that allow us to reason about the occurrence of actions.

The ontologies for activities, states, and time defined in this paper have been implemented on top of C++ using the ROCK knowledge representation tool from Carnegie Group. The successor state axioms and occurrence axioms have been implemented using Quintus Prolog. The formalization of activities is being extended to handle concurrent activities, reasoning about the availability and capacity of resources, and activity-based costing.

This work serves as a testbed for research into enterprise engineering, enterprise integration and integrated supply chain management, as well as opening new areas of application for theoretical work in reasoning about time, actions, and plans.

7.0 ACKNOWLEDGEMENTS

This research is supported, in part, by the Natural Science and Engineering Research Council of Canada, Carnegie Group Inc., Digital Equipment Corp., Micro Electronics and Computer Research Center, Quintus Corp., and Spar Aerospace.

8.0 REFERENCES

[Blackburn 91] Blackburn J. *Time-based Competition.* Business One Irwin, 1991.

[Davenport 93] Davenport, T.H. *Process Innovation: Reengineering Work through Information Technology.* Harvard Business School Press, 1993.

[Davis et al. 83] Davis, B.R., Smith, S., Davies, M., and St. John, W. Integrated Computer-aided Manufacturing (ICAM) Architecture Part III/Volume III: Composite Function Model of "Design Product" (DES0). Technical Report AFWAL-TR-82-4063 Volume III, Materials Laboratory, Air Force Wright Aeronautical Laboratories, Air Force Systems Command, Wright-Patterson Air Force Base, Ohio 45433, 1983.

[Fadel 94] Fadel, F. *Resource Ontology for Enterprise Modelling.* M.A.Sc. thesis, Department of Industrial Engineering, University of Toronto.

[Fadel et al. 94] Fadel, F., Fox, M.S., and Gruninger, M. A resource ontology for enterprise modelling. *Third Workshop on Enabling Technologies-Infrastructures for Collaborative Enterprises*,(West Virginia University 1994), pp. 117-128.

[Fox et al. 93] Fox, M.S., Chionglo, J., Fadel, F. A Common-Sense Model of the Enterprise, *Proceedings of the Industrial Engineering Research Conference 1993,* pp. 425-429.

[Fox et al. 94] Fox, M. S., Gruninger, M., Zhan, Y.. Enterprise engineering: An information systems perspective*Proceedings of the Industrial Engineering Research Conference 1994,* pp. 461-466.

[Fox et al. 95] Fox, M.S., Barbuceanu, M., Gruninger, M. An Organisation Ontology for Enterprise Modelling: Preliminary Concepts for Linking Structure and Behaviour, *Fourth Workshop on Enabling Technologies-Infrastructures for Collaborative Enterprises*,(West Virginia University 1995).

[Gruber 93] Gruber, Thomas R., *Toward Principles for the Design of Ontologies Used for Knowledge Sharing*, KSL 93-4, Computer Science Department, Stanford University, 1993.

[Gruninger & Fox 94] Gruninger, M., and Fox, M.S., The Role of Competency Questions in Enterprise Engineering ,*Proceedings of the IFIP WG5.7 Workshop on Benchmarking - Theory and Practice,* Trondheim, Norway. June 1994.

[Hammer & Champy 93] Hammer, M. and Champy J. *Reengineering the Corporation.* Harper Business, 1993.

[Kim & Fox 94] Kim, H., and Fox, M.S., "Formal Models of Quality and ISO9000 Compliance: An Information Systems Approach", *Proceedings of the 48th Annual Quality Congress*, Milwaukee WI: American Society for Quality Control, pp. 17-23, 1994.

[Kim & Fox 95] Kim, H. and Fox, M.S. An Ontology of Quality for Enterprise Modelling, *Fourth Workshop on Enabling Technologies-Infrastructures for Collaborative Enterprises*,(West Virginia University 1995).

[Lenat & Guha 90] Lenat, D. and Guha, R.V. *Building Large Knowledge-based Systems: Representation and Inference in the CYC Project.* Addison Wesley, 1990.

[Martin & Smith 83] Martin, C., and Smith, S. *Integrated Computer-aided Manufacturing (ICAM) Architecture Part III/Volume IV: Composite Information Model of "Design Product" (DES1).* Technical Report AFWAL-TR-82-4063 Volume IV, Materials Laboratory, Air Force Wright Aeronautical Laboratories, Air Force Systems Command, Wright-Patterson Air Force Base, Ohio 45433, 1983.

[Martin et al. 83] Martin, C., Nowlin, A., St. John, W., Smith, S., Ruegsegger, T., and Small, A. Integrated Computer-aided Manufacturing (ICAM) Architecture Part III/Volume VI: Composite Information Model of "Manufacture Product" (MFG1). Technical Report AFWAL-TR-82-4063 Voluem VI, Materials Laboratory, Air Force Wright Aeronautical Laboratories, Air Force Systems Command, Wright-Patterson Air Force Base, Ohio 45433, 1983.

[Pinto & Reiter 93] Pinto, J. and Reiter, R. Temporal reasoning in logic programming: A case for the situation calculus. In *Proceedings of the Tenth International Conference on Logic Programming* (Budapest, June 1993).

[Reiter 91] Reiter, R. The frame problem in the situation calculus: A simple solution (sometimes) and a completeness result for goal regression. *Artificial Intelligence and Mathematical Theory of Computation: Papers in Honor of John McCarthy.* Academic Press, San Diego, 1991.

[Sathi et al. 85] Sathi, A., Fox, M.S., and Greenberg, M. Representation of activity knowledge for project management. *IEEE Transactions on Pattern Analysis and Machine Intelligence.* PAMI-7:531-552, September, 1985.

[Smith et al. 83] Smith, S., Ruegsegger, T., and St. John, W. *Integrated Computer-aided Manufacturing (ICAM) Architecture Part III/Volume V: Composite Function Model of "Manufacture Product" (MFG0).* Technical Report AFWAL-TR-82-4063 Volume V, Materials Laboratory, Air Force Wright Aeronautical Laboratories, Air Force Systems Command, Wright-Patterson Air Force Base, Ohio 45433, 1983.

[Tham et al. 94] Tham, D., Fox, M.S., and Gruninger, M., A cost ontology for enterprise modelling *Third Workshop on Enabling Technologies-Infrastructures for Collaborative Enterprises,* (West Virginia University 1994).

10

General Concepts of a Manufacturing System Engineering Workbench as a Tool for the Re-engineering of Manufacturing Systems

Garetti M.
professor of Industrial Technology
Dipartimento di Economia e Produzione, Politecnico di Milano
piazza Leonardo Da Vinci 32, 20133 Milano, Italy
phone: 39 2 23994760 fax: 39 2 23994805

Bartolotta A.
research fellow of the National Research Council (CNR)
Dipartimento di Economia e Produzione, Politecnico di Milano
piazza Leonardo Da Vinci 32, 20133 Milano, Italy
phone: 39 2 23994815 fax: 39 2 23994805

Abstract

This paper highlights the main aspects of a research program developed by the Dipartimento di Economia e Produzione of the Politecnico di Milano within the framework of a national research project on the Design and Management of Advanced Production Systems. The study aims at defining the general architecture of a workbench for the design of industrial production systems, identifying appropriate problem-solving tools and developing a prototype-version of such a system. The approach followed by the authors is based on a general descriptive method for generic production system definition, drawn from Zeigler's system entity structure. Based on this general concept, a two-stage design approach is proposed for the design and redesign activities of manufacturing systems. In the authors' opinion this kind of approach could remarkably improve the efficiency of manufacturing systems design and re-engineering tasks, which are major contributors to competitive success in modern industries worldwide.

Keywords

Object-oriented modelling of manufacturing systems, Manufacturing systems modelling, Computer-Aided Engineering, Design of manufacturing systems

1 INTRODUCTION

In the past decade, companies have been forced to cope with many deep changes, most of which were caused by the increasing complexity and instability of processes on the one hand and by stronger market turbulences on the other.

Increased process complexity is linked to the broadening of the technological foundations of products and processes, to the general widening of product ranges and to an increased demand for customization.

The increase in market turbulence has placed strict constraints on the response time of the industrial system and has determined the success of time-based competitive strategies.

These deep changes have forced industries to deploy new manufacturing strategies to adjust to external pressures: hence the fast and dramatic change in the organizational and managerial criteria applied to company and manufacturing strategy. In many cases, industrial engineers have been forced to reengineer corporate manufacturing facilities in order to guarantee their long-term survival. The success of a manufacturing company is more and more closely linked to the speed and efficiency with which it can incorporate new technologies. Since past experiences have shown that information technology is one of the most effective success drivers, when re-designing processes and organizational schemes must take place (Davenport and Short, 1990), the concept of a software system capable of supporting production engineers in the design, analysis and preliminary assessment of the performance of manufacturing systems has started developing. Hereafter this system will be called a *Manufacturing System Engineering Workbench*, or briefly MSEW.

Such a tool appears to be quite useful for manufacturing companies which need to frequently re-engineer plants to preserve their competitiveness: in these companies production engineers must adjust to external changes frequently, or rather their adjustment cycle must be shorter than the external one (Wang *et al.*, 1993).

The process currently used to design or redesign manufacturing systems has been poorly formalized so far. In fact available information tools are highly specialized and difficult to use; moreover they do not allow the sharing of information; thus their use is quite limited. However, considering the costs and resources involved in these activities, it is quite evident that the improvement of the re-design process would be extremely profitable. Like product designers, who have sophisticated and integrated CAD/CAM/CAE tools available, production engineers need advanced computing tools to solve their problems and manage the huge amount of data associated to the design of a manufacturing system more easily.

Therefore, besides automating routine tasks and providing critical technical data to support the necessary decision-making processes, this design environment should allow the use of integrated information tools for the computer-aided design of manufacturing systems (McLean, 1993).

Such a tool could be used to manage the continuous improvement of a manufacturing system, to store information about production resources, to improve production capacity, to design new manufacturing systems etc. In the case of a large manufacturing system, the structured knowledge which would be available in the system database might enable several designers to work simultaneously and to draw up detailed plans and operating models for the whole company in a few days. Many options could rapidly be assessed and implemented, with a significant improvement over current manual methods which may require weeks or months of intensive design activity.

Such a software system is not available yet, but the current level of hardware and software tools will make its implementation viable in a not-distant future.

2 DESIGNING A MSEW

The project of a MSEW first requires the development of a structured knowledge in the domain of manufacturing systems, i.e. a semantically structured definition of the classes of manufacturing systems to be described. Furthermore the need to integrate all the tools to be included in the workbench makes it necessary to share access to common data, i.e. a centralized information system that can be understood and accessed by all modules and tools in the system must be developed. To this purpose a general descriptive method which can be easily loaded into a computer will be suggested for the definition of a generic manufacturing system.

2.1 Object-oriented modelling of manufacturing systems

A manufacturing system can be divided into three different aspects:

- the **structural aspect,** providing a static view of the manufacturing process, is defined by the units making up the system and their relationships (lay-out schemes are a typical method used to highlight the structural aspect of a manufacturing system);
- the **technological aspect,** describing the conversion phases taking place within the manufacturing system, is based on a functional (i.e. dynamic) analysis of the system itself, detecting parts flowing through the system (this aspect is generally illustrated through the use of technological diagrams);
- the **management aspect,** describing the operating steps of the manufacturing system, is provided by the set of functions making up the actual manufacturing process (typically including planning, scheduling and control activities).

Each aspect will only capture some features of a manufacturing system, but the whole system would be exhaustively described by putting the three aspects together.

When following the principles of *object-oriented* programming, each aspect (i.e. the structural, technological and management aspect) of a generic manufacturing system can be described as a set of objects.

More specifically, three different object classes can be defined (see Figure 1):

- plant components
- technological methods
- management methods

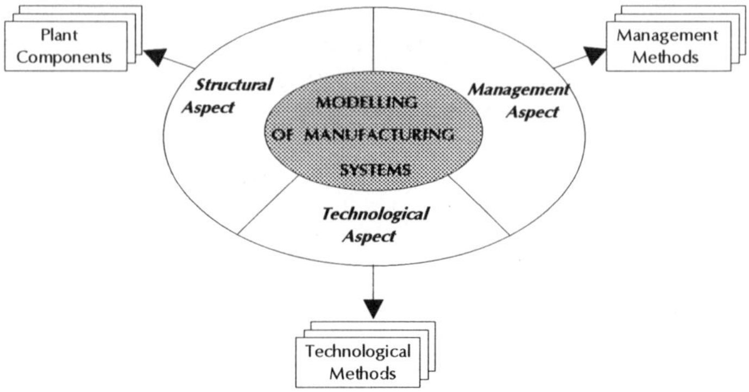

Figure 1 Modelling of a manufacturing system.

Obviously, the identification of components is more intuitive when the structural aspect is examined, since objects defining structural aspects can be quite easily matched to the physical entities making up a manufacturing system. A bigger abstraction effort is required when applying this approach to the other two aspects of the manufacturing system, i.e. the technological and the management aspects, since no matching is possible between objects and physical entities.

The convenience of such an object-oriented modelling system is quite evident: the resulting knowledge is flexible and can be easily changed, objects (plant components, technological and management methods) can be combined freely to create many different models, changes can be restricted to some objects without any need to reconfigure the whole model, etc.

Some modelling detail over this three main aspects is discussed in the following.

Structural aspect

The analysis of the *structural aspect* aims at detecting **Plant Components** and their relationships. Plant components can be divided into:

- permanent objects
- temporary objects.

Permanent objects are resources (i.e. process-shaping objects) always available within the manufacturing system, i.e. components which can be found throughout the whole process (typically machine tools, magazines and handling equipment).

Permanent objects can be divided into:

- stocking agents, that are system resources used to store of temporary objects awaiting to perform certain tasks;
- processing agents, that are system resources (operating stations) dealing with the physical processing of parts or assemblies;

- handling agents, that are system resources used to move parts from an operating station to the next one, linking operating stations to one another, thus generating the part flow.

Temporary objects are entities (process-undergoing objects) which flow through the manufacturing system (typically parts, assemblies and pallets).

Technological aspect

Through the definition of objects called **Technological Methods**, the *technological aspect* of a manufacturing system highlights the dynamic processing behaviour of the system. While the structural aspect focuses on the permanent objects making up the system, the technological aspect focuses on temporary objects, describing the flow of parts, assemblies and pallets inside the system. The following object types are only a few examples within the technological aspect:

- technological cycle of each part type;
- cycle time required for each part type;
- setup time required before a new part type is processed;
- number of scraps for each part type at each inspection station;
- technological lot size, defined as the number of parts which can be processed in parallel by a processing agent.

Management aspect

Through the definition of objects called **Management Methods**, the *management aspect* describes the operating behaviour of the manufacturing process, i.e. the control strategies and management logics governing the interaction between permanent and temporary objects within the manufacturing system.

Generally speaking, information about the management aspect of a manufacturing system fall into the following object groups:

- mechanisms regulating the feeding of parts to the system;
- part routing, if alternative technological routes exist for a part type;
- breakdown of permanent objects;
- dispatching rules (to determine priority levels among parts waiting for a resource);
- loading rules (to determine the feeding sequence of parts to the system);
- management rules of scrap.

2.2 Formal representation of manufacturing systems

Based on the concepts put forth by Zeigler *et al.* (1987, 1989) the semantics developed in the previous chapter for a generic manufacturing system can be formally integrated into a representation diagram which can be easily implemented on a computer, called **System Entity Structure (SES)**.

The System Entity Structure is a knowledge-representing scheme combining decomposition and taxonomy relationships: *decomposition* means explaining how an object can be broken down into its components; *taxonomy* means representing all the possible variants of an object (called *specializations*).

A scheme like the *System Entity Structure* proves very useful in describing large-scale systems, which are extremely difficult to define using conventional, highly-detailed analysis methods. Manufacturing systems definitely fall into this category. Zeigler suggests to break these systems down into several levels, which allows to see the same system from different viewpoints. Such operations are applied on different levels, thereby making it possible to understand how a system can be broken down into its component objects and how such objects can be further broken down into their components.

From a graphical viewpoint, the SES diagram appears as a tree-shaped structure, with *decomposition* relationships being represented as single lines and *taxonomy* relationships as double lines. As the example of Figure 2 describes the root of the mfg systems SES, it only shows decomposition relationships i.e. objects broken into components, this is why taxonomy relationships are present only in deeper SES levels.

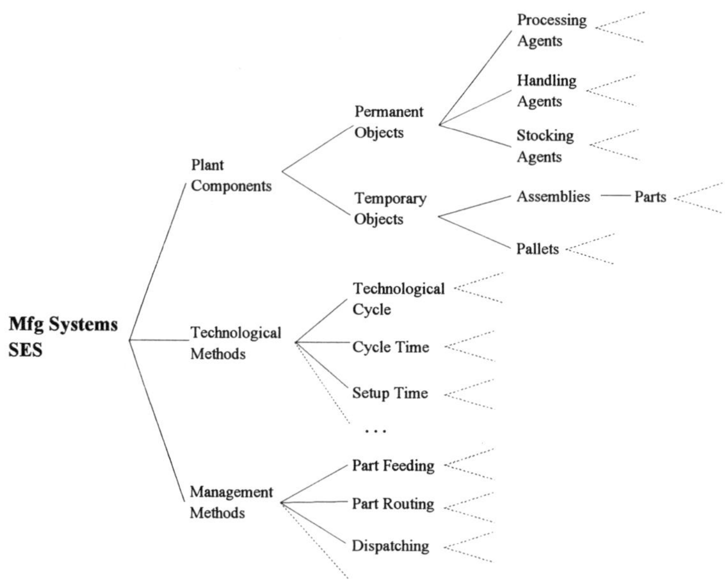

Figure 2 Root of the System Entity Structure of a generic manufacturing system.

The role played by a well-structured knowledge-representing scheme like SES in the MSEW project should not be underestimated. The possibility to store the features of the countless physical and non-physical elements making up a manufacturing system into a single data format and to manage this information in a computer-aided way are an essential pre-requisite for the development of a MSEW.

2.3 General architecture of a MSEW

Defining the general architecture of a Manufacturing System Engineering Workbench means developing an environment of "organized data" (including input, half-processed and output data) and operating tools which can be used by designers to process both previously stored information and new data they have entered themselves, for the computer-aided generation of a manufacturing system model.
The general architecture of the computer-aided system we propose in this paper is roughly shown in Figure 3.

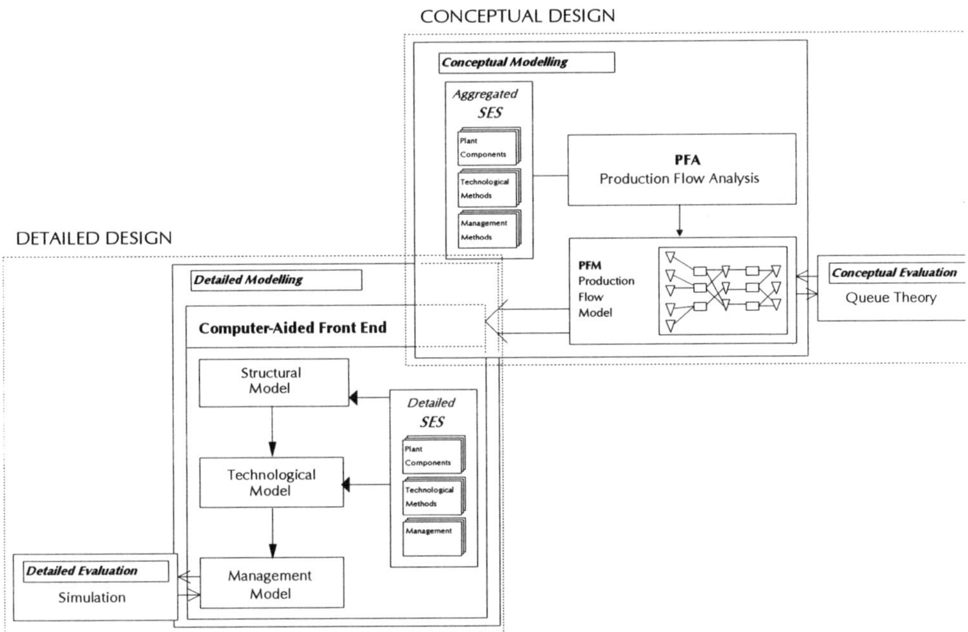

Figure 3 MSEW general architecture.

To meet the need for gradually higher detail levels (Doumeingts *et al.*, 1987), the design process has been divided into two phases:

- a **conceptual design** phase, including a rough design of the manufacturing system. This phase aims at developing a schematic model of the manufacturing system, called *Production Flow Model*, based on the information available in an aggregated version of the mfg system SES, and

- a **detailed design** phase, including a detailed design of the system which builds on the outcome of the previous phase. In this phase, designers use an extended SES version for manufacturing systems, to develop a detailed system model.

Conceptual design phase

The conceptual design phase can be further divided into two steps:

- a modelling step and
- an evaluation step.

During the *modelling step,* a PFA (Production Flow Analysis) module processes general information such as product structure (i.e. a bill of all components making up the end products), gross technological routes of parts and quantitative estimates of the demand for end products to automatically generate a *Production Flow Model* (PFM), i.e. a simplified model of production flows based on the nature and average capacity of processing agents (operating stations), stocking agents and handling agents. Obviously, the detail level of the resulting PFM module will be determined by the detail level of the input data. The PFA module can also generate a simplified graphic representation of the manufacturing system consisting of a scheme similar to the one shown in Figure 3 (where triangles stand for stocking agents, rectangles for processing agents and lines for handling agents).

When filled in with all the relevant quantitative information, the Production Flow Model can provide indications about the most suitable lay-out for the manufacturing system to be designed, since a visual analysis of the flow type (either criss-crossed or linear) can help the designer choose the general configuration which better meets the requirements of the manufacturing system (process-oriented versus product-oriented lay-out).

The *evaluation step* must assess the dynamic behaviour of the PFM, for example by performing a functional model analysis based on queue theory, to ensure that the "static" system configuration resulting from the modelling phase is still valid when the system starts moving.

Detailed design phase

Similarly to the design phase, the detailed design phase can be divided into two further steps:

- a modelling step and
- an evaluation step.

The *modelling step* requires the biggest creativity effort by the designer.

The quantitative and qualitative information collected during the conceptual design phase turn into constraints to be met and into guidelines for the designer's work. The PFM acts as a reference point in this phase.

The system can only act as an interface spurring the designer's creative process as much as possible. Undoubtedly, a graphical interface is best suited to this purpose. The designer needs to have icons representing the objects of a generic manufacturing system and to create a graphic model of the system to be evaluated. In this case, a bottom-up approach is followed to generate the detailed design of the manufacturing system, since the designer has to start from the smallest components making up the system (i.e. the leaves of the SES).

The model resulting from the choice and interconnection of the permanent objects is called a **structural model**, since it represents the structural - i.e. static - aspect of the manufacturing system. In order to build an exhaustive model from a dynamic viewpoint as well, the structural model has to be supplemented with the *technological methods* (i.e. procedures which reflect the processing aspect of the part flow and turn a *structural model* into a **technological model**) and with *management methods* (i.e. procedures reflecting the operating aspects of the manufacturing system which merge into the technological model to create a full **operating model**).

It should be pointed out that it is not reasonable to expect a designer to directly generate a detailed end project starting from the rough outline resulting from the upstream general design phase: several projects with different detail levels will obviously need to be generated, with the most interesting solutions being identified through an iterative trial and error process.

Discrete-event simulation appears to be the most suitable tool for the *detailed evaluation* step. Starting from the operating model of a given manufacturing system, the code for its simulation model has to be directly and automatically generated in a suitable language. This will enable the designer to rapidly evaluate the "goodness" of the project by using the simulator's performance indicators and to make any changes required for refinement purposes.

2.4 The MSEW in a concurrent engineering environment

Unlike traditional design techniques which are sequential, iterative and independent, concurrent engineering (CE) or simultaneous engineering (SE) require a parallel, interactive and cooperative approach to product design. Thus, the CE approach requires that the product and the manufacturing system life-cycles be coordinated and parallelized due to their strong interactions (Kovacs *et al.*, 1991).

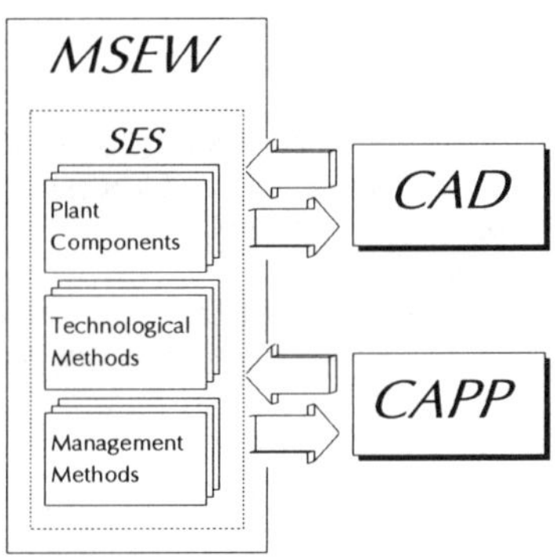

Figure 4 The MSEW in a concurrent-engineering environment.

Hence the need for communication between the MSEW and the CAD (Computer-Aided Design) and CAPP (Computer-Aided Process Planning) systems used by product designers and process designers (see Figure 4). In fact the design specifications and data provided by these systems could be stored into objects making up the System Entity Structure and later gathered by the MSEW user who will interactively assess their feasibility from the manufacturing system point of view.

3 PRELIMINARY RESULTS AND CONCLUSIONS

Some software module of the general architecture described in this paper have been successfully tested in a preliminary and limited implementation of the system:

- a PFA module was developed for the general design phase using an Oracle relational database;
- for a class of automatic assembly lines a SES and a computer-aided front-end device were developed, based on an Gensym Co. G2 shell for expert systems, for the detailed design phase. The automatic generation of the simulation code (based on AT&T Witness simulation software) was also successfully tested starting from the knowledge base of the expert system.

Prototypes for the remaining elements of the suggested architecture need to be developed; the use of an object-oriented simulation software, providing a better integration with the object-based representation of the manufacturing system, is currently under evaluation and more detailed analyses of the semantics of different manufacturing systems need to be implemented and turned into a structured knowledge base.

In conclusion, the potential benefits offered by the development of a *Manufacturing System Engineering Workbench* for the computer-aided design of manufacturing systems have been verified to be well grounded and quite evident, thus calling for a stronger research effort to be performed on this subject in the near future.

4 REFERENCES

Davenport, T.H. and Short, J.E. (1990) The New Industrial Engineering: Information Technology and Business Process Redesign. *Sloan Management Review*, 6, 11-27.

Doumeingts, G., Vallespir, B., Darricau, D. and Roboam, M. (1987) Design methodology for Advanced Manufacturing Systems. *Computers in Industry*, 9, 271-296.

Kovacs, G., Mezgar, I. and Kopsaci, S. (1991) Concurrent Design of Automated Manufacturing Systems using Knowledge Processing Technology. *Computers in Industry*, 17, 205-222.

McLean, C.R. (1993) Computer-Aided Manufacturing System Engineering, in *Advances in Production Management Systems* (eds. I.A. Pappas and I.P. Tatsiopoulos), Netherlands: Elsevier Science.

Wang, W., Popplewell, K. and Bell, R. (1993) An integrated multi-view system description approach to approximate factory modelling. *International Journal of Computer Integrated Manufacturing*, 6(3), 165-174.

Zeigler, B.P. and Rozenblit, J.W. (1987) Design and Modeling Concepts, in *Encyclopedia of Robotics* (eds. R. Dorf and S. Nef), John Wiley & Sons, New York.
Zeigler, B.P.and Zhang, G. (1989) The System Entity Structure: Knowledge Representation for Simulation Modeling and Design, in *Artificial Intelligence, Simulation, and Modeling* (eds. L.E. Widman, K.A. Loparo and N.R. Nielsen), John Wiley & Sons, New York.

5 BIOGRAPHY

MARCO GARETTI is Professor of Industrial Technology at the Dipartimento di Economia e Produzione of the Politecnico di Milano, Italy. After getting a Mechanical Engineering degree from the Polytechnic in 1971, he began his career within Alfa Romeo, where he dealt with mathematical modelling and the computer-aided design of car bodies. In 1974 he joined the Politecnico di Milano and specialized in production systems design and management, with special emphasis on automated production and CIM. He is a senior member of the Computer and Automated Systems Association of the Society of Mechanical Engineering (CASA/SME), a member of the editorial board of PIXEL (an Italian CAD/CAM magazine) and a member of the International Journal of Production Planning and Control.
He is actively involved in the organization of seminars and post-graduate courses on production management and automated manufacturing. His current interests include the design of production systems, production management, intelligent manufacturing systems, simulation and CIM.

ANNA BARTOLOTTA is a research fellow of the National Research Council (Consiglio Nazionale delle Ricerche, CNR) at the Dipartimento di Economia e Produzione of the Politecnico di Milano. She received the degree in Management Engineering in 1994 at the Politecnico di Milano presenting a degree thesis entitled "Framework for a Manufacturing System Engineering Workbench". Her research interests include advanced tools and techniques for the design and management of manufacturing systems.

This study was carried out within the framework of a national research program called "Advanced Techniques for the Design and Management of Industrial Production Systems" funded by the Italian Ministry for University and Scientific Research (MURST 40%).

11

The application of generic process models in Business Process Re-engineering

S J Childe, A M Weaver, R S Maull, P A Smart, J Bennett
University of Plymouth
Drake Circus, Plymouth, PL4 8AA, United Kingdom
Tel +44 1752 232618 e-mail stevec@soc.plym.ac.uk

Abstract

Research work is proceeding on the development of a framework that will help manufacturing businesses identify business processes, process components and links between the processes to form a company-wide view. This paper describes the supporting theory of systems and the structure, development and validation of a model of standard business processes.

Keywords

Business Process Re-engineering; process model; manufacturing; SME; $IDEF_0$

1 INTRODUCTION

The objective of this paper is to describe the development of generic process models for Business Process Re-engineering (BPR) that will encourage companies and participants carrying out BPR projects to take a business process perspective. It will address specifically the application of BPR within Small and Medium Sized Enterprises.

The authors believe that the issue of how to encourage individuals at all levels within a company to think in terms of business processes is critical to the success of a BPR project. This is pointed out by Rummler and Brache (1990) who have found that;

> *"When we ask a manager to draw a picture of his or her business (be it an entire company, a business unit or department), we typically get something that looks like the traditional organisation chart."*

A number of multi-national companies have successfully used generic process models to intervene and change processes within business units, for example Xerox and Shell. The purpose of these generic process models is to encourage individuals within the business units to think in terms of business processes and to provide a starting point for process redesign. The business process view gives the individuals a holistic view of the activities that are carried out within the business units. The authors believe that the use of generic process models could be applied just as successfully by Small and Medium Sized Manufacturing Enterprises (SMEs) to provide a process framework and intervention tool for BPR projects.

2 SME's

The initial problem is to identify which SMEs may benefit from undertaking a BPR project. SMEs have very different characteristics compared to large organisations especially in the area of innovation (Lefebvre et al 1990, Meredith 1987).

Mount et al (1993) provide a framework to deal with this issue which consists of five typical phases of small business development.

1 Owner Operated The owner manages the business and also performs many of the day-to-day productive activities with a small workforce.

2 Transition to owner-managed The owner's role is changing to a state in which the owner is engaged in managing the business full-time, yet the business is small enough not to require a middle level of management.

3 Owner-managed The owner is engaged full-time in the management activities within the business. Supervisory roles may exist but there are no formal functional boundaries.

4 Transition to emergent functional The company is becoming too big to be managed by the owner. Functional boundaries become defined and hence a middle layer of management is required. The addition of specialist middle managers demands substantially more delegation of decision making. In this case the owner is often obliged to screen the viewpoints of senior functional managers and to arbitrate some consensus on a final course of action.

5 Emergent functional organisation A company in which defined functions and managers and a clear organisation structure exists. Middle management is established and functions have frequently established their own objectives, mission statements etc. There may be a conflict of interest between functions, and political manoeuvring may be widespread.

We believe that companies where such conflicts and complexity are emerging are those who may benefit from BPR programmes. The generic models have therefore been developed with emergent functional organisations in mind.

3 GENERIC PROCESSES

In the majority of documented BPR methodologies, including those developed by Coopers & Lybrand, IBM, British Telecom, Xerox and Lucas, one of the initial activities is to identify the core business processes. In identifying the core processes the participants in the BPR project are defining boundaries within their organisation using a process perspective.

By comparing the sets of core processes produced by companies that have undertaken BPR projects, a hierarchy of common processes that are generic across the companies becomes evident. This suggests that a set of standard processes may evolve in process oriented organisations, in the same way that a roughly standardised set of functional divisions (manufacturing, design, sales and marketing, finance, personnel, etc.) developed.

4 AN ARCHITECTURE OF BUSINESS PROCESSES

A manufacturing company can be represented at the most abstract level as a process which transforms inputs into outputs to satisfy the objectives of the various organisational stakeholders. The organisation can be sub-divided into a number of sub-processes that interact to meet these overall objectives. An overall structure or architecture allows each process to be considered without losing the context of its purpose within the whole organisation. The process view of an organisation ensures a strong emphasis on how work gets done and is a "revolutionary change of perspective" from the traditional functional based view of an organisation (Davenport 1993).

There are many examples of organisations identifying a hierarchy of business processes. It is one of the initial activities in the majority of documented BPR including those developed by Coopers & Lybrand (Johansson et al 1993), IBM (Kane 1986), British Telecom (Harvey 1994), Xerox and Lucas (Parnaby 1993). The number of business processes identified at the various levels within the hierarchy varies considerably from organisation to organisation. Davenport (1993) gives a number of reasons for this variation:

- *Processes within organisations are almost infinitely divisible.*
- *The identification of processes can be exploratory and iterative.*
- *An organisation seeking to carry out incremental changes is likely to focus on improvements in sub-divisions of processes whereas for radical changes an organisation should attempt to define processes as broadly as possible.*

Examples of process identification by organisations can be found in Davenport (1993) and the Business Intelligence report on BPR (Harvey 1994) and many case studies in journal articles, for example Shapiro et al (1992), Davenport and Short (1990).

Two activity types "primary" and "support" activities are identified by Porter in his "value chain" concept (1985). The "primary activities" are those activities that interface with the external customer and add value to a product either by designing, manufacturing or by selling the product. The "support activities" are those activities that enable the primary activities to function.

"Management" activities represent a third type of process, including activities which do not directly add value to the customer, the direction setting, enabling change or managing performance activities. For example Veasey (1994) refers to "Management, Support and Value Adding" processes; Royal Mail have "External Customer, Support and Management" processes; Lucas have "Development, Delivery Operations and Support" processes; Pagoda (1993) have "Manage, Operate and Support" processes. The CIM-OSA standard (AMICE ESPRIT 1989) also groups processes into "Manage, Operate and Support".

The grouping of the processes under "Manage, Operate and Support" emphasises some of the general characteristics of the processes and the approaches to redesigning the different types of processes may be different. For example, the concept of value-added must be applied differently in the Operate and Manage areas. Paradoxically, the grouping of processes is a functionally based analysis rather than a process analysis and must be seen as less important than the analysis of the processes themselves.

4.1 The "Operate" processes

The "Operate" processes are those processes which directly produce value for customers. Value is added if activities lead directly to the fulfilment of a customer's requirements. The core operational processes identified by Champy (1995) and Meyer (1993) for a business are "customer service", "product development", and "order fulfilment". The "customer service" process transforms knowledge of customer requirements and the market into customer orders. The "product development" process transforms the actual or perceived requirements of a customer into a design that can be manufactured. The "order fulfilment" process takes the order, manufactures and delivers the product to the customer.

The focus of the work in developing a set of generic processes has been on the "Operate" processes because these are the processes where greatest gains in competitive advantage can be made (Hammer and Champy 1993, Meyer 1993, Johannson et al 1993). Analysis of these processes will also

> *illuminate the most important support process impediments and do so within the context of meeting customer needs* (Meyer 1993)

A recent survey (Harvey 1994) also showed that the most commonly cited processes that organisations were targeting for re-engineering included customer service, logistics and new product development.

From our discussions with companies and our comparison of the lists of core processes developed by a number of organisations including Xerox, IBM and Rover, many companies further divide the "customer service" process into two parts. The two parts are the process of getting an order from a customer and the process of providing support to the customer after the order has been fulfilled. We have called these processes the "Get Order" process and the "Support Product" process.

We have thus identified a set of four "Operate" processes within a manufacturing company. We have named each one with an imperative verb so that the process names are consistent with the $IDEF_0$ models. The four "Operate" processes are

- Get Order
- Develop Product
- Fulfil Order
- Support Product

4.2 Process definition

There are many different views of what should be included or excluded within the boundaries of each process. Each organisation is likely to have a different view. To describe a consensus view of the "Operate" processes we are developing a precise description using a *root definition* and an $IDEF_0$ model of each of the processes showing activities and flows in each process and between the four processes. These are intended to provide what Wilson (1984) terms a "Consensus Primary Task Model".

To develop a rigorous definition of each process, a "root definition" of the process was defined. The concept of a "root definition" is part of the Soft Systems Methodology (SSM) described by Checkland (1981). A root definition should be a "concise description of a human activity system which captures a particular view of it" (Checkland 1981). Checkland also developed a mnemonic CATWOE by which the six elements that should be covered in a root definition can be remembered. The six elements paraphrased from Checkland are;

- *Customers* of the process, beneficiaries or victims affected by the processes activities.
- *Actors* or agents who carry out or cause to be carried out the main activities of the process.
- *Transformation*, the means by which defined inputs are transformed into defined outputs.
- *Weltanschauung*, the outlook or framework that makes the root definition meaningful.
- *Ownership*, the agency having a prime concern for the system and the ultimate power to cause the system to cease to exist.
- *Environment*, features of the environment of the process that must be taken as given.

Since the generic process models stem form the same work, the Actors, Weltanschauung and Ownership for each are the same. The Actors in each process are the people and machines within the manufacturing company under consideration. These cannot be defined more precisely, as the model has to preserve its generic nature. The Weltanschauung for each model is the same, that is to say they are all intended to be more helpful than a neutral model which would be acceptable to *all* manufacturing companies, but which in its theoretically wide application would lose all meaning. Rather it is intended to produce a consensus model which will accommodate the Weltanschauung of the majority of manufacturing companies. Ownership can only be expressed as the owner of the manufacturing company. In some specific cases, process owners may be created which provide the owner role for a particular process, but this can not be seen as a general concept until the process architecture is generally accepted, thus, it can not be part of it.

The root definitions that capture the view of the authors with respect to the "Operate" processes of any manufacturing company is as follows;

The "get order" process contains activities performed by humans and machines. Its principal transformations are to transform a product or concept of a product into a customer order, to translate customer requirements into a form meaningful to the other processes and to use market data to identify potential requirements for new products. It includes the flow of information that is required to satisfy a customer by providing information to the customer and to the other "Operate" processes. The process constantly seeks to ensure that customers' requirements are met and that there are sufficient orders to meet the stakeholder requirements.

The "develop product" process contains activities performed by humans and machines. Its principal transformation is from knowledge into the specification of a product that can be produced to meet customer requirements. It includes the flow of information to enable development of the specification of a product that can be manufactured and the development of product concepts that may fulfil future customer requirements. The process constantly seeks to provide specifications for products that will meet the requirements of customers whilst balancing stakeholder requirements.

The "fulfil order" process contains activities performed by humans and machines. Its principal transformations are product orders into products and enquiries into specifications. It includes the flow of both the material and the information that result in the fulfilment of the external customer order or enquiry. The process constantly seeks to fulfil customer requirements whilst balancing stakeholder requirements.

The "support product" process contains activities performed by humans and machines. Its principal transformation is a need for support into a product that continues to meet the requirements of a customer. It includes the flow of the resources and information that are required to meet the customer's support requirements. The process constantly seeks to fulfil the customer's support requirements whilst balancing stakeholder requirements.

In the tradition of Checkland's Soft Systems Methodology, the root definitions are being revised as more knowledge about the processes is gained.

5 DEVELOPMENT AND VALIDATION OF THE PROCESS MODELS

5.1 Modelling technique

The model of the "Operate" processes has been developed using $IDEF_0$ (CAM-I 1980). $IDEF_0$ is widely used in the manufacturing sector for modelling processes. $IDEF_0$ comprises:

- A set of methods that assist in understanding a complex subject;
- A graphical language for communicating that understanding;
- A set of management and human-factor considerations for guiding and controlling the use of the technique.

$IDEF_0$ uses top-down decomposition to break-up complex topics into small pieces which can be more readily understood and which are set in their proper context with respect to other system elements. An $IDEF_0$ model is an ordered collection of diagrams, related in a precise manner to form a coherent model of the subject. The number of diagrams in a model is determined by the breadth and depth of analysis required for the purpose of that particular model. At all times the relationship of any part to the rest of the whole remains visible.

In summary $IDEF_0$ provides the ability to show what is being done within a process, what connects the activities and what constrains activities. It uses a structured set of guidelines based around hierarchical decomposition, with excellent guidance on abstraction at higher levels. If used well this ensures good communication and a systemic perspective.

5.2 Level of analysis

The level of analysis is critical when developing a generic model. For the generic model to be of any use it must contain elements which are at a level of detail that allows meaningful discussion within a particular company. Conversely, too much detail would restrict its application. A very detailed model would become specific to a particular company. Thus an attempt is being made to judge the appropriate level of detail.

Using $IDEF_0$ as a modelling technique ensures that the context for any part of a process model under analysis in relation to the whole of the process model is always known. Therefore a company can focus on the part of a process model it is particularly interested in and develop a further levels of detail without losing its context within the whole process.

5.3 Information sources

The models have been developed with the involvement of a number of manufacturing companies varying in size from Times 1000 companies to Small and Medium Manufacturing Enterprises (SMEs) with under 500 employees.

The information used to develop the process models has been extracted and assimilated from a number of sources including literature (especially Harrington 1984, CAM-I 1984, Porter 1985), previous work (Childe 1991), generic models described in other modelling methods and individual models of company processes.

The $IDEF_0$ models of the "Operate" processes will cover all four types of manufacturing companies defined by Wortmann (1990); Make-to-stock, Assemble-to-order, Make-to-order and Engineer-to-order.

5.4 $IDEF_0$ standard process models

The "Operate" processes are represented in a single $IDEF_0$ model that shows the interactions between each of the processes and external customers, suppliers and other parts of the organisation that are outside the boundaries of the model. $IDEF_0$ has allowed us to develop a model of each process separately and then combine the $IDEF_0$ models into an integrated model of the "Operate" processes. The complete model includes a set of $IDEF_0$ diagrams and a glossary of terms.

5.5 Validation

The validation of the process models is currently being carried out. Validation methods include criticism and comment by academic colleagues and industrial practitioners experienced in BPR and manufacturing management and a comparison by third parties to their own process models.

6 APPLICATION

In the introduction the critical issue of getting employees to think in terms of business processes was identified. The generic process models are intended to be used as an intervention tool to encourage the participants of a BPR project within a manufacturing company to take a business process perspective. The participants in a BPR project would generally be individuals from the functions who currently perform activities within the process, guided by objectives set by senior management.

In the initial stages of the BPR project, following the identification of a core process to be redesigned, the participants would be presented with the generic process model and glossary of terms and asked to compare the generic process model against the activities within the company. These activities would be carried out under the guidance of an internal or external facilitator.

In carrying out a comparison the model encourages the participants to;

- Take a business process perspective as the generic model provides an existing process framework.
- Develop a consensus view of their own company's process by debating the differences between the generic model and each participants perceived view of the company's process.
- Identify and change the generic model to represent their company's process.
- Identify immediate changes that could be made to the company's process as differences between the model and reality are found.
- Consider the systemic relationship of all parts of the process as model provides a structured medium where inconsistencies in the changed model can be identified easily.

In comparison with current BPR approaches where the participants are encouraged to develop a process model of the existing business process, the use of generic models reduce the danger of participants reverting to tradition functional thinking by providing a process focused framework. It also provides greater momentum to the project than a "blank sheet of paper" and the generic process model is non-political having being produced externally. The non-political nature of the generic process model should enable participants to more freely criticise the model and in doing so generate debate and understanding amongst the group.

7 CONCLUSION

This paper has described the development of a set of generic process models for business process re-engineering in small and medium sized manufacturing companies. Initial validation of the models has supported the view that generic models would be useful in the re-engineering of SME's and the models have raised considerable interest. Further development and validation of the models is proceeding.

Acknowledgements
The research work described in this paper has been jointly funded by the Engineering and Physical Science Research Council under a grant (GR/J/95010) entitled "Specification of a Business Process Re-engineering methodology for Small and Medium Sized Manufacturing Enterprises" and British Aerospace (Systems and Engineering) Ltd.

REFERENCES

AMICE ESPRIT (1989) : CIM-OSA Reference Architecture
CAM-I (1980), *Architect's Manual ICAM Definition Method "IDEF-0"*, Arlington, Texas
CAM-I (1984), *A study program for defining a computer integrated manufacturing (CIM) generic concept for electronic products*, Arlington, Texas
Champy J (1995), *Re-engineering Management*, HarperBusiness, NY
Checkland P (1981), Systems Thinking, Systems Practice, J Wiley & Sons
Childe S J (1991), *The design and implementation of manufacturing infrastructures*, PhD Thesis, Polytechnic South West, UK
Culverhouse P F, 1993, Four design routes in electronics engineering product development, *Journal of Design and Manufacturing*, Vol 3
Davenport T H and Short J E (1990), The new industrial engineering: information technology and Business Process Redesign, *Sloan Management Review*, Summer
Davenport T H (1993), *Process Innovation*, Harvard Business School Press
Hammer M, Champy J (1993), *Re-engineering the corporation*, HarperCollins, NY
Harrington J (1984), Understanding the manufacturing process, Marcel Dekker Inc.
Harvey D (1994), *Re-engineering: The Critical Success Factors*, Business Intelligence
Johansson H J, McHugh P, Pendlebury A J, Wheeler W A (1993), *Business Process Re-engineering- BreakPoint Strategies for market dominance*, J Wiley & Sons
Kane E J (1986), IBM's Quality Focus on the Business Process, *Quality Progress*, April
Lefebvre L A, Lefebvre E, Poupart R (1990), The shape of the new winner: Innovativeness and strategic edge in small firms, *National Productivity Review*, Vol 9 No 3
Meredith J (1987), The strategic advantages of new manufacturing technologies for small firms, *Strategic Management Journal*, Vol 8, pp249-258
Meyer C (1993), *Fast Cycle Time*, The Free Press, NY
Mount J, Zinger J T, Forsyth G R (1993), Organising for development in the small business, *Long Range Planning*, Vol 26 No 5, pp 111-120
Pagoda Associates Ltd (1993), *Pagoda Process Management Report*, London
Parnaby J (1993), *Business Process Systems Engineering*, Lucas Industries plc, November
Porter M E (1985), *Competitive Advantage: Creating and Sustaining Superior Performance*,

The Free Press, NY
Rummler G A & Brache A P (1990), *Improving Performance*, Jossey-Bass
Shapiro B P, Rangan V K, Sviokla J J (1992), Staple yourself to an order, *Harvard Business Review*, July-August
Veasey P W (1994) Managing a programme of business re-engineering projects in a diversified business, *Long Range Planning*, Vol 27 No5
Wilson B (1984), *Systems: Concepts, Methodologies and Applications*, J Wiley
Wortmann J C (1990), Towards One-of-A-Kind Production: The Future of European Industry, *Advances in Production Management Systems*, Elsevier Science Publishers, Holland

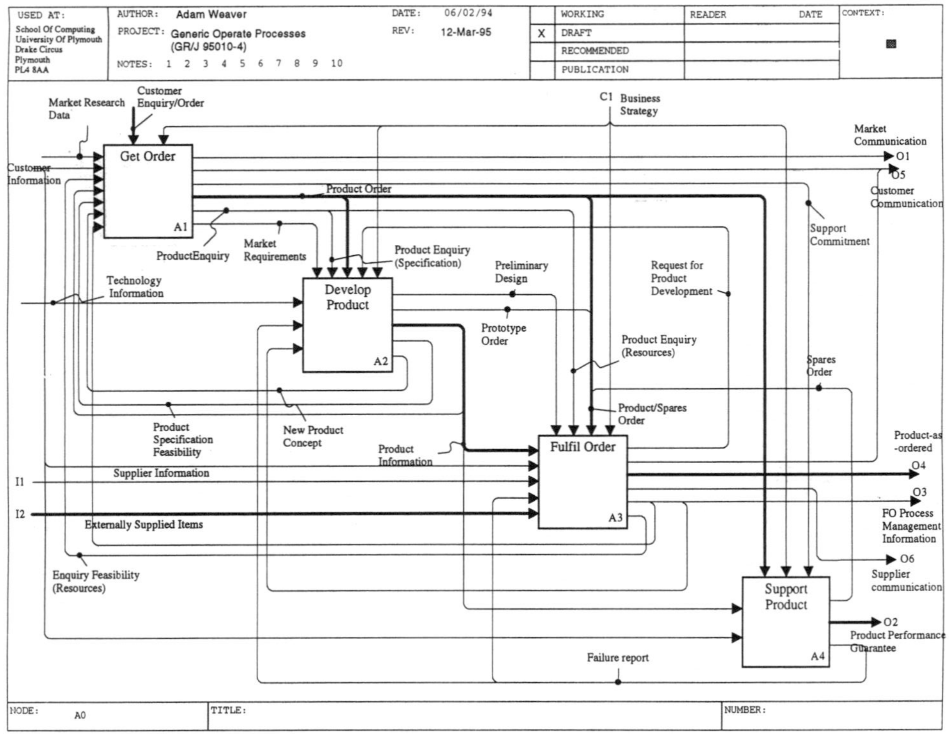

USED AT:
School Of Computing
University Of Plymouth
Drake Circus
Plymouth
PL4 8AA
AUTHOR: Adam Weaver
PROJECT: Generic Operate Processes
(GR/J 95010-4)
NOTES: 1 2 3 4 5 6 7 8 9 10
DATE: 06/02/94
REV: 12-Mar-95
WORKING
X DRAFT
RECOMMENDED
PUBLICATION
READER
DATE
CONTEXT:
Market Research Data
Customer Enquiry/Order
Customer Information
Get Order
A1
C1 Business Strategy
Market Communication
O1
O5
Customer Communication
Product Order
ProductEnquiry
Market Requirements
Product Enquiry (Specification)
Support Commitment
Technology Information
Develop Product
A2
Preliminary Design
Request for Product Development
Prototype Order
Product Enquiry (Resources)
Spares Order
Product/Spares Order
Product Specification Feasibility
New Product Concept
Product Information
Fulfil Order
A3
Product-as -ordered
O4
Supplier Information
I1
O3
I2
Externally Supplied Items
FO Process Management Information
O6
Enquiry Feasibility (Resources)
Supplier communication
Support Product
A4
O2
Product Performance Guarantee
Failure report
NODE: A0
TITLE:
NUMBER:

The authors are members of a research team at the University of Plymouth engaged upon various research projects in the field of systems modelling and design in manufacturing and service industries. Current projects are in the areas of business process re-engineering in SME's, production management, concurrent engineering and manufacturing strategy. Current collaborators include manufacturing companies, software companies, financial institutions and hospitals.

Dr Steve Childe is a Postdoctoral Research Fellow in the School of Computing, University of Plymouth. His research interests lie in the area of manufacturing and business systems, including Business Process Re-engineering and Computer Aided Production Management. Dr Childe is a grant holder on EPSRC Grant GR/J95010 "Specification of a BPR Methodology for SMEs". He is a founder member of the European Operations Management Association and a member of IFIP's Working Group 5.7 (Computer Aided Production Management).

12

Library Based Modeling of Process Chains

G. Zülch and A. R. Fischer

ifab - Institute of Human and Industrial Engineering (Institut für Arbeitswissenschaft und Betriebsorganisation) University of Karlsruhe, D-76128 Karlsruhe, Kaiserstraße 12, Germany, phone: +49 721 608-4250, Email: gert.zuelch@mach.uni-karlsruhe.de

Abstract

This contribution gives an overview of the method Object-Oriented Modeling of Process Chains OMP. It is based on a library of process segments and allows an efficient configuration of complex process networks. The presented library represents a reference model, which is prepared to model the technical processing of customer oriented orders in a small series and one-of-a-kind production system.

After an introduction into the poblem an object oriented concept will be discussed. The important features of OMP will be described. Finally, the influence of the level of detail in modeling will be investigated with the help of simulation.

Keywords

Object-oriented modeling method, reference model for processes, system of predetermined times, modeling on different levels of detail, simulation.

1. INTRODUCTION

While on one hand, modeling means a simplification of the real system, on the other hand it is necessary to include all facts which are necessary to describe the dependencies of the system's entities. As a result, models of very complex systems are still complex themselves, especially if a good model quality is claimed. Thus, modeling means not only simplification and finding the main points of the system but as well requires the ability to handle the complexity of the model. In consequence it should be possible to choose any view on any level of detail which is included in a model representation. As a first step there is the task of creating a (detailed) model.

This paper focuses on the modeling of process networks, that represent the technical work-off of customers orders (see figure 1). The number of tasks in a process chain relates directly to the structure of the requested product, i.e. to the components and parts which have to be adapted to the special requirements.

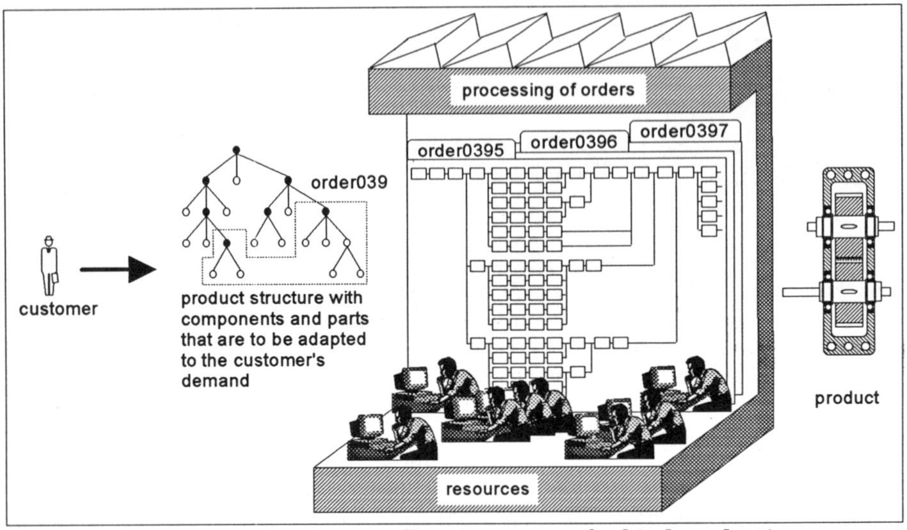

Figure 1 Order processing in a small series or one-of-a-kind production.

Besides creating the process network, i.e. finding all tasks that are necessary to fullfil the order and setting them in a logical relation to each other, it is necessary to guess the task's efforts, especially its operations durations. Both, configuration of the process network and the qualified guess of durations will be supported by the method Object-Oriented Modeling of Process Chains (OMP).

2. OBJECT-ORIENTED MODELING OF PROCESS CHAINS (OMP)

OMP is a general method to configurate complex processes based on a library. The latter includes a number of connected process segments and represents a reference model for processes. The concept of OMP is done under a strong object-oriented paradigm. Its main characteristics are abstraction, encapsulation, combination of data and behavior, inheritance and emphasis of object structure in contrast to procedure structure (Rumbaugh et al., 1991).

The objects of OMP are provided in figure 2. Its upper half includes all objects of the reference model, i.e. the objects for the process-library. The lower half of it includes all objects for the specific process-chains. They will be populated on basis of the reference processes. The symbol diagram (symbols following Rumbaugh, 1991 and Grobel, Kilger and Rude, 1992) shows all associations between the objects, e.g. the connection between *process* and its origin - the *ref-process*.

Modeling a specific order will be done by configuration of process-chains from the library. This task will be realized by the object's features. The procedure of modeling is specified by the complete specification of all objects, i.e. their attributes and their methods. While modeling, the objects call each other by their interfaces. As an example, the *detail* method of *process* is roughly described.

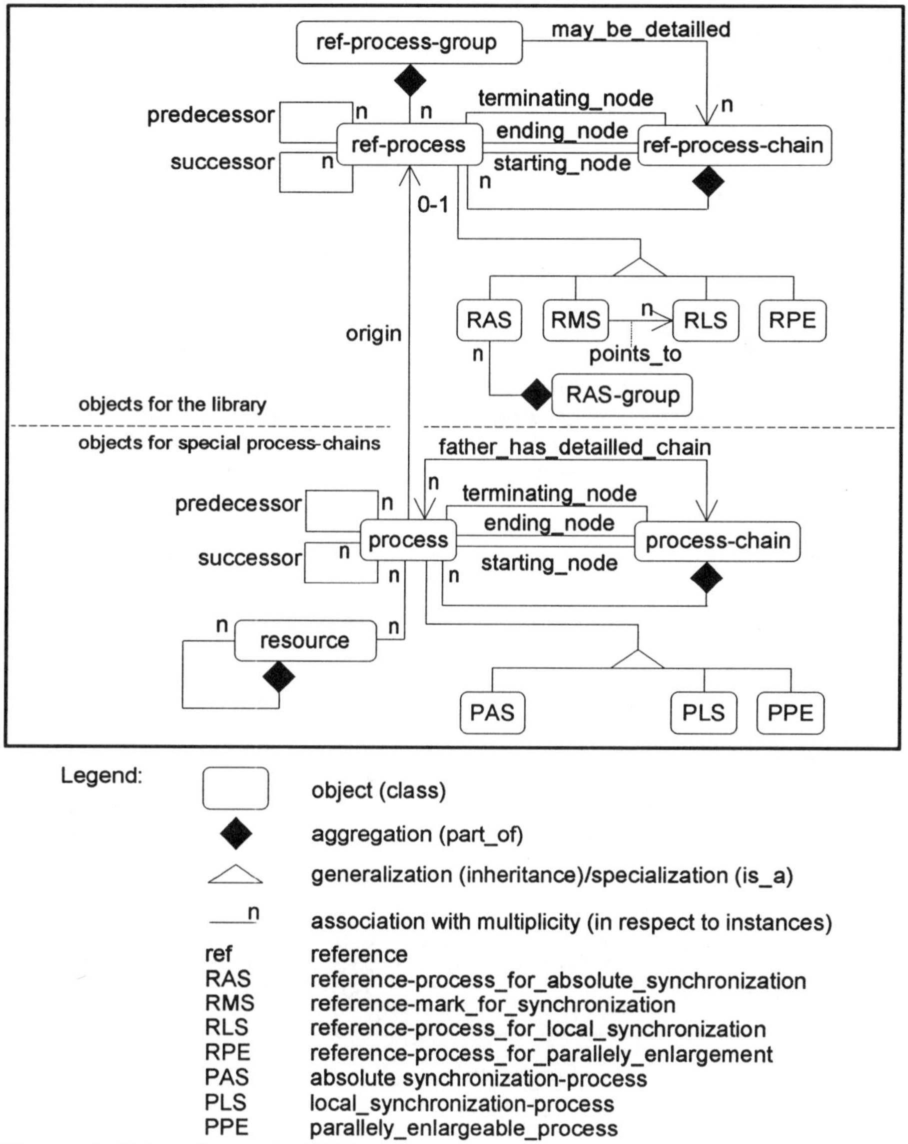

Figure 2 Object classes in OMP.

A *ref-process-chain* and a time level (high, middel, low) are given as input. OMP provides possible chains, which could be choosen to detail the *process*: its origin *ref-process* belongs to a *ref-process_group* and this may be detailled by a couple of *ref-process-chains* (see figure 2). The information which chains are allowed to detail an actual process is represented in the library. Thus, the method works context sensitive.

3. THE LIBRARY - A REFERENCE MODEL FOR ORDER PROCESSING

As shown OMP is based on a library. Its population includes process segments and knowledge of their configuration possibilities. The reference model encloses process chains for the technical processing of customer oriented orders in a small series or one-of-a-kind production system for goods.

A process dictionary is worked out as an elementary basis. Its ref-processes belong to the 32 ref-process-chains of the library LIB. Some of them for example represent the procedure of designing a component C or part P. Modeling a complex process-chain means specifying the customer variable part of a product structure. Table 1 shows an extract of the library. The grey marked ref-processes

Table 1 LIB-C: an extract of the library of process chains.

Processes in Process-chains	*H*	*M*	*L*	*ref-process-group*
LIB-C		R		
determining principle variant for C	30	30	30	
choosing solution principle for C	90	60	30	
determining DWS	90	60	30	
defining claims for adaptable sub-design-working spaces	120	60	30	
defining interfaces for the SDW	240	120	60	
determining purchasing parts	360	240	120	
designing a component	R	R	R	C-SOL, C-ALT, C-REC-SOL, C-REC-ALT;
designing a part	1050	780	285	P-ALT, P-SOL;
designing alternative DWS	R	R	R	DWS-ALT;
integrating moduls in DWS	360	180	30	
checking conformity	660	360	210	CONF_CHECK;
choosing alternative	120	60	30	
...	..	..	..	

Legend:		
	ALT	(design) alternative
	DWS	design-working-space
	H, M, L	plan time high, middle, low
	R	recursive structure is possible; plan time will be dependent from the component structure
	REC	Recursion
	SDW	sub-design-working-space
	SOL	(design) solution

are detaillable - also by alternative chains. The latter ones are fixed by the ref-process-group. There exist as well ref-process-chains which can be detailled recursively (REC), because the process-chain structure occurs similar to the recursive product structure (a component includes again components). Basically, each process has three plan times. Thus, a synthetical determination of planning times goes along with the configuation procedure of a special order.

4. MODELING ON DIFFERENT LEVELS OF DETAIL - INVESTIGATION WITH SIMULATION -

With regards to the level of detail there are two aspects. The first one refers to the level of detail which should be modeled. The second one considers the level of detail which should be used for an application.

OMP allows a very efficient modeling, even if the level of detail is very high. Nevertheless, the developed library includes ref-process-chains for a rough modeling as well. It allows to stop the detailled configuration procedure at any level. Then, specifying the tasks for a component simply include functions instead of processes, e.g. rough designing, detail designing and operations planning. A function includes the tasks of all sub-components and parts which are included with the actual component. The single tasks, e.g. operations planning for part 4717 and for part 0814, can no longer be handled separately. Subsequently, the process orientation will be lost and replaced by a function oriented modeling. A parallel view of the tasks will become impossible as well as a realization of simultaneous engineering. An aggregated process only can be assigned to one organizational unit and an organizational distortion may occur. In addition, it becomes difficult to guess the plan time for a complex task or function.

As a first result it can be summarized that a high level of modeling brings advantages if there exists a method which allows an efficient modeling procedure. The idea is not to reduce the complexity but to give support to handle it with regards to modeling.

The second aspect refers to the application of a process-chain model. There exist a lot of topics. One could be to plan a suitable organizational structure (Zülch and Grobel, 1992; Grobel 1993). Then, the level of detail should be high, because the tasks should be assigned separately and under several aspects to the different organizational units. Only detailled process-chains enable to model a product or process oriented organization in contrast to a functional one.

Another application refers to resources coordination or order control. This means a dynamical processing or simulation of modeled process-chains. As an example an order with the adaption of 2 components is modeled. Each of them contains 1 sub-component and 2 respectively 3 parts. In addition also the design of alternatives is considered and modeled. To investigate the influence of the level of detail a model of the same thing, the customer order, was created on four different ways: (a) with OMP and use of the full specification of the product structure; (b) with OMP and selecting a more aggregated level of detail but with information of the complete specification; that is coming up from (a); (c) with

OMP and the same level of detail as (b) but with chains for rough modeling and therefore, without full information; that is direct specification from the top to a certain specification of the product structure; (d) using (a) and aggregating its processes with respect to the given organizational structure; organization oriented modelling.

A good impression of the different modeling results is depicted in figure 3. It shows how the aggregation of the processes reduces the possibilities of realizing a parallel processing. The models (b) and (c) in relation to (a) aggregate the product structure. The sub-components are no longer broken down. Here begins the transition to a function oriented modeling. For example, the operations plans of the first sub-component's 3 parts can no longer worked out separately. Case (d) is fully aggregated with respect to the organizational structure. The organizational structure consists of 9 departments and is the same for all four process-chain models. Function oriented modeling in this regard means modeling with respect to the given organizational structure. In figure 3, all tasks of stability calculation are done for all parts in a single department (see vertically framed processes). The processes just left to it in (b) and (c) can be done in two different departments: One creates operations plans for rotary parts the other one for cubic parts.

Aggregation can be done under several aspects, e.g. product structure oriented or with respect to a given organizational structure. For dynamical handling and simulation the latter one is useful and mostly necessary. Why that...?

... it is because of a probem with the full detailled process plan. It provides too much freedom for dynamical processing. A simulation program tries processing as parallel as possible (see figure 4). There is no restriction or dynamical rule to solve the problem. Defining rules would bring simulation more and more to a determined processing. This would be the end of simulation.

Processing an order means changing the view from the product structure to the resources. This makes it useful to build packages of the process-chain's tasks which can be made in direct sequence (or parallely) at one department. The problem is how to aggregate the tasks. The criterian may not only depend on the product structure. There is no hint how large a package of tasks should be. Thus, a method for useful aggregation remains as an open point.

The four ways of modeling lead to different system loads: (a), (b) and (d) have the same total load. They all are modeled with full information by the detailled structure from (a), i.e. a synthetic modeling with pre-determined durations is practiced.

An optical impression of the orderprocessing, done with the simulation program FEMOS (Grobel, 1992) is illustrated in figure 4. It is obvious that a higher level of detail goes along with more parallely processing.

A real and valid processing is positioned somewhere in between (a) and (d). The process chain from type (a) allows to process any degree of parallelity and sequency. The remaining problem is the dynamic rule how to control resources and orders.

The level of detail may influence directly and massively absolute characteristics and even twice the goal accomplishments.

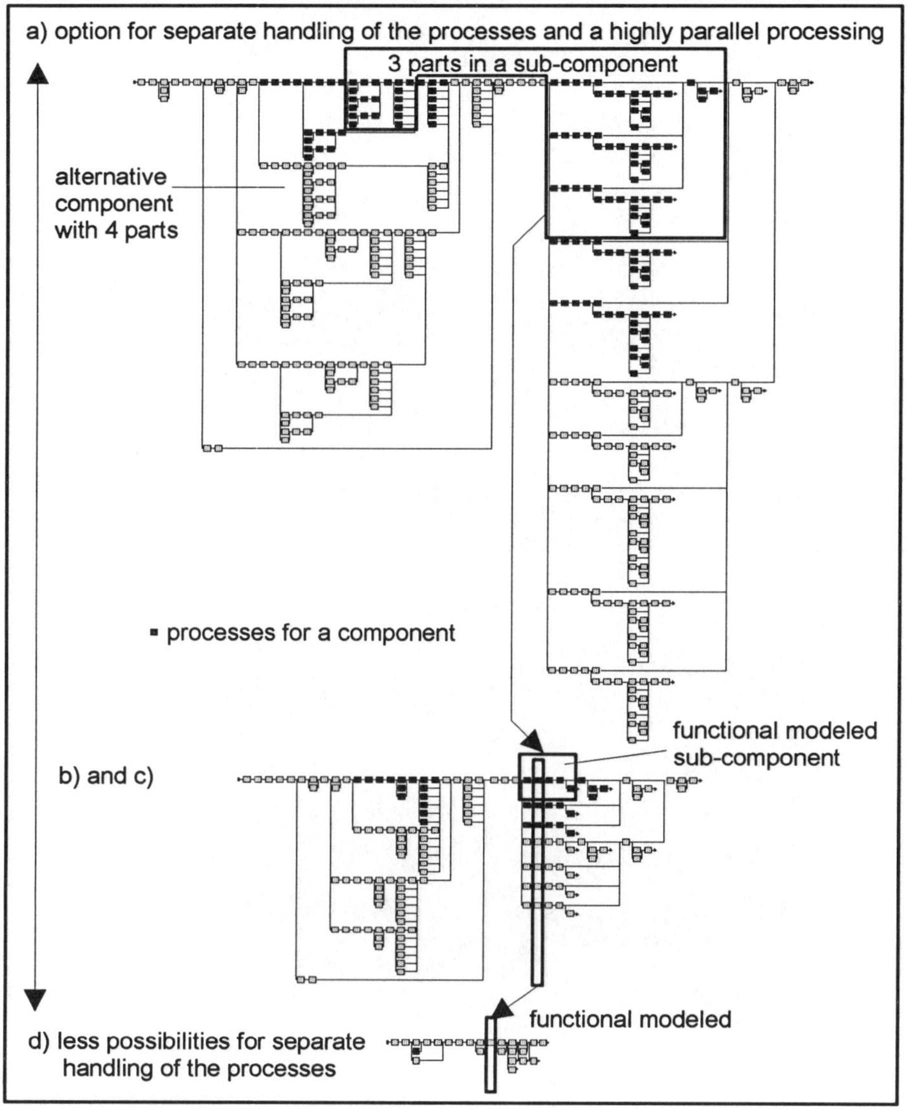

Figure 3 Modeled process chains for the same order

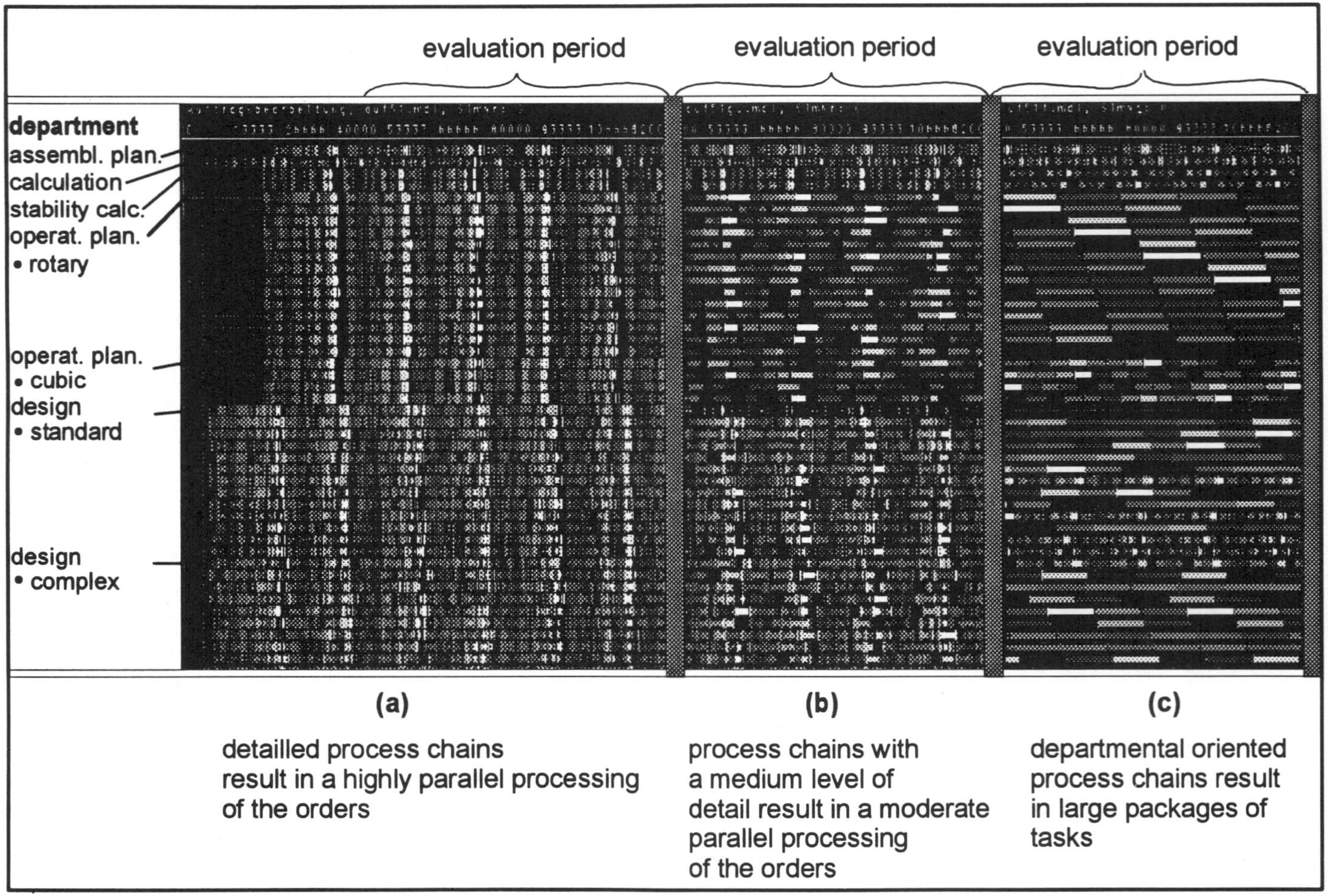
evaluation period
evaluation period
evaluation period
department
assembl. plan.
calculation
stability calc.
operat. plan.
• rotary
operat. plan.
• cubic
design
• standard
design
• complex
(a)
detailled process chains
result in a highly parallel processing
of the orders
(b)
process chains with
a medium level of
detail result in a moderate
parallel processing
of the orders
(c)
departmental oriented
process chains result
in large packages of
tasks

5. SUMMARY AND OUTLOOK

OMP allows a truely process or task oriented modeling of even complex processes. As the example has shown configurating a process e.g. for a component encloses rearly all tasks which are in connection with it. In contrast, function oriented modeling means an aggregation of tasks, mostly with respect to the orgnizational structure.

One of the major advantages of OMP is its context sensitive support for modelling. Thus, it enables a very efficient configuration procedure and provides a representation which includes several levels of detail with respect to the product structure. It supports as well a synthetic determination of plan times with predicted durations of tasks for higher aggregated processes.

The level of detail influences the modeling result and as well the dynamic order processing heavily. The rules for dynamic processing and simulation have to consider this influence. To assure predictability of system's behavior some more investigations are necessary. Simulation systems of today should be improved with respect to a valid dynamic processing of complex process chains.

6. REFERENCES

1. GROBEL, T. (1992). *Simulation der Organisation rechnerintegrierter Produktionssysteme.* Karlsruhe, Institut für Arbeitswissenschaft und Betriebsorganisation, Universität Karlsruhe, ISSN 0940 - 0559.
2. GROBEL, T; KILGER, C. and RUDE, S. (1992). Objektorientierte Modellierung der Produktionsorganisation, in *Information als Produktionsfaktor* (ed. W. Görke; H. Rininsland; M. Syrbe), Springer-Verlag, Berlin et al., 411-419. (Informatik aktuell)
3. GROBEL, T. (1993). *Analyse der Einflüsse auf die Aufbauorganisation von Produktionssystemen.* Karlsruhe, Universität Diss., Institut für Arbeitswissenschaft und Betriebsorganisation, Universität Karlsruhe, ISSN 0940 - 0559.
4. RUMBAUGH, J.; BLAHA, M.; PREMERLANI, W.; EDDY, F. and LORENSEN, W. (1991) Prentice Hall. *Objectoriented modeling and design.* Englewood Cliffs, NJ.
5. ZÜLCH, G. and GROBEL, T. (1992) Simulating the Departmental Organization for Production to Order, in *'One-of-a-Kind' Production: New Approaches* (ed. B. E. Hirsch; K.-D. Thoben), IFIP Transactions B: Applications in Technology, B-2, Amsterdam et al: North-Holland, 177-193.

7. BIOGRAPHY

G. Zülch and A. R. Fischer, Institute of Human and Industrial Engineering (Institut für Arbeitswissenschaft und Betriebsorganisation - ifab), University of Karlsruhe, Germany.

Prof. Dr.-Ing. Dipl.-Wirtsch.-Ing. G. Zülch, born in 1946, studied mechanical engineering at the University of Technology at Brunswich, and Industrial Engineering at the Superior School of Technology at Aix-la-Chapell, Germany. After 10 years of experience in research and industry, he became head of the then newly founded ifab-Institute at Karlruhe University in 1985.

Dipl.-Ing. A. R. Fischer, born in 1963, studied mechanical engineering (production technology) at the University of Karlsruhe, Germany, from 1983 to 1989. Since his graduation in 1989 he has been working as researcher at the ifab-Institute of Karlsruhe University. His research interests mainly concern modeling methods with respect to production systems and its analysis with support of simulation.

13

High Level Process Modelling: A Key Feature of Any BPR Project

A. W. Chan P. Bradley***
** National Research Council (NRC), Ottawa,Ontario, Canada,*
Tel: +1-613-993-9437 E-mail: A_Chan@syslab.nrc.ca
*** CIMRU, University College Galway, Ireland.*
Tel: +353-91-750414 E-mail: Padraig.Bradley@ucg.ie

Abstract

A wide range of software tools are available for use in BPR programmes. This paper debates some of the issues involved in using enterprise modelling tools. Additionally, a specific enterprise modelling tool, and its use in two case studies, is described.

Keywords

Business process reengineering, enterprise modelling, EMS, BPR

1 INTRODUCTION

Business Process Reengineering (BPR) is an approach for achieving dramatic improvements in a company's performance in a relatively short period of time. BPR has helped companies to understand how the different functions or processes in their business are related. Many approaches exist for improving performance in manufacturing (JIT, OPT, etc.) but few approaches offer the opportunity to make dramatic improvements in the non-manufacturing or "white-collar" areas of a company.

Some of the goals of BPR include (Classe, 1994):

- achieving step changes in performance,
- moving from a function to a process based capability,
- emphasising customer focus,
- integrating work,
- developing a process management culture.

High level process modelling is a very helpful, perhaps necessary, first step in any Business Process Reengineering (BPR) project. It provides a global view of how key processes in an enterprise are interrelated, and predicts the potential impact of reengineering lower level processes. Also, it provides a framework for incorporating new process modules and facilitates in the migration from 'As Is' to 'To Be' scenarios as processes are reengineered in an organisation.

2 BUSINESS PROCESS REENGINEERING

Reengineering was a term coined by Michael Hammer in 1990 to describe the process of change that certain organisations were undertaking in order to achieve dramatic performance improvements. The improvements being sought were dramatic and radical in nature and these companies were able to attain significant performance improvements while also totally changing the focus of the organisation. Hammer himself says, "... I never say that I invented reengineering. What I say is that I may have discovered it, which is an entirely different thing" (Barrier, 1994).

Hammer & Champy define reengineering as "the fundamental rethinking and radical redesign of business processes to achieve dramatic improvements in critical, contemporary measures of performance, such as cost, quality, service and speed" (Hammer & Champy, 1993). The key words in this definition, which form the basis of the reengineering approach are:

- fundamental,
- radical,
- dramatic,
- business processes.

Reengineering begins with no assumptions, and the most fundamental (or basic) questions about the companies, and how they operate, need to be asked. Reengineering programs should concentrate on what the company should be doing and should ignore what the company is currently doing. Reengineering is about redesigning the way that the company does business. This involves becoming process oriented. In reengineering, radical redesign means disregarding existing structures and procedures and inventing completely new ways of accomplishing work. This is done by starting with a 'clean slate' and redesigning the processes to reflect these new approaches. Hammer argues for the 'neutron bomb' approach of obliteration and argues strongly against automating the current ways of doing business (Hammer, 1990).

Reengineering is also about achieving dramatic performance improvements (Hammer & Champy, 1993). The levels of improvement sought need to be multiplicative and not fractional (10X rather than 10%) (Davenport, 1993). The dramatic improvements required means that reengineering must be undertaken as an all or nothing proposition. The processes should be redesigned in such a way as to allow these dramatic improvements to be achieved.

Reengineering must focus on redesigning fundamental business processes and not on departments or other organisational units. Most businesses are not 'process oriented'; they are focused on tasks, on jobs, on people and on structures, but not on processes. A business process can be viewed as a set of logically related tasks performed to achieve a

definite business outcome (Davenport & Short, 1990). These business processes have two important characteristics:
1. they have customers, and:
2. they cross organisational boundaries.

As one of the primary concerns of reengineering are business processes, reengineering is often known as business process reengineering, or simply BPR.

Customer order fulfilment is a good example of a business process as the output of the process (i.e. product/service) is required by a customer and the process itself spans several departments in the organisation. BPR involves reengineering these processes to achieve radical improvements in cost, time, responsiveness, performance, quality, etc. These reengineered processes should provide the company with dramatic improvements in cost, response time, and performance, as well as reflecting the company's strategy. The focus in BPR is on 'why' a particular process activity is undertaken.

Processes, not organisations, are the object of re-engineering. All work is performed as part of a process, but often these processes tend to be invisible, fragmented and unmanaged. This is due to the fact that these processes are spread across different organisations with many people working on small pieces of the overall process. No one person manages the overall process and all personnel working on a process identify themselves as working in a particular department but not on a particular process.

3 HIGH LEVEL PROCESS MODELLING

The general requirement that simulation models be specified and validated at a very basic level of detail has been a source of frustration in many modelling exercises. It is often desirable to develop 'quick and dirty' models to obtain ballpark results and to do sensitivity analysis without undertaking a detailed study leading to detailed specifications of the processes in the system to be modelled. Alternative actions can be evaluated at a coarse level, and a small set of selected actions may then be investigated in greater detail.

High level process modelling implies modelling the key processes in an enterprise, including business and manufacturing and/or service processes, by taking a holistic view of these processes to include their interactions with one another and their overall impact on key measures of effectiveness for the enterprise. Hierarchical decomposition, which allows the specification of processes and activities in successively greater detail, is a useful concept in process modelling, and high level process modelling is a useful first step in this procedure. In enterprise modelling studies, the objectives may be to assess the overall performance of the system, to examine the balance between work groups, to study material supply and logistics issues, or to look at overall plant decision making, such as planning and scheduling procedures. The interactions with the outside world and various measures of effectiveness may have to be considered. High level process modelling provides the basis for some activities carried out within the system to be modelled in great detail while others are left in broad terms.

Whereas lower level modelling is often concerned with material flows, high level process modelling is more likely to emphasise the modelling of information flows. High level process modelling should be planned so as to allow progressive refinement of the model over time. This implies assuring that the consistency in the interactions between different segments in the model can be maintained. Boundaries and interfaces must be carefully defined so that one part of the model can be modified without invalidating other parts of the model. In defining the boundary of an object, the consistency of information should be a concern. For example, a department modelled at a detailed level may require data from the same department modelled at an abstract level, or vice versa.

High level process modelling can be part of a top-down or bottom-up approach to process modelling. In the former case, it is part of a process of improved definition of a model. Until a process, activity or function is decomposed into elements whose behaviour can be defined in detail, there will be some uncertainty about the accuracy with which it should be represented in the model. Sensitivity analysis may reveal which parts must be broken down into greater detail to ensure reliability of the results. In the latter case, subsystems which have been modelled in detail and already validated could be aggregated to form a larger model. For example, cell models might be incorporated into a shop floor model; shop floor models could be combined to form a plant model, etc. The details within the sub-model might be hidden in the higher level model by aggregating the summary statistics generated while running the detailed models.

High level process modelling may also involve approximation by treating activities in less detail than in real life. For example, an activity such as *Process Workpiece on Machine X* consists of at least three sub-activities, loading the workpiece into the machining position, machining it, and unloading it from the machining position. Where detailed statistics on resource utilisation are not being sought, the simplification is adequate for modelling the start and finish of processing the workpiece.

4 BPR AND HIGH LEVEL PROCESS MODELLING

Reengineering (or BPR) must begin with a survey of the process landscape to identify processes that are candidates for reengineering (Davenport, 1993). The key activities in identifying a process for re-engineering are as follows:

- Enumerate major processes,
- Determine process boundaries,
- Assess strategic relevance of each process,
- Render high-level judgements of the "health" of each process,
- Qualify the culture and politics of each process.

When the objective is radical process change, a process should be defined as broadly as possible (Davenport, 1993). The fewer and broader the processes, the greater the scope for reengineering success, while the problems of understanding, measuring and changing the process are also greater. Once the processes have been identified at a high level, the boundaries between these processes need to be managed. The following questions may help in defining these boundaries:

- When should the process owner's concern with the process begin and end?
- When should process customers' involvement begin and end?

- Where do processes begin and end?
- Is the process fully embedded within another process?
- Are performance benefits likely to result from combining the process with other processes or subprocesses?

What processes should be chosen for re-engineering? Processes that may be suited for re-engineering can be grouped into three types (Hammer & Champy, 1993), namely:

1. Broken processes,
2. Important processes,
3. Feasible processes.

Broken processes are generally well known and have some common symptoms, which include:

- extensive information exchange, data redundancy and rekeying,
- inventory, buffers and other assets,
- high ratio of checking and control to value adding,
- rework and iteration,
- complexity, exceptions and special cases.

Important processes tend to impact on outside customers. These processes contain points where the customer (on whom the organisation depends) has contact with the organisation. These processes should be customer oriented and capable of delivering the performance required by the customer. Feasible processes are determined by sets of factors that determine the likelihood of success of a particular re-engineering effort. One such factor is cost, as high cost reduces feasibility. Once a process has been selected, it should be properly understood through careful analysis.

Hammer has identified seven principles that should guide any business process reengineering exercise undertaken (Hammer, 1990). These principles, shown below, can be used as a guide in developing high quality 'To-be' process configurations.

1. Organise about outcomes, not tasks.
2. Have those who use the output of the process perform the process.
3. Subsume information processing work into the real work that produces the information.
4. Treat geographically dispersed resources as though they were centralised.
5. Link parallel activities instead of integrating their results.
6. Put the decision point where the work is performed, and build control into the process.
7. Capture information once and at the source.

In any BPR project, it is important to gain an overall understanding of how the key processes in an enterprise relate to one another before focusing in on one or a few processes to reengineer. This is important both for the 'obliterate' approach or the 'migrate' approach commonly used in reengineering exercises. In the former case, the high level process modelling will provide a 'birdseye' view of the philosophy and culture of the company, underscoring any obvious areas where new technology, such as information technology, can play a significant role in reengineering. In the latter case, the high level, enterprise-wide process model will help to highlight the key decision points and bottlenecks in the system, thereby pinpointing processes and process steps which should receive the most attention. In both cases, the value-add of each process step will be demarcated. Not only can non-value-adding steps be identified for elimination, but it can also be predicted where the most impact can be expected when value-adding steps are re-engineered.

Many companies have started to look beyond the four walls of their organisation, and are attempting to integrate their customers and suppliers into their value chains and systems. The interactions between the processes in an enterprise and external agencies, such as customers and suppliers, can be much more appropriately captured with high level process modelling. With the insight gained from these models, BPR can be carried out in such a way that customer viewpoints and supplier system capabilities will be explicitly considered rather than as an afterthought.

5 USE OF A PROCESS MODELLING TOOL

The development of a high level model of the business process of an organisation can help in understanding how the processes relate to each other. The activities in the process, the process boundaries, the resources used in the process all need to be understood. Due to the amount of detail, and the number of iterations involved in developing a reasonably detailed high level model, the use of a modelling tool offers specific benefits. A properly chosen process modelling tool satisfies the following criteria (Davenport, 1993):

1. It is fast and easy to use at a high level;
2. It is applicable to the portrayal and analysis of the new process, enabling new and old processes to be compared in the same formats and perhaps even driven by the same set of simulation variables;
3. It provides not only a descriptive, but also an analytical, model of the process, facilitating an understanding of such factors as time, cost, and other resources consumed by the process; and
4. It supports the addition of successive levels of systems and data-oriented detail, enabling it ultimately (and seamlessly) to serve a useful purpose during the systems design and/or prototyping stages.

An integrated modelling tool that will permit modelling the processes in an organisation at various levels with flexibility is thus very desirable. Ideally, the same tool can be used to model processes at a very high level, giving a global view of how processes relate to one another, as well as to model low level processes in great detail, allowing these processes to be finetuned for maximum operational efficiency.

The Enterprise Modelling System (EMS), jointly developed by the National Research Council of Canada and a consortium of companies, is an example of a modelling tool available to build multi-level models of processes found in an enterprise (Chan, 1994). It can facilitate the establishment of a business or manufacturing strategy for an enterprise. This should be a dynamic process which allows periodic updates to the strategy by taking into account the changing conditions of external variables, such as competition, consumer taste, economic environment, and unpredictable occurrences in the marketplace. The EMS provides special features to model processes from an overall enterprise-wide level, to processes distributed in multiple plants at different geographical locations, to processes spanning different departments in a plant, to manufacturing processes found in various manufacturing cells in a department, and all the way to control processes for manufacturing and material handling equipment within a manufacturing cell. Not only can process models be created at the various levels in the enterprise hierarchy, but a combination of process models of varying degrees of details can be set up within the same framework, and made to run together by passing the output of one model as input to another model. This is particularly useful when the user would like to home in on the details of one process, while maintaining a high level view of the overall picture.

The EMS permits a range of levels of abstraction in model building by providing high level constructs (templates and modules) rather than restricting the user to basic block commands. A single organisation unit may represent a plant or warehouse in a high level model, and may be exploded into a whole set of resources and activities at a lower level. Common measures of performance which are tracked by the EMS are time, cost and value. A breakdown of the time, cost and value for each subprocess or activity in the definition of a high level process is available to the user. Whereas *time* is based on the estimated duration of an activity specified by the user, *cost* is computed based on a fixed cost and a variable cost, which in turn is derived from the unit cost of a resource. *Value* is assigned by the user to reflect the value an activity is deemed to be worth to the customer, providing a basis to examine the value added by each component in a value chain. In all cases, either fixed values, values within a range, or values based on statistical distributions can be used. These measures of performance provide the user with a high level view of the expected impact of process changes, facilitating decisions to focus in on specific subprocesses or activities.

6 TWO EXAMPLES

The Enterprise Modelling System has been used in the case study of a multi-plant manufacturer for the high level process modelling of the order fulfilment process. Key processes corresponding to functions in marketing, operations, receiving, shipping, production and stores have been modelled at a coarse level. This has provided not only a useful documentation for the current processes in the organisation, but also a basis for Business Process Reengineering studies in the future, and for selecting processes for in-depth review.

An example of a high level process in the *Order Fulfilment* process definition is the *Fill Materials Order* process. This process is broadly broken down into five subprocesses as shown in Figure 1. The *Receive Materials Request* subprocess has the resource Stores Supervisor receiving a Materials Request from either the production shop floor or from another plant in the enterprise. The next subprocess, *Generate Move Documentation*, has the Stores Supervisor taking the input Materials Request and transforming it into the output, Materials Move Documentation. The next subprocess, *Retrieve Materials*, involves taking the Materials Move Documentation to the Storeman and deciding whether

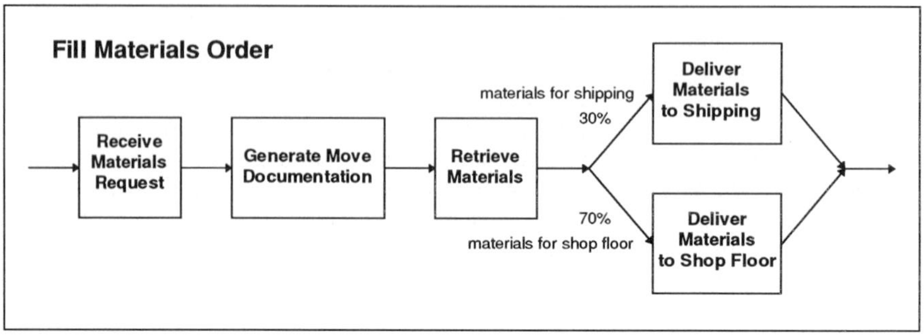

Figure 1 Example of a High Level Process.

the material requested is for the shop floor or for another plant. Depending on the nature of the request, the Storeman transfers the materials requested to the proper destination, either the shipping department or the production shop floor. Rather than modelling the attributes of each Materials Request in detail to determine which branch in the process definition should be taken, the high level process model simply assigns a probability for each of the ensuing branches (see Figure 1). Subsequent refinement of the model may warrant modelling the branching decision in greater detail.

While resources employed in the execution of the *Fill Materials Order* process, such as Stores Supervisor, Storeman, Shipping Staff, etc. have been specified with their respective position in the organisation structure, unit cost, capabilities, etc., the high level process model is not designed to track the utilisation of any individual resource. In other words, these resources are treated in an aggregate manner, with one designation representing possibly a whole group of similar resources, and their individual identities are not explicitly modelled. With the help of overall summary information such as time and cost charts (Figure 2), the user will get a high level view of the relative impact of changes made to each subprocess. This will form the basis of re-engineering alternatives, as well as decisions to go to lower levels of detail for selected subprocesses, such as the Retrieve Materials subprocess. In such a case, the user would be concerned with the actual retrieval of the requested materials, and the logistics of transporting to the proper location, possibly including the use of materials handling devices, and so on.

FirstSTEP, a commercially available business process re-engineering tool based on the EMS technology, has been adopted by a major financial institution to model high level processes in the Global Custody Services it offers to its clients in both mature and emerging markets throughout the world. Key components of these services include asset safekeeping, trade settlement, income collection, corporate action processing and accounting. High levels of customisation and exception processing are involved, which in turn lead to high levels of customer contact and inquiry processing, requiring the definition of complex support processes.

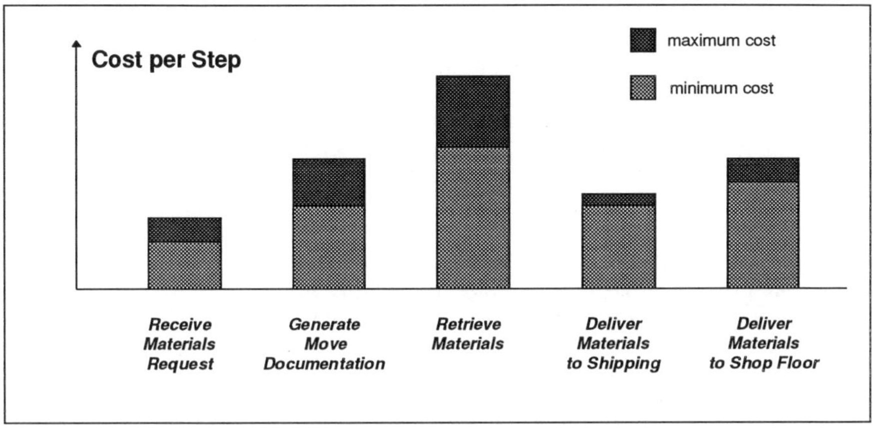

Figure 2 Example of a Process Cost Chart.

A high level model of the current processes is being constructed using FirstSTEP, facilitating the forecast of capacity requirements and the identification of process bottlenecks and streamlining opportunities. Impact analysis of re-engineered processes and possible enhancements to the system are being carried out within the safety of a 'virtual' environment. Other areas of investigation planned using the model include bottlenecks caused by redundant controls, backlogs resulting from rigid work distribution rules, and the elimination of 'black holes' or unnecessary hand-offs. It is expected that the high level process model will provide a valuable resource for future re-engineering and capacity modelling activities within the Customer Service Unit in the company.

7 Conclusion

In general, modelling the processes found in an enterprise at a high level will provide useful insight into the relationships between the key processes. As well, it will help to highlight the bottlenecks so that they can be the focus of attention in process redesign. The cost effectiveness of reengineering certain processes can also be assessed, based on the relative cost and time impact of various processes on the organisation. The high level process model developed can provide a framework whereby modules of individual processes with much greater detail can be plugged in, and the overall impact on the company can be studied. The use of a flexible and integrated process modelling tool will greatly facilitate this exercise.

8 References

Barrier, M.(1994), Re-engineering Your Company, *Nation's Business*, pp: 16-22.
Chan, A.W.(1994), Enterprise Modelling, *Modern Manufacturing: Information Control and Technology*, (ed. M. Zaremba, B. Prasad), Springer-Verlag, London.
Classe, A. (1994), Software Tools for Re-engineering, *Business Intelligence*, London.
Davenport, T. H. (1993). *Process Innovation: Reengineering Work through Information Technology*, Harvard Business School Press, Boston.
Davenport, T. H., Short, J. E. (1990), The New Industrial Engineering: Information Technology and Business Process Redesign, *Sloan Management Review*, pp: 11-27.
Hammer, M. (1993), Reengineering Work: Don't Automate, Obliterate, *Harvard Business Review*, pp: 104-112, Boston.
Hammer, M., Champy, J.(1993), *Reengineering the Corporation: A Manifesto for Business Revolution*, Harper Business Press, New York..

9 Biography

Albert W. Chan is a Senior Research Officer in the Institute for Advanced Manufacturing Technology at the National Research Council of Canada. He received a Bachelor of Science in Mathematics from the University of Minnesota, and a MSc and PhD in Industrial Engineering from the University of Florida. His research interests include simulation and computer modelling, decision support systems, and process modelling and re-engineering of business and industrial systems.

Padraig Bradley is a Senior Research Engineer in the Computer Integrated Manufacturing Research Unit (CIMRU) at University College Galway. He has obtained a Bachelor and Masters degree in Engineering from University College Galway. His research interests lie in the areas of manufacturing systems, knowledge based systems and business process reengineering.

14

Investment-based Analysis of IT-enabled Business Process Reengineering:

A Case Study within Engineering Data Management

Theo J.W. Renkema & Tom Dolan
Department of Information and Technology
Graduate School of Industrial Engineering and Management Science
Eindhoven University of Technology
The Netherlands
tel/fax+31-40-472290/451275
tre@bdk.tue.nl; or *tdl@bdk.tue.nl*

Abstract

IT investments are being increasingly undertaken as part of larger business process reengineering (BPR) exercises. The reported failure rates of such BPR projects are typically high. Effective management of the associated investment decision is, therefore, even more important for the success of the project and the business itself. A framework for the evaluation of such investments is presented, which extends existing appraisal techniques to account for important organisational factors. An industrial case study concerning the introduction of an Engineering Data Management (EDM) system is presented and analysed in terms of this framework. It is shown that the framework is amenable to the structuring of such an investment decision. The paper concludes with some recommendations for future extensions to the research.

Keywords

Business Process Reengineering, IT investments, Investment evaluation, Engineering Data Management (EDM)

1 INTRODUCTION

Business Process Reengineering (BPR) has emerged as an influential theme in the industrial engineering and information systems disciplines. Many organisations are embarking on BPR initiatives or have an interest in doing so. Recent surveys show for instance, that around 60% to 70% of the responding organisations are already involved in BPR programs (Klein, 1993; Butler 1994). The reported failure rates of such BPR projects are, however, typically high. Hammer and Champy (1993) estimate that up to 70% of BPR projects fail. Belmonte and Murray (1993) state that 'statistics show that less than 45% of companies that try business process redesign are successful at achieving their intended goals'. The situation that many firms face is perhaps best expressed by Douglass (1993) when arguing that 'Business reengineering is a long, expensive process; most organisations underestimate the time and cost involved. But the payoffs can be staggering. In fact, reengineering may mean the difference between survival and extinction down the road.'

This paper contends that a more rigorous investment-based analysis of IT-enabled BPR projects will provide one step forward in ensuring that BPR delivers value for money. A framework for the analysis and management of BPR-investment evaluation is proposed and applied in the analysis of a case study in Engineering Data Management (EDM). The paper is structured as follows. Section 2 introduces and delineates the concept BPR, and provides our view of a BPR investment as a specific type of IT investment. The third section then goes on to sketch the problems and methods in evaluating IT investments. Building on these two sections, the fourth section proposes a framework for investment-based analysis of BPR. This framework is then applied in a case study of EDM in section 5. The final section concludes with an assessment of the expected contribution of the proposed framework and some future directions of research in the realm of investment-based analysis of BPR projects.

2 BUSINESS PROCESS REENGINEERING AND BPR INVESTMENTS

2.1 BPR: its content and dimensions

There is little agreement in the literature of what BPR exactly is or should be. Even the name it is given differs from author to author. BPR has, for instance, also been called Business Engineering (Meel et al. 1994), Process Innovation (Davenport 1993) and Core Process Redesign (Kaplan and Murdock 1991). Grover et al. (1993) identify several common elements in the numerous definitions which are typical characteristics of BPR:

- The radical redesign of business processes;
- The typical employment of IT as an enabler;
- The attempts to achieve organisational level strategic outcomes.

In order to get more in-depth insight into the characteristics of different BPR approaches, Jones (1994) distinguishes between five dimensions of BPR (summarised in Table 1), based on:

- **The scope of the main processes** (the organisational level of the analysis focus);
- **The scale of change** (incremental or radical);
- **The means of achieving change** (systematic or inspirational);
- **The source of the change model** (the underlying concepts of the approach);
- **The role of IT** (the contribution of IT)

Author	Scope of main processes	Scale of change	Means of achieving change	Source of change model	Role of IT
Hammer	Organisation	Radical	Inspiration	Novel	Essential enabler
Davenport and Short	Organisation /dept./task	Incremental	Systematic	Industrial Engineering	Recursive relationship
Harrington	Department/ task	Incremental	Systematic	Kaizen	Automation (secondary
Kaplan and Murdock	Organisation	Radical	Systematic	TQM	Key enabler
Business Intelligence	Department/ task	Incremental	Systematic	Kaizen	In theory, not necessary
Davenport	Organisation	Radical	Systematic	TQM + IE + 3 others	Primary enabler
Johannson et al.	Organisation	Radical	Systematic	TQM	One of several 'Break-Points'
Morris and Brandon	Organisation /dept./task	Incremental	Systematic	Industrial Engineering	Not necessarily based on IT

Table 1: Characteristics of different BPR approaches (taken from Jones 1994)

2.2 What is a BPR investment?

From the latter section it can be concluded that the currently available BPR approaches still exhibit a large variety on several dimensions. Essential common elements however emerge in the proposed approaches. These common elements provide the foundation on which this paper's view of a BPR investment is based.

A BPR investment is regarded as a long-term commitment of organisational resources in order to achieve ambitious business goals through (radical or incremental) transformation of business processes with IT as an enabling technology.

As such we see a BPR investment as a specific type of IT investment. Traditionally, IT investments were used to simply automate existing business processes, in order to increase efficiency by internal cost savings and cost avoidance ('automating IT investments'). A BPR investment however stresses the recursive relationship between IT and business processes (see Figure 1) with ambitious business goals in order to improve organisational effectiveness and to gain competitive advantage ('transformating IT investments')[1].

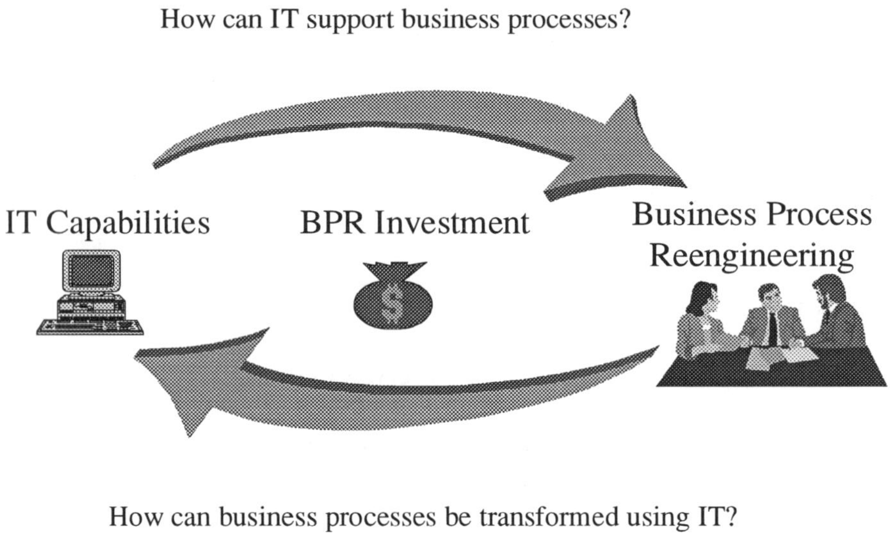

Figure 1: BPR investments and the recursive relationship between IT and business processes (adapted from Davenport and Short, 1993)

3 EVALUATING AND MANAGING IT INVESTMENTS

3.1 IT investment evaluation in organisations

Our view of BPR as a 'transformating' type of IT investment allows us to use existing insights and theories with respect to the evaluation of IT investments. Evaluation of IT investments is not a new problem area in the information systems discipline. Already in 1961 the IFIP devoted

---

[1] IT investments aimed at the automation of existing business processes can also yield benefits beyond efficiency. Improved efficiency is, however, the initially expressed aim of these investments.

its first conference to evaluation issues (Frielink, 1961) and in 1968 Joslin wrote his book on computer selection (Joslin, 1968).

The evaluation and management of IT investments has, however, gained the renewed interest of both management and academics. Throughout the last thirty years, IT has come to play a central role in organisations and consequently the sums of money invested in IT have increased considerably. It is estimated that large organizations spend up to 50% of their annual capital expenditures on IT (Farbey et al., 1993). Senior management are, therefore, unreceptive towards 'act of faith' investment decisions, and seek more robust evidence for the cost-effectiveness of proposed investments. Several empirical studies of IT investment decision making practices have recently been conducted, (e.g. Farbey et al. 1993; Willcocks and Lester, 1993). These studies indicate that although many organizations have formal finance-based justification procedures in place, actual decision-making practice is often unstructured and of an ad-hoc character. Also the many non-financial methods for IT evaluation do not seem to be employed in decision-making practice.

4 INVESTMENT-BASED ANALYSIS OF BPR

This section proposes a framework for the analysis and evaluation of IT-driven BPR investments that draws on and extends existing methods. The framework is grounded on our contention that the ultimate aim of any BPR investment evaluation is to improve and facilitate decision-making on BPR investments. This has lead us to believe that we need an explicit view of BPR investment evaluation as a decision-making and largely communicative process. To develop this view a model of BPR investment evaluation was designed. Four aspects are distinguished that can all four be used to improve the investment evaluation and to manage the underlying decision-making process. Figure 3 summarises the four aspects, while the remainder of the section elaborates them further.

4.1 The product of the BPR investment evaluation

At the heart of the framework lies the product of the investment evaluation, i.e. the set of arguments (evaluation or decision criteria) on the basis of which the decision whether to invest or not is made. Every investment decision is made against the background and judgement of advantages, disadvantages and risks. It is best to make these as explicit and debatable as possible. These investment arguments can either be financial or non-financial advantages and disadvantages, in addition to possible risks involved (see table 2). These arguments also offer a language to communicate the implications of the investment and to increase commitment to the decision made.

The many existing methods for IT investment evaluation (see section 3) focus on the *product* dimension of investment decision-making in order to arrive at a 'go/no-go' decision. Generally they prescribe a rigid set of investment arguments, without proper recognition of the many local business goals and priorities in an organisation. Table 2 provides a more general structure for a set of investment arguments. It can be used to derive locally based evaluation criteria, using the methods for IT investment evaluation as an input. These criteria can then be graphically presented to the different stakeholders in the evaluation by means of a multi-criteria, ratio or portfolio approach. A further treatment of this process is given by (Kusters and Renkema 1994)

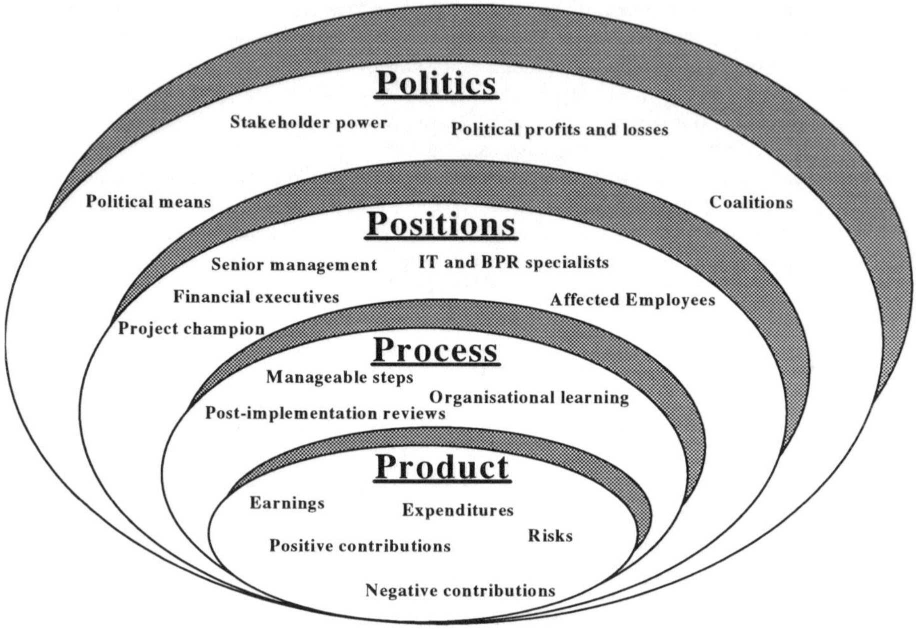

Figure 2: Framework for BPR investment evaluation.

Arguments	**Positive**	**Negative**	**Risks**
Financial	Earnings: - unique - recurring	Expenditures: - internal - external	Cost of capital Hurdle rate
Non-financial	Positive contribution	Negative contribution	Development risks External risks

Table 2: The product of the investment evaluation (adapted from Kusters and Renkema, 1994)

4.2 The process of the BPR investment evaluation

The second aspect of the framework shown in Figure 2 refers to the process of the BPR investment evaluation. This process considers the different phases the evaluation goes through;

both prior to, and during, project execution. Important recommendations for decision support lie in:

❑ Decomposing investment decision-making into manageable steps, analogous to well known decision-making models (Harrisson 1987):

- problem statement, formulation of project goals;
- evaluation of alternatives;
- choice of investment alternative;
- implementation of the chosen solution.

This subdivision of steps is not meant as a linear and rigid procedure, but more as a pattern of thought, with possible loops.

❑ Performing post-implementation reviews of the investment decision to monitor and control the investment across its life cycle. These reviews provide valuable information on whether the investment actually delivers value for money and to what extent there is still room for improvement. These post-implementation reviews can also be used to establish an investment climate in which organisational learning is encouraged. Many investment decisions are made by 'jumping from one project to the other'. Explicit knowledge of prior investments and their realised value can very much contribute to improved decision-making. Regular reviews of the investment also minimise the phenomenon of 'investment entrapment' (Dinther, 1993); a situation in which ever-greater resource commitments are made because of too much emotional involvement, without sound evaluations of increased investments.

4.3 The positions of the different BPR project stakeholders

The third aspect that receives special attention in Figure 3 concerns the positions of the different BPR project-evaluation stakeholders. It is advisable to involve all appropriate people in decision-making with respect to the BPR investment. These include:

❑ Senior management
As discussed in section 2.2, BPR investments are aimed at achieving ambitious business goals. Such goals cannot and should-not be striven for without the support of senior management.

❑ IT and BPR specialists
BPR investments are highly dependent on the expertise of specialists in the field of IT in general, and in the field of BPR in particular. Many BPR projects are, in fact, initiated by the IT department (Douglass 1993).

❑ Financial executives
Although BPR initiatives have many consequences that cannot easily be expressed in financial terms, it is worthwhile to attempt the estimation of financial consequences as accurately as possible. Financial executives typically have experience in this area, and in addition their analytical rigor can also be used for the estimation of non-financial consequences and for risk analysis.

❑ Employees whose work processes are reengineered
Every BPR investment is aimed at the reengineering of some work process, which may often be performed by highly professional employees. Their knowledge of the candidate processes is

indispensable in deciding what the precise consequences and risks of the BPR project are. Also, their commitment to performing the reengineered work is a prerequisite for achieving value for money from BPR.

❑ Project champion

It has been shown (see e.g. Farbey et al., 1993) that the likelihood of success of investment projects is considerably improved when there is a 'project champion' involved. This championship refers to the special effort that is made by some involved stakeholder to make the BPR effort a success. This stakeholder does not necessarily have a formal role that implies such an effort. The more powerful this champion's position in the organisation is the better.

4.4 The politics of the BPR project

The previously discussed aspects of the framework sketched a homogeneous, rational picture of an organisation. This view implies for instance that the different stakeholders in the investment evaluation share the same intentions, goals and priorities (in terms of Lammers (1983) the 'system model'). A more realistic view is that of an organisation in which different stakeholder groups have their own wishes and preferences (the 'coalition model'; Lammers, 1983). Such a view allows for the recognition of conflicting interests and the use of political means to safeguards one's interests (Peffer, 1981; Mintzberg, 1983).

Decision support with respect to the politics of the BPR evaluation lies in what has been called 'stakeholder analysis'. Boonstra (1991) suggests the following steps in such an analysis:

1. Listing of stakeholders, their estimated power and impacts of the proposed investment;
2. Assessment of possible 'winners' and 'losers' and their possible (political) 'profits' and 'losses';
3. Establishment of feasible strategies (e.g. financial compensation) to influence the political account of profits and losses.

5 THE INDUSTRIAL CASE STUDY

We have applied the investment framework described above to a real-life industrial project (an Engineering Data Management project) in order to:

- verify its applicability in an industrial context;
- investigate limitations which will become the subject of further research.

Engineering Data Management (EDM) was chosen as it represents a typically industrial, large-scale, IT investment which has possibilities to not only automate, but also transform the underlying business process. These possibilities exist because EDM provides an organisation-wide communications infrastructure. The infrastructure is not enough, however, to ensure improved development and production, that requires complementary process reengineering.

EDM implementation entails formalism (of procedures) and structure (of documents). It demands that previously ad hoc work practices must be examined and perhaps re-designed, that informal codes of behaviour regarding documentation must be analysed and modified to fully avail of the EDM functionality. This close relationship between IT and the underlying process to be supported is critical to the successful implementation of EDM (or indeed any major IT investment). It is widely accepted that the most serious barriers to successful IT implementation are not technological but organisational in nature (Scott Morton, 1991) and any serious IT/BPR project should account for this fact.

Unless the entire development process is re-examined in the light of the new possibilities, and unless management and users are committed to re-engineering those processes which hinder optimal IT exploitation; the project will fail. Similarly, the IT product (in this case an EDM System) must support good design in all its organisational aspects, in an integrated way.

This realisation is the cornerstone of IT-enabled BPR, and also of the investment appraisal framework presented in this paper, which regards the IT investment from both a technical and organisational perspective. These themes will be expanded in the case description which follows in section 5.2, after a brief description of the case-study site.

5.1 Introduction to the site:

The site for the case in the headquarters of a world-wide manufacturer of high technology medical diagnostic equipment employing 10 000 people. The site is the main centre for both design and manufacturing, and the development organisation is composed of business units[2] related to major technical application areas. The company has a waterfall-based development model (see Figure 3), supported by a manual documentation system for project management and review.

phase 1	phase 2	phase 3	phase 4	phase 5
Feasibility	Overall Design	Detail Design	Integration - Test	Start Production

Figure 3: the phases in the product development process

This company is the leader in the highest end of the market, but recognises that the industry is entering a very dynamic period and wants to prepare its development activities to meet the competitive demands. In particular; timing and costs will become much more critical as success factors, as will the provision of software to support the entire application environment. These pressures are resulting in a development process which must deliver a more complex product in a shorter timescale, this was the impetus for the EDM investment. It is also worth noting that one of the 'selling points' of the project was the fact that EDM would *enable* an improved development process (there were even estimated cost savings presented). In fact the view held by the project proposers (the IT department) was that EDM was an infrastructure product which would facilitate the re-design of the development process (*i.e.* IT-induced BPR) to improve productivity.

2 They are not business units in the strict sense as sales responsibility is held at regional level.

5.2 Applying the framework to the EDM case study

The aim of this paper is to investigate the suitability of the evaluation framework described, to an industrial, BPR scenario. To do this we have taken a current EDM implementation project; investigated the evaluation process; and structured it in the context of the framework (the four P's).

❑ **The product of the BPR investment evaluation**

As discussed in section 4.1, the product of the investment evaluation refers the 'go/no-go' decision and the set of investment arguments that is used to arrive at this decision. This EDM investment project was justified on the need to improve the product development process in order to meet increased competitive demands. Improvement in this respect lies in delivering a more complex product in a shorter timescale, against lower costs. The more efficient management of product and process data was regarded an important prerequisite for this. One of the important 'selling points' of the EDM project was that it would enable an improved development process, where improvement was particularly visible in the estimated cost savings. EDM was considered an infrastructure product which would facilitate the redesign of the development process.

The estimated cost savings were projected in the report of a large industrial consultancy firm, commissioned to assess the feasibility of implementing the EDM system. These consultants organised a workshop with several developers, which investigated current practice and identified the core problems with the development process. Estimates were made of the cost savings possible from the removal of these problems by EDM, and these cost savings outweighed the financial expenditures considerably. As such, the report sketched a picture of an investment with only limited risks. The more qualitative, strategic benefits emerged later on in the investment evaluation.

There was also the choice of the software package in order to be able to implement the proposed EDM system. A 'buyers guide' of available EDM software packages (which was commercially available) was used the get an initial idea of the available packages. The choices between the three final candidates (two external packages, one package developed by a sister company) was made on the basis of functional and technical requirement and intuitions concerning the packages.

❑ **The process of the BPR investment evaluation**

As with many BPR projects, the EDM investment idea was launched from the IT department. The department foresaw major advantages in the use of a uniform EDM software tool. Internal triggers within the IT department were the limited capacity of the existing data-management tools and their lack of uniformity. Several years before, a comparable investment initiative was proposed by the IT department. This project was far from successful. Perceived causes for this lie in: technological immaturity, organisational immaturity and the fact that the then proposed system was installed on a mainframe computing environment. This centralised structure highly contrasted with the professional, autonomous, working environment of the product developers.

The IT department put great effort into 'selling' the investment idea to development and management. The projected improvements in the development process gradually turned their efforts into a success. The management team established an investment budget and decided to install a 'study committee' to investigate the possibilities of EDM implementation. This committee contracted a large industrial consultancy firm to explore the feasibility of EDM. In the light of the consultants' findings, the committee then proposed to invest in the system and management decided to go ahead with the EDM investment. A project committee was installed to guide the implementation of the system. It was also decided to implement the system gradually, through pilot projects. At this early stage of the project, no post-implementation of the investment decision has taken place.

❑ **The positions of the BPR investment evaluation stakeholders**
The IT department played a central role throughout the investment decision-making process. It proposed the EDM investment, and its expertise was used in the continuation of the investment. The support and attention of the management team was also seen as an important factor. One member of this team put a lot of personal effort into the success of the EDM project. He is seen as a good candidate for the title 'project champion'. The study committee consisted of representatives from the several involved departments. This committee collaborated with the aforementioned consultancy firm. The project committee that is responsible for the ongoing system implementation phase comprises a representative of the IT department, the engineering department and product development. This committee has also consulted another plant of the case study organisation (in Germany) to ensure uniformity in their respective approaches to EDM.

❑ **The politics of the BPR investment evaluation**
Several issues can be mentioned that relate to the politics of the investment evaluation. The political atmosphere was very much influenced by the different backgrounds and perceptions of the IT department versus the product developers. This can be regarded as a typical example of the classical conflict between 'IT' and 'users'. Political issues resulted from the rather long distance between the management of the IT department and the management team. Their communication mainly took place via separated hierarchical channels. A final point worth mentioning is that the focus on the justification through cost savings can be traced back to the cost consciousness of the Manufacturing Engineering department in the 'study committee'.

5.4 Analysis of the case study

The EDM case raises some diagnostic questions with respects the way BPR investment evaluation was carried out. These can be seen as points for improvement with respect to future BPR initiatives in the case study organisation.

❑ The **product** of the investment evaluation, i.e. the 'go/no-go' decision and its constituting investment arguments, was chiefly oriented towards projected cost savings. The more strategic, and possibly more essential, investment arguments emerged later on in the evaluation. This concentration on cost savings more or less stood in the way of a well-founded and explicitly communicated strategic investment orientation, accounting for both financial and non-financial opportunities, drawbacks and risks.

- The **process** and **positions** of the BPR investment evaluation were fairly straightforward, including the project committee's and senior management's support (the role of the 'project champion' in the latter group was seen to be very beneficial). An important role was played by an external consultancy firm, who predicted many efficiency gains. The focus of external legitimatisation of the investments might, however, have been inhibiting internal commitment to the decision made. It is not fully clear to what extent organisational stakeholders felt that were the investment would indeed be worthwhile or were told so.

- The **politics** of the investment evaluation reflects the often reported divergence between the work climates of the IT department and end-users. As such, this observation implies that within the case study organisation, room for improvement lies in removing cultural barriers between the two groups. This claim, however, also holds for the information systems discipline in general. A further observation regarding the politics of the investment evaluation is the seemingly over-involvement of the people that were particularly cost-consciousness. This fuelled the already cost-driven investment justification, possibly at the cost of the interest of other stakeholders who would benefit most, i.e. the product developers.

6 CONCLUSION

IT investments are increasingly undertaken as part of larger BPR exercises. The reported failure rates of BPR programs are however typically high. The management of the associated investment decision has therefore become more important in ensuring that the BPR investment delivers value for money. This paper argued that a more rigorous, investment-based analysis of BPR projects will provide one step forward in achieving this. A framework for the analysis and management of BPR investment evaluation was proposed and used to examine a case study in Engineering Data Management.

It has been shown that the proposed framework can be used to analyse BPR investment evaluation both in terms of the arguments put forward in the course of the decision-making process (the product of the evaluation), as well as other important organisational factors (the process, positions and politics of the evaluation).

The analysis of the case convinces us that the research reported here brings a new and challenging research field to light. The many claims made with respect to the value to be gained from BPR investment highly contrast with the finding that BPR programs are often the subject of disappointed expectations. These expectations can only be judged on their merits if they are debated as part of a well-structured investment evaluation. Important recommendations for future work, therefore, lie in:

- Extending the analysis of BPR investments in terms of the proposed framework to other organisational settings and other research. This will ensure that the generic framework is free from the excess influence of local circumstances and the researchers involved. Such analyses give the investing organisation a means to diagnose their current way of managing BPR investment evaluation as a first step towards encouraging organisational learning.

- A second and more important recommendations concerns the use of the framework in a more prescriptive, action-oriented manner. For this, more detailed prescriptions on all aspects of the framework should be given. This implies more research on the distinctive characteristics of BPR investment evaluation and the way it is carried out in practice.

7 REFERENCES

Belmonte R.W. and Murray J. (1993), Getting ready for strategic change: surviving business process redesign, *Information Systems Management*, Summer, 23-39

Boonstra, A. (1991), Political aspects of information systems development (in Dutch), *Informatie*, nr.12, 857-864

Butler, C. (1994), The role of I/T in facilitating BPR: observations from the literature, in: *Business process re-engineering: information systems opportunities and challenges* (Glasson, B.C. eds.), IFIP TC8, North-Holland, Amsterdam, 147-160.

Checkland P. (1989), Soft systems methodology, in: *Rational analysis for a problematic world* (Rosenhead J. ed.), John Wiley, Chicester. 71-100.

Davenport, T.E. and Short, J.E. (1990), The new industrial engineering: information technology and business process redesign, *Sloan Management Review*, Summer, 11-27.

Davenport T.E., (1993), *Process innovation: reengineering work trough information technology*, Harvard Business School Press, Boston.

Dinther, M. van. (1992), *Caught in a web, psychological backgrounds of investment decisions* (in Dutch), AXA, Equity and Law.

Douglass, D.P. (ed.) (1993), The role of IT in business reengineering, *I/S Analyzer,* August, 1-16.

Farbey, B.; Land, F. and Targett, D. (1993), *How to assess your IT investment: a study of methods and practice*, Butterworth-Heinemann, Oxford.

FinTech (1992), Spotlight: business process reengineeering brings big benefits, FinTech Electronic Office, December 2, 4.

Frielink, A.B. (ed.,) (1961), *Auditing Automatic Data Processing*, Elsevier, Amsterdam.

Grover, V.; Teng, J.T.C. and Fiedler; K.D. (1993), Information technology enabled business redesign, *Omega*, nr. 4, 433-447.

Hammer M. and Champy, J. (1993), *Reengineering the corporation: a manifesto for business revolution*, Nicholas Brealey, London.

Harrington H.J. (1991), *Business Process Improvement*, McGraw Hill, London.

Harrison, E.F.(1987), *The Managerial Decision Making Process*, Houghton Mifflin, Boston.

Jones (1994), M., Don't emancipate, exaggerate: rhetoric, reality and reengineering, in: *Transforming organizations with information technology* (Baskerville, R.; Smithson, S.; Ngwenyama,O.; and DeGross, J.I. eds), IFIP 8.2, North-Holland, Amsterdam, 1994, 357-378.

Joslin, E.O.(1968), Computer selection, London, Addison-Wesley, 1968.

Kaplan, R.B. and Murdock L. (1991), Core process redesign, *McKinsey Quarterly*, 27-49

Klein, M.M. (1993), IE's fill facilitator role in benchmarking operations to improve performance, *Industrial Engineering*, September, 40-42

Kusters, R.J. and Renkema, T.J.W. (1994), Designing local evaluation models for IT investments, in: *Proceedings fifth European Software Cost Modelling Conference*, (ed. P. Rook), Ivrea (I),

Lammers, C.J. (1983), *Comparing Organisations*, Spectrum, Utrecht.

Meel, J.W. van; Bots, P.W.G. and Sol, H.G., Towards a research framework for business engineering, in: *Business process re-engineering: information systems opportunities and challenges* (Glasson, B.C. eds.), IFIP TC8, North-Holland, Amsterdam, 581-592.

Mintzberg, H. (1983), *Power in and around organizations*, Prentice Hall, Englewood Cliffs.

Morris D. and Brandon J. (1993), *Reengineering your business*, McGrawHill, London.

Pfeffer, J. (1981), *Power in organizations*, Pitman, Boston.

Renkema, T.J.W. (1994), Investing in the information technology infrastructure, *in: Proceedings Doctoral Consortium of the ICIS*, Vancouver Island, University of Victoria, DD 1-13.

Scott Morton, M. (ed.) (1991), *The corporation of the 1990's, information technology and organizational transformation*, Oxford University Press, Oxford.

Willcocks L. and Lester S. (1993), How do organizations evaluate and control information systems investments? Recent UK survey evidence, In: *Human, Organizational, and Social Dimensions of Information Systems Development* (Avison D. Kendall J.E. and DeGross J.G. Eds.), IFIP 8.2, North-Holland, Amsterdam, 15-40

8 BIOGRAPHY

Theo J. W. Renkema holds a masters degree (1992) in Economics from the University of Groningen, the Netherlands, which he received with the distinction 'cum laude'. He was employed as a research assistant in the Business Economics department of Groningen University ('89 - '92), investigating potential decision support for the organisational diagnosis tasks of interim-managers. Currently he is conducting doctoral research in the Faculty of Industrial Engineering and Management Science, Eindhoven University of Technology. The central theme of this research is the organisational decision-making process and evaluation of investments in the information-infrastructure. The research is conducted in close co-operation with a large financial services organisation.

Tom J. Dolan obtained his primary and masters degrees in Industrial Engineering at University College Galway, Ireland, in 1990 and 1992 respectively. He then worked as a research engineer at the CIM Research Unit in Galway for 2 years, which included a 9 month placement at the Eindhoven University of Technology. In November 1994 he began his doctoral studies in Eindhoven. His current research interests include: requirements management, product data modelling, and design process improvement. His research is being supported by the medical equipment manufacturer referred to in this paper.

15

Simultaneous Change

R. Feurer
Hewlett-Packard GmbH
Herrenberger Straße 130, 71004 Böblingen, Germany,
Tel.: (49) 7031 14 6395, Fax (49) 7031 14 3888.

K. Chaharbaghi
Cranfield University
Cranfield, Bedford, MK43 0AL, United Kingdom,
Tel.: (44) 234 750 111, Fax (44) 234 750 875.

Abstract

With increasing levels of uncertainty in competitive environments organisations need to become more dynamic in their strategy development capabilities. Any change in the environmental conditions has to be matched by a simultaneous change in all the dimensions relating to strategy development in order to ensure the organisation's survival and growth. This paper presents the results of a study which has been carried out to explore the relationship that exists between the concept of simultaneous change and business performance. Based on the findings a mapping process has been developed which helps organisations in identifying areas that need to be addressed to make simultaneous change possible.

Keywords

Strategy development, change, competitiveness, business performance, dynamics, strategy process, communication, value system.

1. INTRODUCTION

There is a general cognizance that competitive environments are changing at an accelerating rate and that the critical success factors are constantly being challenged. This implies that organisations must continuously change their competitive strategies with implemented strategies being in place only for a short period of time. Given these considerations, organisations can no longer afford to adopt a sequential approach to strategy development which entails several stages including data collection, development of strategic options, evaluation, selection and implementation. In the face of accelerating change there is a need for a dynamic approach in which opportunities are identified and evaluated simultaneously in the light of the organisation's existing and potential future competencies together with the level of resource commitment necessary. Figure 1 summarises the differences between the sequential and simultaneous approaches to strategy formulation and implementation.

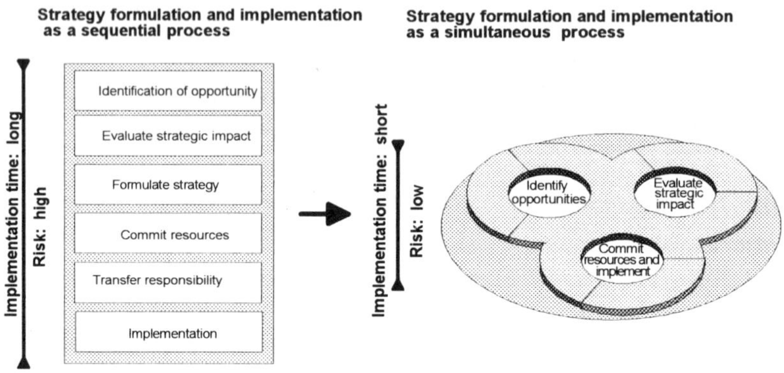

Figure 1 Sequential versus dynamic strategy formulation and implementation.

The simultaneous approach to strategy formulation and implementation brings about three major benefits when compared with the sequential approach. First, it reduces the time between the identification and exploitation of business opportunities which is one of the most important considerations when operating in a highly dynamic environment. Second, it reduces risk. This is because the sequential approach requires, at a very early stage, commitment of investments in terms of capital, people and time whereas the simultaneous approach constantly evaluates and dynamically adjusts resource commitments. Third, the simultaneous process continuously adjusts strategies according to the changes in the competitive environment. A dynamic approach to strategy formulation and implementation requires organisations to apply a structured framework thereby enabling the realisation of the following:

- Goal setting.
- Goal communication and negotiation.
- Dissemination and application of strategic knowledge which can be both internal and external.
- Formulation of strategies at that point in the organisation where the most appropriate strategic knowledge exists.
- Performance measures that are constantly aligned with the organisation's value system, goals and objectives.
- Alignment of strategies to:
  - Eliminate or reduce goal conflicts between entities.
  - Minimise strategy overlaps and redundant efforts.
  - Co-ordinate activities that span over several business entities or regions.
  - Maintain same overall direction and focus.

The requirements concerning speed, flexibility and maintenance of overall direction can only be achieved if organisations are able to simultaneously change the level of dynamics in each of their main strategy development capabilities. The main strategy development capabilities that have been identified in earlier research include (Feurer, Chaharbaghi, 1995):

- Timeframe for strategy formulation and implementation.
- Strategy formulation and implementation process.
- Organisation structure.
- Communication.
- Organisation values.

A study involving 40 multinational organisations has recently been carried out in order to establish the relationship that exists between the level of dynamics in each of these dimensions, the level of dynamics in the competitive environment and business performance. The aim of this paper is to present the results of this study and to introduce the concept of simultaneous change as a prerequisite for the realisation of dynamic strategy formulation and implementation. A mapping process is also introduced which measures the level of dynamics in each of the dimensions related to strategy formulation and implementation (i.e. timeframe for strategy formulation and implementation, strategy process, organisation structure, communication and organisation values). This mapping process can be used to identify actions which must be taken in order to make simultaneous change possible.

2. THE NEED FOR SIMULTANEOUS CHANGE

The application of a dynamic approach requires strategy formulation and implementation to be treated as a cognitive process rather than a process of conception ((DeGeus, 1988) (Senge, 1990). However, existing organisation strategy development capabilities, such as the process of strategy development, timeframe for strategy development, organisation structure, communication and organisation values, do not often lend themselves to such a dynamic approach. This is because these capabilities have been based on maximising the efficiency and effectiveness of a given strategy development approach rather than flexibility. In highly dynamic competitive environments, however, internal and external conditions constantly alter and as a result superior performance will depend on the ability of the organisation to change its capabilities accordingly. This lack of flexibility is reflected in the growing number of change management project failures. A recent study conducted in the U.S.A. revealed that 85% of top-managers were dissatisfied with the outcome of change management projects. The major factors contributing to these failures include poor co-ordination and lack of commitment; poor co-ordination is caused by different functions having different levels of dynamics in their strategy development capabilities whereas lack of commitment is the result of organisation values not being compatible with the speed of change within the environment. It follows that for achieving superior business performance the following conditions must be satisfied:

1. The strategy development capabilities of an organisation should be aligned with the level of dynamics in its competitive environment.
2. All strategy development capabilities should exhibit the same level of dynamics.

Several approaches have been proposed in order to evaluate the level of dynamics in a given competitive environment. These approaches are generally based on the speed of change, level of complexity and uncertainty. Ansoff and Sullivan (1993), for example, describe the dynamics of the environment in the form of environmental turbulence comprising :

- **Complexity of events** which occur in the environment.
- **Familiarity** of successive events.
- **Rapidity** with which events evolve after they are first perceived.
- **Visibility** of the consequences of these events.

Other approaches assume that the attractiveness of and the level of competitiveness within a given environment play an equally important role in the level of dynamics as well as business performance. These are expressed in the form of either environmental munificence (Dess, Beard, 1984) or environmental hostility (Miller, Friesen, 1983). Regardless of the approach used, different levels of dynamics in competitive environments necessitate different approaches to strategy development. The existence of such a relationship has been reported by a number of studies (Bourgeois, Eisenhard,1988) (Ansoff, Sullivan, 1993). However, these studies do not take into consideration that the application of different strategy development approaches will require a simultaneous change in all dimensions of strategy development capabilities.

3. SIMULTANEOUS CHANGE AND BUSINESS PERFORMANCE

In order to explore the relationship between simultaneous change and business performance, a survey was carried out involving 40 multinational organisations. Primary data was collected in the form of a questionnaire that was mailed to the Chief Executive Officer, a member of the board or a dedicated executive who had been identified as being responsible for strategy development. The mailing was preceded and followed up by telephone calls. In addition, secondary data was collected concerning financial performance.

The level of dynamics in the competitive environment was conceptualised through a number of questions considering speed of change, the increase in dynamics over time as well as the level of competitive pressures in the environment concerned. The level of dynamics in each of the strategy development capabilities (i.e. timeframe of strategy development, strategy development process, organisation structure, communications and values) was operationalised through ten questions and a total of 40 items in the questionnaire. Ratings for each of the above dimensions were obtained through a combination of qualitative and quantitative answers to both open and closed questions. Where possible five point scales were employed to minimise the error potential resulting from an assessment of qualitative responses.

The performance of the organisation was conceived as a multidimensional construct based on a holistic understanding of organisation competitiveness (Feurer, Chaharbaghi, 1994). The dimensions considered include performance in the eyes of the shareholders, performance in the eyes of the customers, performance in the eyes of the stakeholders (such as employees) and long-term ability to sustain performance. Financial performance data was collected using a timeframe of five years and included measures such as ROI (Return on Investment), ROS (Return on Sales) and sales growth. Primary data on the dimensions of customer satisfaction, investment in people, investment in technology and investment in society was obtained through the questionnaire using a five point scale. Based on the collected data, a number of ratings for each organisation were obtained and normalised using a five point scale. These ratings include:

1. Dynamics rating of the competitive environment in which the organisation operates.

2. Dynamics rating of the organisation computed as the average of the individual levels of dynamics in the strategy development capabilities.
3. Difference between the lowest and highest levels of dynamics in strategy development capabilities (thereafter referred to as Δ1).
4. Difference between the level of dynamics in the environment and the average level of dynamics in the organisation's strategy development capabilities (thereafter referred to as Δ2).
5. Business performance rating.

The first step of the analysis involved a comparison of the dynamics ratings and levels of dynamics in the strategy development capabilities for the four highest and four lowest performing organisations (see figures 2 and 3). The analysis of figures 2 and 3 reveals that no single strategy development capability accounts for superior performance. However, the high performing organisations exhibit a more uniform pattern in all dimensions relating to strategy development capabilities (this is characterised by Δ1 having a small value). They also feature a greater degree of alignment between the overall dynamics rating and the dynamics rating of the environment (this is characterised by Δ2 having a small value). On the other hand, low performing organisations have a higher value for Δ1 (which implies non-uniformity in the level of dynamics of different strategy development capabilities) as well as a higher value for Δ2 (which indicates a high degree of misalignment between the level of dynamics in the organisation's strategy development capabilities and the level of dynamics in the competitive environment).

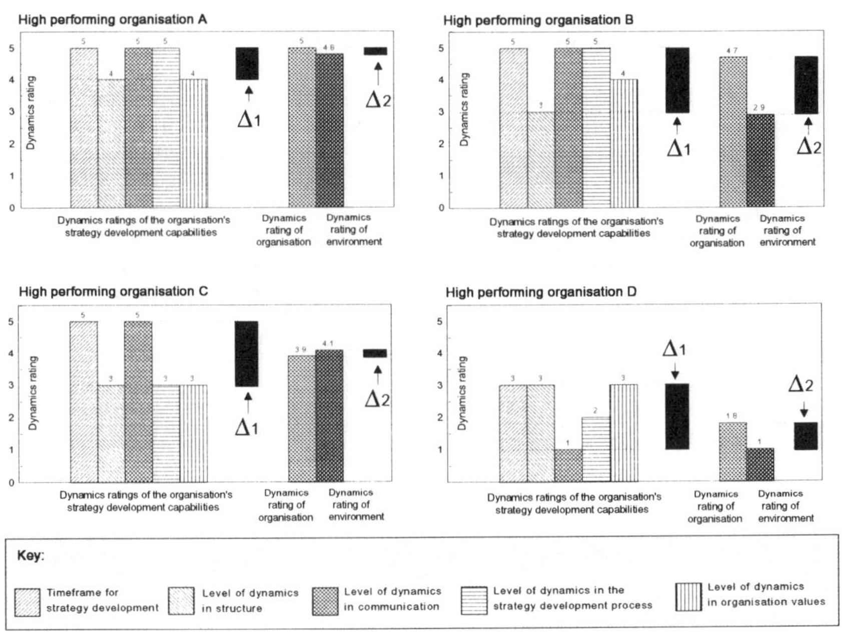

Figure 2 Dynamic ratings of high performing organisations.

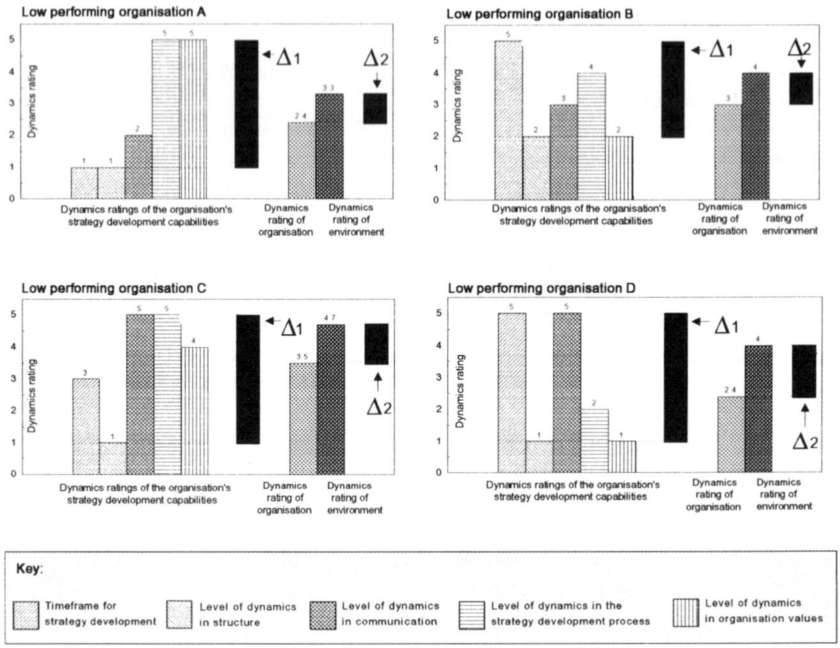

Figure 3 Dynamics ratings of low performing organisations.

These findings are further supported by the three dimensional scatter diagram given in figure 4. This figure clearly suggests an inverse correlation between business performance and Δ1 values. This demonstrates that organisations with small differences in their level of dynamics in strategy development capabilities have a superior performance when compared to those with larger differences. This figure also reveals that the high performing organisations are able to adjust the level of dynamics in their strategy development capabilities to the level of dynamics in the competitive environment.

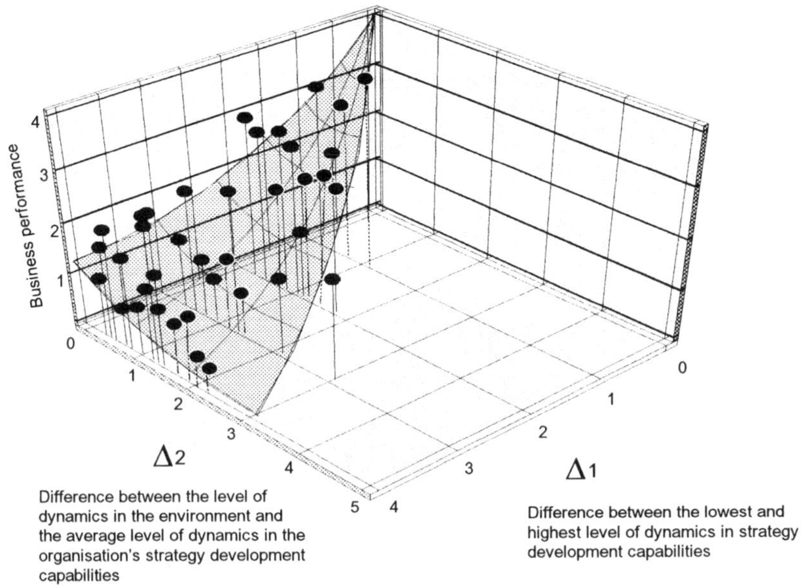

Figure 4 Relationship between Δ1, Δ2 and business performance.

5. MAPPING THE LEVEL OF DYNAMICS OF STRATEGY DEVELOPMENT CAPABILITIES

A mapping process is presented below which can be used by organisations to realise simultaneous change. The mapping process first requires the selection of criteria for measuring the level of dynamics for each of the strategy development capabilities as well as the competitive environment. The next stage involves collecting data from all individuals who may potentially be affected by the change process. This data is then cumulated and can be visualised using a radar chart. The radar chart will identify potential internal and external misalignments or gaps. The mapping process outlined above will be particularly useful if it is repeated over time or if it is used to compare the perceptions of different groups or individuals within an organisation.

The following provides examples of knowledge that can be generated using the mapping process:

- Change in strategy development capabilities over time.
- Perception of dynamics levels by different groups within the organisation.
- Perception of dynamics levels by different members in the same group.

- Comparison of the required and actual levels of dynamics as perceived by a given project team (i.e. perceived gap).
- Comparison of the perceived gaps as determined by the project team and the project sponsor.

The resulting radar charts can be used to identify problem areas which should be addressed before initiating a change project. Such a mapping process ensures that internal misalignments are minimised while improving communication, motivation and commitment. Figure 5 summarises the mapping process and provides an example of a radar chart.

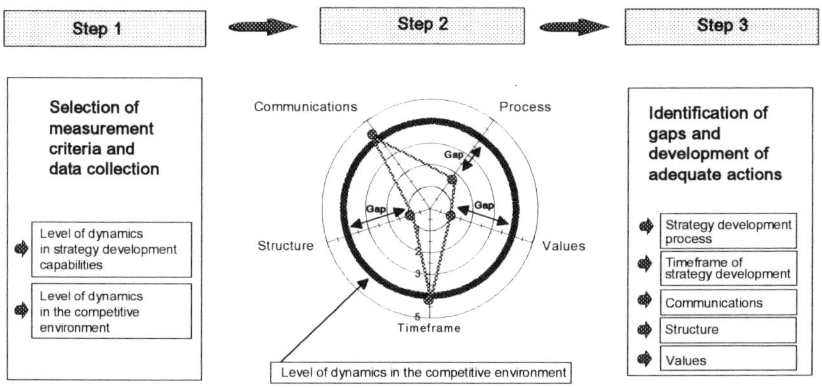

Figure 5 Mapping process.

6. CONCLUSION

In the face of increasing levels of uncertainty in competitive environments organisations must take a dynamic approach to strategy formulation and implementation in order to survive and grow. Such a dynamic approach represents a continuous and simultaneous process in which the strategy development capabilities are constantly aligned with changes in the competitive environment. In dynamic environments, high performance directly depends on the ability of an organisation to minimise internal and external misalignments in the level of dynamics in their strategy development capabilities. Using a mapping process, organisations can analyse the level of dynamics in each of their strategy development capabilities and take appropriate actions in order to minimise potential gaps. Such a process ensures that any change in the environment can be matched by a simultaneous change in all the strategy development capabilities of the organisation.

REFERENCES:

Ansoff, H.I., Sullivan, P.A. (1993) Optimising Profitability in Turbulent Environments: A Formula for Strategic Success. *Long Range Planning*. **26,5,** 11-23.

Bourgeois, L.J., Eisenhard, K. (1988) Strategic Decision Processes in High Velocity Environments: Four Cases in the Microcomputer Industry. *Management Science*. **14.** 816-835.

DeGeus, A.P.(1988) Planning as Learning. *Harvard Business Review*. March-April, 70-74.

Dess, G., Beard, D. (1984) Dimensions of Organisational Task Environments. *Administrative Science Quarterly*. **29**, 52-73.

Feurer, R., Chaharbaghi, K. (1994) Defining Competitiveness: A Holistic Approach. *Management Decision*. **32, 2**. 49-58.

Feurer, R., Chaharbaghi, K. (1995) Dynamic Strategy Formulation and Implementation. *Journal of General Management,* **20, 3**.

Feurer, R., Chaharbaghi, K. (1995) Strategy Formulation: A Learning Methodology. *International Journal of Benchmarking for Quality Management and Technology*. **2,1,** 35-52.

Feurer, R., Chaharbaghi, K. (1995) Performance Measurement in Strategic Change. *International Journal of benchmarking for Quality Management and Technology*. **2,2.**

Feurer, R., Chaharbaghi, K., Wargin, J, Weber, M. (1995) Hewlett-Packard's Business Alignment Framework. *Strategy, Structure and Technology*. Auerbach Publications, New York.

Miller, D., Friesen, P. (1983) Strategy-making and Environment: The Third Link. *Strategic Management Journal*. **4**. 221-235.

Senge, P. M. (1990) The Fifth Discipline - The Art and Practice of the Learning Organisation. Century Business, Bantam Doubleday Dell Publishing Group Inc., London.

BIOGRAPHIES:

Dr. Kazem Chaharbaghi lectures at Cranfield University and his special interests encompass strategy development and management learning.

Rainer Feurer works as a strategic consultant in the Professional Services Organisation of Hewlett-Packard in Germany.

16

Business Process Reengineering for Small and Medium Sized Manufacturing Enterprises

J. McSwiney
Business Development Manager, Cascade Designs, Dwyer Road, Midleton, Co. Cork, Ireland. Tel: +353 21 632399, Fax: +353 21 632583

Abstract
This paper examines a range of manufacturing performance programmes. Specific features from each of these programmes that are found in BPR are identified. The differences between a BPR implementation and a WCM implementation for three small and medium manufacturing enterprises are also described in some detail.

Keywords
BPR, WCM, SMME, performance improvement

1 INTRODUCTION

This paper presents the results of research carried out by the author between January and October 1994. The research was facilitated by both the Computer Integrated Manufacturing Research Unit (CIMRU) and AMT Ireland. The objectives of this research were as follows:

1. To explore, in the context of manufacturing companies, the novelty of Business Process Reengineering (BPR) and preceding performance programmes, in particular World Class Manufacturing (WCM).
2. To explore the relevance of Business Process Reengineering programmes to small and medium sized manufacturing enterprises (SMMEs).

2 A VIEW ON BUSINESS PROCESS REENGINEERING

The first task of the research was to choose a view of Business Process Reengineering. There is no recognised single authority or single view of BPR. In particular, authorities differ along two lines: radical versus incremental change, and the level of information technology (IT) dominance.

In relation to radical versus incremental change issues, both Hammer and Davenport come down on the side of radical change. Hammer et al (1990) emphasise that continuous improvement is not BPR and therefore the simple achievements of continuous improvement is tantamount to failure. They point out that: "Reengineering strives to break away from the old rules...At the heart of business reengineering is the notion of discontinuous thinking - identifying and abandoning the outdated rules that underlie current business operations". Similarly Davenport and Short (1990) are of the view that "Objectives of 5% or 10% in all business processes each year must give way to efforts to achieve 50%, 100% or even higher levels in a few key processes...the level of change targeted should be radical". Johansson et al (1993) also accord with the radical view and quote Paul O'Neill, Chairman of ALCOA to support this. "I believe that we have made a major mistake in our advocacy of continuous improvement....Continuous improvement is exactly the right idea if you are the world leader in everything you do. It is a terrible idea if you are lagging...It is probably a disasterous idea if you are far behind. ...we need rapid quantum leap improvements". Childe and Maull (1994) quote an IEEE expert, J. Atkins, as being at the other extreme. Atkins places no requirements on the scale of improvement required. He describes BPR as: "..the current popular term for examining an organisation's business processes and recommending automation or changes to achieve strategic goals".

In relation to the issue of information technology (IT) dominance, all BPR advocates agree on the enabling power of information technology. Hammer (1993) views IT to be "..an essential enabler..". Similarly Davenport (1993) proposes that: "...information technology should be viewed as a force to fundamentally reshape the way a business is done. ... Business process redesign and information technology are natural partners, yet ... their relationship ... has barely been exploited at all. But the organisations that have used IT to redesign boundary crossing, customer driven processes have benefitted enormously". Gant (1992) sees BPR as simply: "the redesign of processes to take advantage of the enormous potential of information technology". Clearly Gant views IT as both a means and an end. However, not all views are dominated by the IT factor as the above. For example, Touche Ross (1993) provide the following definition of BPR in the promotional literature for their reengineering program facilitation service: "Reengineering: A multi-disciplinary approach to implementing fundamental change in the way work is performed across the organisation with the goal of dramatically improving performance and stockholder value". Notice that there is no mention of information technology.

For the purposes of the research associated with this paper the radical improvement attribute of BPR was retained as an essential element. Information technology was considered as both a key element of the reengineered business processes and also as a tool for use during the business process redesign activity. However, IT was viewed strictly as a means and not as an end to BPR. It was recognised that valid non-information technology BPR solutions exist.

3 THE NOVELTY OF BPR

Hammer and Champy (1993) argue that Business Process Reengineering is fundamentally different from anything that has gone before. "Nor is reengineering the same as ...Total Quality Management...To be sure quality programmes and reengineering share a number of common themes ... However the two programmes also differ fundamentally. ... Fundamentally reengineering is about reversing the industrial revolution. Reengineering rejects the assumptions inherent in Adam Smith's industrial paradigm - the division of labour, economies of scale, hierarchical control and all other appurtenances of an early stage developing economy". At the other extreme Libby (1994) introduces a BPR implementation at Cochrane Furniture in North Carolina as follows: "...came across the book 'The Goal' by Eliyahu M. Goldratt that introduced him to a REENGINEERING approach called SYNCHRONOUS MANUFACTURING. Consistent with TOTAL QUALITY MANAGEMENT thinking synchronous manufacturing is a way to speed material flow through a process using techniques for reducing inventories while improving delivery performance. Cochrane Furniture has achieved genuine LEAN PRODUCTION by establishing a synchronous manufacturing environment". The capitals serve to highlight the variety of improvement programmes referenced.

Decade	**Approach**	**Pioneer or Early Authority**
1690(-)	Division of Labour	Adam Smith
1890	Scientific Management	Frederick Taylor
1900	Mass Production	Henry Ford
1920	Industrial Engineering	F. Gilbreth & F. Taylor
1930	Human Relations Movement	Elton Mayo
1950	Japanese Quality Revolution	J. M. Juran & W. E. Demming
1960	Materials Requirements Planning	William Orlicky
1970	Manufacturing Resource Planning	Oliver Wright
1970	Focused Factory	Wickham Skinner
1980	Total Quality	Philip Crosby
1980	Just In Time	Taiicho Ohno
1980	Computer Integrated Manufacturing	
1980	Optimised Production Technology	Eliyahu Goldratt
1980	ISO 9000	NASI
1980	World Class Manufacturing	Richard Schonberger
1980	Benchmarking	Rank Xerox
1990	Lean Manufacturing	Jones & Roos
1990	Business Process Reengineering	M. Hammer & T. Davenport

Table 1 Major approaches to manufacturing performance improvement

The Goal was first published in 1986, certainly prior to any popular use of the term 'reengineering'. Clearly Libby is of the view that BPR is not new and that it is intimately related to what this paper considers to be previous approaches to manufacturing performance improvement.

The following methodology was used to explore the novelty of BPR.

1. Construct a list of the manufacturing performance improvement programmes or approaches to date.
2. Analyse this list in chronological order comparing and contrasting with previous ideas.
3. Finally analyse Business Process Reengineering, showing clearly how it compares and contrasts with preceding programmes.

A chronological list of what were identified as the major approaches to manufacturing performance improvement are shown in table 1.

The approaches presented in table 2 were found to have pre-empted elements of BPR. The final row (of table 2) identifies the features attributed as being original to BPR.

This research was particularly concerned with the uniqueness of Business Process Reengineering in relation to World Class Manufacturing. It is useful to consider how a BPR implementation would differ from a WCM implementation. The differences can be categorised under the headings of methodology and solution space.

3.1 Methodology

A BPR company analysis will usually begin with the customer and then look at entire business processes including the administrative activities. The majority of WCM implementations will focus on the manufacturing function. Administrative functions may or may not be affected. All BPR projects will be initiated from the top down. WCM projects will be initiated by a forum of direct level employees, i.e. from the bottom up. Such projects will centre around the implementation of WCM techniques or will focus on specific manufacturing activities.

3.2 Solution Space

BPR embodies a prior expectation that there will be significant restructuring. There is no such prior expectation in a WCM implementation. People will expect the overall departmental structure to remain intact, although reorganisation within the manufacturing department may be anticipated. The significance of this lack of expectation is that if major restructuring is attempted it may meet with huge resistance.

Within BPR radical performance improvement is viewed as an essential outcome. Consider the following axiom of performance measurement: whatever is measured will improve and if it is not measured it will not improve (loosely attributed to Kaplan 1987). The relevance in the current context is that if performance of a change initiative is not measured in terms of radical improvement then the chances of achieving radical improvement are diminished. There is a danger in WCM initiatives that every reasonably attractive project identified will be pursued, possibly to the detriment of overall achievement. The BPR approach is to ignore the 'small fry in the hope of a big catch'.

Approach to Manufacturing Performance Improvement	Contribution to Business Process Reengineering
Division of Labour	Recognition of a relationship between structure and performance. Achievement of radical performance improvement through restructuring.
Scientific Management	Recognition of the possibility of radical performance improvement. Process analysis and design.
Mass Production	A specific case of very effective reengineering of the order fulfilment business process which was subsequently taken as a general solution.
Industrial Engineering	A hierarchical top down systematic approach to process design. The professional process redesigner (i.e. the I.E.)
Manufacturing Resource Planning	Recognition of information technology as a potential process improvement enabler. A programme style implementation as a potential process improvement enabler. A programme style implementation methodology, a separate chain of command for implementation.
The Focused Factory	Recognition of the benefit of organisation around a clearly defined customer focused task (not quite a business process focus).
World Class Manufacturing (includes the ideas associated with JIT, TQM, Lean Manufacturing and the human relations movement)	Cellular manufacturing, i.e. organisation around a product, was a fully fledged precursor to business process orientation. People empowerment and the team approach to performance improvement. Radical performance improvement through redesign of a stable manufacturing process. However, radical improvement was not considered essential for success.
Business Process Reengineering	**Novelty:** The elevation and total focus on the business process. The prior expectation that there will be significant restructuring a "blank state" approach to planning the next state of an existing system. Radical performance improvement as an essential requirement, the exclusion of continuous improvement as a successful outcome. The elevation, though not origination, of the idea that IT must be viewed as much more than an automation tool (Pava (1983) and Browne et al (1988) both emphasised this view).

Table 2 BPR elements found in manufacturing performance improvement approaches

Information technology occupies a central role as an enabler of BPR solutions. WCM programmes on the other hand will in some cases be ill-disposed to IT-driven solutions. This attitude is a reaction to the poor impact on overall company performance of many early Manufacturing Resource Planning (MRPII) implementations. These implementations suffered from attempting to automate un-streamlined systems (Browne et al, 1988). World Class Manufacturing arrived on the manufacturing performance improvement programme scene hot on the heels of such misadventures. At one time advocates of WCM saw themselves in opposition to advocates of MRPII. Although such opposition is now mostly in the past, WCM practitioners tend not to jump to IT as an early enabler of performance improvement. It is important to point out that IT solutions have advanced considerably since the days when MRPII was giving IT a bad name.

4 THE RELEVANCE OF BPR TO SMMEs

BPR is undoubtedly relevant to manufacturing companies. Hammer et al (1993) present overviews of manufacturing related BPR implementations at Hallmark and Kodak. Davenport and Short (1990) see BPR as an application to business processes of approaches pioneered by Industrial Engineers in manufacturing operations. This has given rise to the term white collar Industrial Engineering. However the question of the relevance of BPR to smaller manufacturing organisations is not so clear cut. Very few overt references to BPR implementations in a small company were discovered. Even a cursory analysis of BPR will raise fundamental questions about the relevance of BPR to small companies.

The most effective way to investigate the relevance of BPR to SMMEs would have been to analyse a number of smaller company BPR programme implementations. Such implementations were not available. An alternative approach would have been to create models of existing small or medium sized manufacturing enterprises and to carry out virtual implementations on the models. The weakness with the virtual BPR approach is that at best it would indicate the relevance of the concept of process redesign. It would give no indication of a small company's willingness to undertake the BPR challenge or a small company's ability to carry out a process redesign and implement the outcome. A survey of SMMEs would have been yet another possible way of exploring the relevance of BPR programmes. The problem with such an approach is that BPR is a new and is not yet a widely understood concept. A survey questionnaire or interviewer would both have had to introduce the concept of BPR and obtain useful views on its relevance. This was considered impractical. Faced with this dilemma it was necessary to improvise. The approach taken was to examine a number of World Class Manufacturing implementations. Although WCM and BPR are not the same they do share common principles and similar approaches to implementation. The basic assumption is that an examination of a WCM implementation will give a valid insight into the viability of a potential BPR implementation. The overall methodology employed to test the relevance of BPR to SMMEs is presented as:

1. Define the concept of a small to medium sized manufacturing enterprise.
2. Formulate a number of hypothesis about the relevance of BPR programmes to SMMEs.
3. Test the hypotheses on three SMMEs using WCM implementations as indicators of potential BPR implementations.
4. Reject or reformulate the hypotheses to reflect the empirical evidence.

The Task force on Small Business Report (1994) defines SMEs (small and medium-sized enterprises) as having less than 250 employees. The Irish Companies Act specifies that the average number of employees of a small company should not exceed 50. The balance sheet total should not exceed £1.25 million and turnover should not be in excess of £2.50 million. The figures for a medium company are 250 employees, £5 million balance sheet and £10 million turnover (Doolan 1991). The companies chosen for the research all had less than 100 employees.

Two sets of hypothesis are presented below. The first set are positive in relation to the relevance of Business Process Reengineering programmes to SMMEs. The second set are negative. The hypotheses are derived from Barrier M. (1994), Bannock G. (1981), Barber M. et al (1994), Barber J. et al, (1989), McGee J. (1987), and The Task force on Small Business Report (1994).

Positive

1. The continued survival or growth of a small or medium size manufacturing enterprise may depend on re-engineering style radical change and performance improvement.
2. SMMEs are capable of implementing fast change and therefore the risk normally associated with a re-engineering programme may be reduced.

Negative

3. Because of their simple structure SMMEs will not need a 'full blown' BPR programme to exploit lucrative opportunities for restructuring.
4. Managers of SMMEs may not be open to formal improvement programmes.
5. Managers of SMMEs may not be open to the idea of employee empowerment (employee empowerment is an essential element of most BPR solutions).
6. SMMEs may lack the internal expertise to take advantage of BPR style information technology solutions.
7. An SMME may not be able to dedicate sufficient internal resources to achieve successful implementation of a BPR programme.
8. The challenges facing SMMEs may reside mostly in the finance and marketing arenas and therefore may not warrant a business process focus.
9. It is likely, particularly in the area of product development, that an SMME's business process will reside partially outside the organisation. This may have a detrimental impact on the SMME's ability to re-engineer such processes.

The above hypothesis were tested on three case companies each of which are briefly described in table 3.

Company Designation	Number of Employees	Sales Turnover (£)	Ownership	WCM Initiation	BPR Feature
Company A	34	n/a	United States	October 1993	IT enabled solution
Company B	85	£6,400,000	Irish	January 1994	Major restructuring
Company C	80	£4,500,000	United States	March 1994	Radical improvement

Table 3 Companies used for case studies

Company A specifically requested that no indication of turnover be given, even in a disguised document. All WCM implementations were on-going at the time of the research and had progressed sufficiently to support analysis. The above three companies were selected from a range of alternatives because in each case the WCM programme solution included a particular BPR programme feature. This feature was not to be the sole focus of the analysis, but it was expected to add validity to the findings. The author was actively participating in each implementation. Some personnel in the companies were aware of the research, but it is certain that their behaviour was not affected directly by the research goals. The small sample size may have limited the overall population validity of the research conclusions, i.e. the degree to which it is possible to generalise from the sample observed to the wider population of the same types. However, the participative ethnological style approach ensured particularly strong ecological validity, i.e. the degree to which it is possible to generalise from the context in which the research has taken place to other contexts and settings. In other words, the research was carried out in a natural situation which was very representative of the natural situation about which knowledge was sought.

The following table is a summary of the case study verdicts on each hypothesis. The verdicts were based on analysis of the following aspects of each case:

- The chosen WCM solution.
- The chosen approach to implementation.
- The implementation experience.

The verdicts from each case for each hypothesis were combined to yield a conclusion (see table 4). A conclusion may be to support, oppose or remain inconclusive on the validity of a hypothesis. A conclusion is strong if all results agree. It is weak if one disagrees. It is neither strong nor weak if two results agree and the other is neutral. The conclusion is inconclusive otherwise.

The final step was to reformulate the tested hypotheses into a view on the relevance of BPR to SMMEs. There are potentially many varieties of BPR programmes, even within the restricted view adopted by the research. Similarly there are numerous varieties of SMMEs. Therefore, a simple yes or no answer will not suffice. A qualified answer is presented. The style is emphatic rather than suggestive.

A Business Process Reengineering programme is relevant to a small or medium sized manufacturing enterprise under the following conditions:

1. The SMME is faced with a situation where survival or growth depends on radical performance improvement or a radical change in the way business is carried out.
2. The SMME commits in advance to dedication of sufficient internal resources to the BPR effort. This is not simply an allocation of personnel. Some of the existing duties of the chosen personnel must be reassigned for the duration of the project.
3. The SMME must be either in control of the major portion of the business process or interest, or they must have the cooperation of the other organisations involved.
4. The BPR effort is facilitated by external BPR experts.
5. Information technology is viewed strictly as a means and not an end. It may potentially be a very powerful enabler, but equally powerful non-IT related BPR solutions may exist.
6. The external BPR experts are able to provide appropriate support for IT solution, design, implementation and maintenance. The maintenance may be contracted elsewhere, but is an essential aspect of a viable IT solution.
7. The implementation pays particular attention to the difficulty of changing inappropriate management styles. Training may not be enough. Visits to other organisations and pilot exercises may be required. Some managers within the SMME may not make the transition.

	Hypothesis	**Co. A**	**Co. B**	**Co. C**	**Conclusion**
1.	An SMME may need to undergo radical change to continue growth.	Neutral	Supported	Supported	Supported
2.	An SMME's ability to achieve fast change reduces the potential risk of a BPR implementation.	Supported	Opposed	Supported	Weakly Supported
3.	SMMEs will not need a formal improvement programme in identifying opportunities within their simple structure.	Opposed	Supported	Opposed	Weakly Opposed
4.	Managers of SMMEs may not be open to formal improvement programmes in general.	Opposed	Opposed	Supported	Weakly Opposed
5.	Managers of SMMEs may not be open to employee empowerment.	Opposed	Neutral	Supported	Inconclusive
6.	SMMEs lack the internal expertise to take advantage of IT solutions.	Supported	Supported	Supported	Strongly Supported
7.	An SMME may not dedicate sufficient internal resources to BPR.	Neutral	Supported	Supported	Supported
8.	SMME challenges are in finance and marketing and therefore do not warrant a business process focus.	Opposed	Opposed	Opposed	Strongly Opposed
9.	SMME business processes reside partially outside the organisation and may be difficult to reengineer.	Supported	Supported	Opposed	Weakly Supported

Table 4 Hypotheses and their conclusions

5 REFERENCES

Bannock, G. (1981) *The Economics of Small Business: Return from the Wilderness.* Basil Blackwell, Oxford.

Barber, J., Metcalfe, J.S., Porteous, M. (1989) Barriers to Growth: The ACARD Study, in *Barriers to Growth in Small Firms* (ed. J. Barber et al), Routledge, London.

Barrier, M. (1994) Re-engineering Your Company, *Nations Business.*

Browne, J., Bradley, P. (1994) The Generic Model of Manufacturing and The Extended Enterprise. Unpublished, CIMRU, UCG, Ireland.

Childe, S.J., Maull, R.S. (1994) Frameworks for Understanding Business Process Re-engineering, Submitted to *International Journal of Production Management*, Vol 14 No. 12.

Davenport, T.H. (1993) *Process Innovation : Re-engineering Work Through Information Technology.* Harvard Business School Press, Boston, Massachusetts.

Davenport, T.H., Short, J.E. (1990) The New Industrial Engineering: Information Technology and Business Process Redesign, *Sloan Management Review*, Summer.

Doolan, B. (1991) *Principles of Irish law.* Gill and Macmillan, Dublin.

Ford, H. (1926) Mass Production, *Encyclopaedia Britannia*, 13th Edition, Suppl. Vol2, pp. 821-823.

Gant, J.G., (1992) Work Management: the next step in imaging, *Chief Information Officer Journal*, Fall, pp. 60-64.

Goldratt, E., Cox, J. (1986) *The Goal.* Creative Output Books, USA.

Hammer M., Champy, J. (1993), *Re-engineering the Corporation : A Manifesto for Business Revolution.* Nicholas Brealey, London.

Hammer, M. (1990) Re-engineering Work: Don't Automate, Obliterate, *Harvard Business Review*, July-August , pp 104 - 112.

Johansson, J.H., McHugh, P., Pendlebury, A.J., Wheeler , W.A. III (1993) *Business Process Re-engineering: Breakpoint Strategies for Market Dominance.* John Wiley & Sons, Chichester, West Sussex, England.

Kaplan, R.S., Johnson, H.T. (1987) *Relevance Lost: The Rise and Fall of Management Accounting.* Harvard Business School Press, Boston, Massachusetts.

Libby, B. (1994) Re-engineering, By the Book, Manufacturing Systems (MFS), Vol 12, Issue 4, April, pp 52-53

Mayo, E. (1933) *The Human Problems of an Industrial Civilisation.* Macmillan, USA.

McGee, J., (1987) Barriers to Growth: The Effects of Market Structure, in *Barriers to Growth in Small Firms* (ed. J. Barber et al.), Routledge, London.

Mullee, W.B., Porter, D.B. (1963b) Process Chart Procedures, *Industrial Engineering Handbook*, McGraw-Hill Book Company, Inc., New York.

Pava, C. (1983) *Managing New Office Technology: An Organisational Strategy*, The Free Press, New York.

Schonberger, R. (1986) *World Class Manufacturing: The Lessons of Simplicity Applied.* Free Press, New York.

Skinner, W. (1984) The Focused Factory, *Harvard Business Review*, May-June.

Smith A. (1776) *The Wealth of Nations.* Glasgow.

Task force on Small Business (1994) Report, Government Publications, Dublin.
Taylor, F.W. (1911) *The Principles of Scientific Management.* Harper and Row, New York.
Touche Ross (1993), *Re-engineering For Results*, TM Management Consultancy Services Brochure, Touche Ross, London.
Wight, O. (1981) *MRP II: Unlocking America's Productivity Potential.* CBI Publishing, Boston MA.

6 BIOGRAPHY

Joe Mc Swiney graduated from the Cork Regional Technical College in 1986 with a Bachelor of Technology Degree in Electronics Engineering. In that year he joined Concurrent Computers Irl. Ltd., initially as a Manufacturing Engineer and later as Manufacturing Supervisor. In 1989 he joined the CIM Applications Group of AMT Ireland as a Manufacturing Consultant. He spent over five years with AMT Ireland assisting companies improve their manufacturing performance, primarily through the implementation of World Class Manufacturing. In 1994 he completed a two year part time Masters of Business Administration (MBA) degree at University College Galway. In January of 1995 he joined Cascade Designs Ltd. as Business Development Manager.

17

Self assessment - A tool for business process reengineering

A. Rolstadås
Norwegian Institute of Technology, Department of Production and Quality Engineering
R. Birkelands vei 2, 7034 Trondheim, Norway
Telephone: + 47 73 59 38 00
Telefax: +47 73 59 71 17
E-mail: arolst@protek.unit.no

Abstract

Industry experiences a changed competitive environment characterised by customer focus and global competition. To meet this situation manufacturing and development will have to be done globally as well. This is referred to as the virtual enterprise.

In the virtual enterprise performance becomes a key issue. Measurement of performance must be based on an extended productivity definition where productivity is the ratio between value added and resources consumed. This definition requires a process oriented approach. In this the definition and identification of business processes is essential. These processes must be studied and reengineered.

The study of business processes will include performance measurement. This can be done using the TOPP approach.

TOPP is a research programme where amongst other a model for productivity studies is developed. This model comprises system variables, functions, cycles and management philosophies.

Productivity studies can be done using the self assessment approach. This is either based on a questionnaire or on the definition and continuous evaluation of standard business processes.

Keywords

Production management, enterprise modelling, productivity, benchmarking.

1 THE GLOBAL COMPETITION

Today industry places a major focus on manufacturing excellence and quality. The competitive environment has changed. It is not sufficient to be a cost champion for an enterprise to survive. Competition is experienced in several dimensions such as cost, time, quality and flexibility.

This changed competitive environment is of course based on a changed marked requirement. Some key characteristics for this new market situation are:

- Declining markets
- Global competition
- Customer in focus
- Life cycle requirements
- Environment protection restrictions.

The answer to this is the virtual enterprise (Kimura, 1993, Rolstadås, 1994).

This future enterprise is «lean» or «agile». The customer is in focus. All activities in the company must add value for the customer. Otherwise they represent a waste of resources. The customer worries about price, quality, service, and delivery. Total quality management has become a new topic addressing all these problems. Actually, it goes far beyond product quality. It looks at quality in every link and every activity. The customer's expectations must be exceeded.

Lean manufacturing is only possible with an efficient and lean production management. High quality must be secured at minimum cost and with the shortest possible delivery time.

The market is international. Each business process in the company must be benchmarked against the very best world wide (Rolstadås, 1993). But not only the market is international. The same is true for the company. A successful competitive company may place its development activities in region A and its manufacturing in regions B and C. These activities may be moved at any time to the region offering the most favourable conditions. The company will focus on its core business processes and may outsource the rest to more competitive suppliers.

These suppliers may have long term contracts including technological development. The company will establish a strategic alliance with its suppliers. This may even extend to the customers or vendors. This is what is meant by the *virtual* enterprise. It consists of a number of units geographically distributed but managed as one total unit although the subunits may be under separate management.

The competition leading up to the virtual enterprise can be understood by studying the stakeholders model depicted in Figure 1 (Rolstadås 1995). This model emphasises that the company is competing in several marketplaces. Long term survival and competitiveness is not limited to customers, but depends heavily on attractiveness towards the different stakeholders. In fact, most companies are competing about the best suppliers, the best lenders, the best alliance partner, the best employees etc. This competition is mainly a matter of position and terms in the relationship between company and stakeholder.

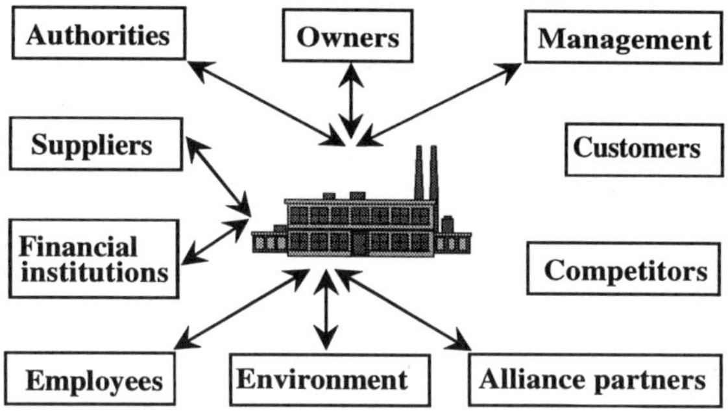

Figure 1 Stakeholders Model.

The new approach focuses on measuring performance rather than efficiency. However, this requires a process oriented view. The enterprise is considered as a set of interdependent business processes rather than a set of functions. To develop and change a company to this new philosophy requires a new way of thinking and some tools. Methods of enterprise modelling represent a foundation for implementing such a change and enterprise modelling techniques therefore becomes a crucial tool.

2 PRODUCTIVITY VERSUS PERFORMANCE MEASUREMENT

The classical definition of productivity is the ratio between output and input. Based on this definition, the classical approach to performance measurement has been published by Sink and Tuttle (Sink, 1985, Sink and Tuttle, 1989). The model claims that the performance of an organisational system is a complex interrelationship between the following seven performance criteria:

1. Effectiveness, doing the right things, at the right time, with the right quality, etc. Defining the criterion as a ratio, effectiveness can be defined as Actual Output/Expected Output. Figure 2 illustrates this.

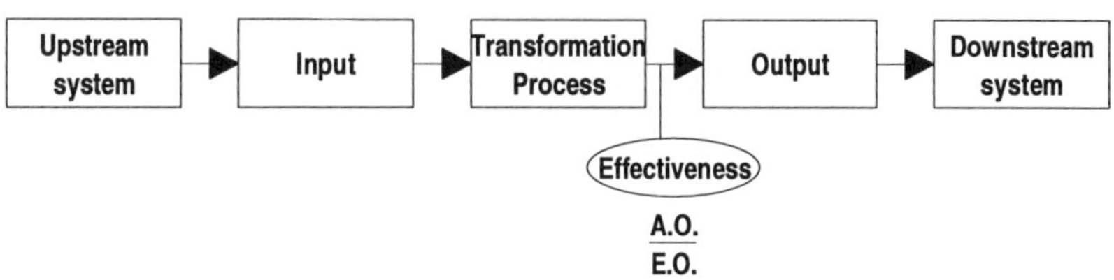

Figure 2 Operational Definition of Effectiveness.

2. Efficiency, this is an input- and transformation process-question, defined as Resource Expected to Be Consumed/Resources Actually Consumed, as shown in Figure 3.

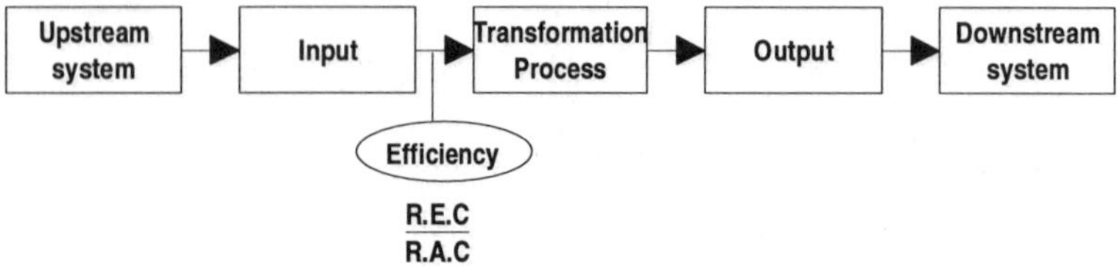

Figure 3 Operational Definition of Efficiency.

3. Quality, where quality is an extremely wide concept. To make things more tangible, quality could be measured at six checkpoints:

   - Upstream systems
   - Inputs
   - Transformation value adding process
   - Outputs
   - Downstream systems
   - Quality management process.

4. Productivity, this is the traditional ratio of Output/Input, but it appears as just one of several criteria. For an illustration, see Figure 4.

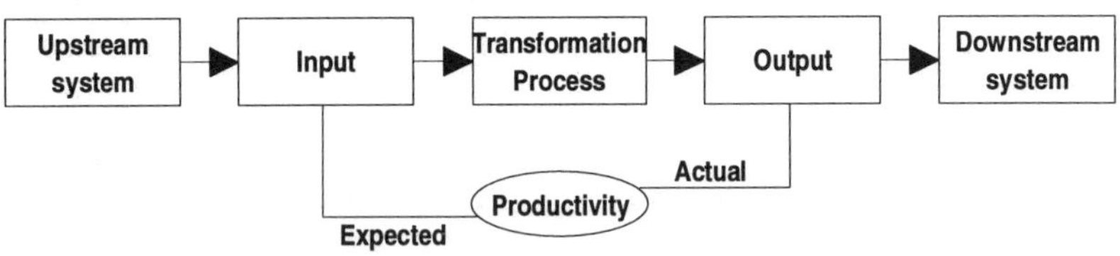

Figure 4 Operational Definition of Productivity.

5. Quality of work life, one essential, but often forgotten element contributing to a well performing system.

6. Innovation, a key element in sustaining and improving performance.

7. Profitability/Budgetability, the ultimate goal for any organisation.

The traditional definition of productivity is insufficient to deal with the challenges the virtual enterprise will meet in the global market. For this purpose a new definition has to be applied. This will define productivity as the ratio between value added and resources consumed. This definition is much wider and includes all aspects of performance both inside the enterprise and in the supplier and customer markets. It is therefore more correct to talk about performance rather than productivity.

The performance oriented view requires a process oriented approach rather than a functional oriented approach. The process oriented thinking covers the interface between functional units in the enterprise in addition to the functions themselves. The process oriented approach therefore represents a more holistic thinking than the functional oriented approach.

3 BUSINESS PROCESS ORIENTATION

Process oriented thinking is by many considered as the new possibility for competitiveness improvement. By functional oriented thinking, only marginal improvement seems to be possible. The process oriented thinking includes two aspects (Rolstadås 1995):

- Definition of business processes
- Reengineering of business processes.

It could be questioned whether reengineering represents anything new. Some argue that the techniques behind reengineering are known form JIT and Time Based Management. Process flow analysis is a common denominator for several of these approaches. Actually, techniques and methods in reengineering are known from current management philosophies. Reengineering is not a streamlined concept with own tools, but a creative approach where the result is the only matter of concern. Tools and methods are adapted to suit this situation.

Reengineering is then simply a popular description for a strategy with a set of characteristics rather than a philosophy. However, people often tend to extend new buzzwords into independent philosophies.

Reengineering can be defined as (Camp, 1989):

Reengineering is fundamental rethinking and radical redesign of business processes to achieve dramatic improvements in critical, contemporary measures of performance, such as cost, quality, service, and speed.

The definition states that reengineering means search for dramatic change and productivity breakthrough. Reengineering is not the kind of tool a company applies if its performance needs a 10 % boost. Business process reengineering is a strategic tool useful for achieving radical breakthroughs on business performance. Typical examples could be reduction of delivery times by 75 %, reduction of cycle times by 75 %, reduction of product development times by 50 %, reducing the number of components by 50 % and so forth. Change in this magnitude is impossible by simply improving the individual performance. That is why strong adjectives as fundamental, radical and dramatic are used to describe the magnitude of change in the processes. This may be illustrated from the following list of bullet items (Rolstadås, 1995):

- Customer orientation. Implementations should be justified from a customer perspective.
- Process orientation. Reengineering goes beyond predefined organisational boundaries and looks at the entire process to consider all tasks involved in transforming the input into output.
- Focus on core businesses. Reengineering is reserved for the main processes that could translate into value and give a competitive advantage.
- High ambition. Applying reengineering implicates an aim for breakthroughs, not evolution (60 % or higher).
- Rule-breaking. No traditional or established truth is allowed to be kept out from the reengineering process. Paradigms like specialisation, sequentiality, and timing have to be abandoned.
- Devotion for simplification. Complexity hides waste and simplification is necessary to increase speed and reduce costs.
- Creative use of information technology. Creative use of information technology in different ways. Reengineering is not the same as automation. The process is the core and information technology is just a tool to reinforce improvements.
- Rapid payback. Reengineering involves high risk due to the comprehensiveness and rapid payback is necessary to handle this.

It is not possible to describe a universal reengineered process. However, there exist some typical sources for action when reengineering a process (Camp, 1989, Carr, 1993).

- Several jobs are combined into one. Benefits from combining jobs could be fewer errors and misunderstanding, less handling, increased speed, and less administration.
- Decentralisation of decisions. Decisions should be made where the need for them arises.
- The steps in the process are performed in a natural order. Sequentalisation forced by the organisational structure is abandoned. Concurrent methods are applied when appropriate.
- Designing flexibility into all parts of new processes. Multiple versions of processes could be a solution.
- Work is performed where it makes the most sense. Reengineering boosts the traditional organisational structure and focuses on doing the activities as close to where they are required as possible. Support activities like detailed planning, quality assurance, purchasing etc. are adapted to enable decentralisation.
- Checks and controls are reduced. This is nonvalue-adding work and should be minimised.
- Reconciliation is minimised. This is done by cutting back the number of external contact points a process has in order to reduce the possibility that inconsistent data are received.
- A case manager provides a single point of contact. They act like a buffer between the process under reengineering and the customer.
- Hybrid centralised/decentralised operations are prevalent. Advantages of centralisation and decentralisation are combined in the same process. Information technology and empowered operators could make hybrid organisations possible.

As it is clear from the preceding, there is no general definition of business processes. They are all adapted to the actual case.

The process oriented approach is ideal for studying the performance of an enterprise. Such performance studies can be done at three levels:

- Self assessment
- External evaluation
- Benchmarking.

Self assessment means that the enterprise evaluates itself over time. It must be based on a model of the enterprise, defining business processes.

External evaluation involves external experts to review the performance of the enterprise compared to what the expert perceives as best practice. Again this will have to be based on a similar enterprise model as for self assessment.

Benchmarking is extending the external evaluation to compare selected business processes with such processes in other enterprises that can demonstrate best practice or clear performance excellence.

4 THE TOPP PROGRAMME

TOPP is a research programme sponsored by the Norwegian Research Council (Moseng, Bredrup, 1993). The overall goal of TOPP is to *«Focus on the total productivity for the whole enterprise and stimulate an industrial climate that improves competitiveness»*.

Important key objectives and key issues are:

- Time to market
- Quality
- Flexibility
- Total cost.

The programme involves co-operation between

- The Federation of Norwegian Engineering Industries
- The Norwegian Institute of Technology
- Industrial enterprises,

and is sponsored by

- The Research Council of Norway.

The TOPP programme is planned for the period 1992-95. The following tasks are included:

- Analysing company productivity and competitiveness
- Implementing actions for industrial productivity improvements (industrial projects, seminars, courses, industrial networks, etc.)
- Generating new knowledge (research projects, analysing productivity data, etc.)
- Long term competence programme.

TOPP is bases on an enterprise model. This is again based on the FOF model (Wortmann, 1989, Falster, 1991) indicating performance by effect on performance indicators as design choices are changed. In TOPP the design choices are called system variables.

The following system variables are considered:

- Products
- Facilities
- Equipment
- Personnel
- Organisation and systems.

To study an enterprise, it is necessary to look at functions performed on the system variables. To represent these functions, the value adding chain defined by Porter may be suitable. However, since TOPP addresses the manufacturing industry, it can be made more specific as shown in Figure 5.

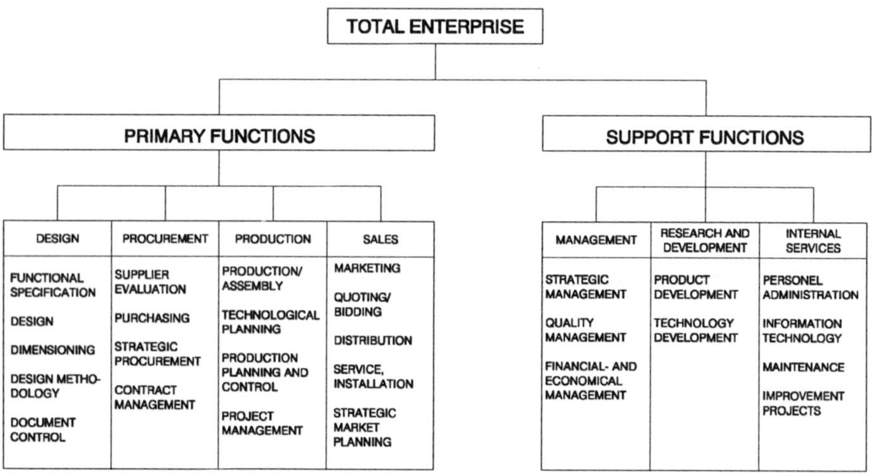

Figure 5 Functions in the Value Adding Chain (Moseng, Bredrup, 1993).

The description of the functions along the value adding chain (Figure 5) represents a static picture of the company. For a study of productivity performance, this may be insufficient. An enterprise may be «world champion» within each function, but it may still be less competitive. The reason is, of course, what is happening on the transfer from one function to another and how functions may be run concurrently. This more dynamic aspect may be taken into account by following the various flows in the company. These flows are referred to as cycles. A possible set of cycles could be:

- Material cycle (follow materials through the enterprise)
- Order cycle (follow a customer order through the company)
- Product cycle (follow a product through its whole life from idea to destruction)
- Supplier cycle (follow all steps through the interface between the enterprise and a supplier for a purchase)
- Customer cycle (follow all steps through the interface between the enterprise and a customer for a sales order).

It may be sufficient to study system variables, functions and cycles. However, TOPP has added a fourth dimension to this: management philosophies. A management philosophy is a way of thinking and an object of focus that will affect the total performance of the company. Examples of such management philosophies are:

- Total Quality Management
- Just in Time
- Continuous Improvement
- Customer Satisfaction
- Time Based Management
- Zero Defects
- Zero Inventories
- Concurrent Engineering
- Management by Objectives
- Partnership and Strategic Alliances.

5 TOPP SELF ASSESSMENT

The TOPP self assessment is based on the TOPP model described in the preceding and a set of 22 general business processes shown in table 1.

There are two different approaches to self assessment:

- Questionnaire
- Process evaluation.

The first approach uses a questionnaire in standard format to be answered by the enterprise. This questionnaire consists of three parts (TOPP, 1992):

- Part 1
  Facts about the company such as products, equipment, human capital, finance, costs, customers, etc.

- Part 2

Overall evaluation of different functions and system variables to be answered confidentially by a limited number of employees at different levels in the organisation.

Process no.	Types of process	Process
1		Sales and marketing
2		Procurement
3		Technological planning
4		Design and engineering
5	Primary	Production planning and control
6		Production and assembly
7		Distribution
8		Order processing
9		Invoicing and reimbursement
10		Strategic management
11		Financial management
12	Support	Personnel administration
13		Information management
14		Maintenance
15		Internal control of health, environment and safety
16		Continuous improvement
17		Market development
18		Product development
19	Development	Technology development
20		Human resource development
21		Supplier development
22		Development of external relations

Table 1 General Business Processes in the TOPP Self Assessment Model.

- Part 3
  Detailed evaluation of primary and support functions and system variables. This part is answered by expert groups in the company. Management is represented in all groups.

The scale of assessment comprises an interval from 1 to 7 with 7 as «best practice».
Based on the response to the questionnaires, four types of reports are provided:

1. Individual report based on data from own company
2. Report based on data from all participating companies
3. Report based on data from groups of companies. Criteria for selection of groups could be competitors, products, production technology, etc.
4. Results from research. An R&D programme is established to analyse the total database to search for success factors, etc.

The second approach to self assessment uses the defined business processes in table 1. The enterprise will carry out this evaluation without any external assistance. The objective is to follow and monitor trends in performance to obtain a continuous improvement.

The methodology consists of the following steps (TOPP, 1994):

a) Identification of critical and important business processes in the company
b) Selection of analysing areas and indicators to measure the business processes
c) How to organise the self assessment
d) Collection of data
e) Presentation of results
f) Evaluation of results, actions.

6 REFERENCES

Camp, R.C. (1989) *Benchmarking-The Search for Industry Best Practices That Lead to Superior Performance.* ASQC Quality Press.

Carr, D.K. (1993) *Managing for Effective Business Process Redesign.* Journal of Cost Management, 7 (3).

Falster, P., Rolstadås, A., Wortmann, H. (January 1991) *FOF Production Theory, WP2 Report: Design of a Conceptual Model.* Trondheim.

Kimura, F. (1993) *Virtual Manufacturing Environment.* IMS Globemann 21 Meeting, Kyoto.

Moseng, B., Bredrup, H. (1993) *A Methodology for Industrial Studies of Productivity Performance.* Production Planning & Control, vol. 4, no. 3.

Rolstadås, A. (1993) *Manufacturing Industry Benchmarking.* APMS'93. Elsevier Science Publishers.

Rolstadås, A. (1994) *Beyond Year 2000-Production Management in the Virtual Company.* IFIP WG5.7 Conference on Evaluation of Production Management Methods, Gramado, Brazil, March 1993. Elsevier Science Publishers.

Rolstadås, A. (Ed) (1995) *Performance Management-A Business Process Benchmarking Approach.* Chapman & Hall.

Sink, D.S. (1985) *Productivity Management: Planning, Measurement, and Evaluation, Control, and Improvement.* John Wiley & Sons, New York.

Sink, D.S. and Tuttle, T.C. (1989). *Planning and Measurement in Your Organisation of the Future.* Industrial Engineering and Management Press, Norcross.

TOPP, (1992) *TOPP-A Productivity Program for Manufacturing Industries, Questionnaire for Analyzing Competitiveness.* University of Trondheim, Trondheim, Norway.

TOPP, (1994) Handbook for Self Assessment, University of Trondheim, Trondheim, Norway.

Wortmann, J.C. (1989) *Towards an Integrated Theory for Design, Production, and Production Management of Complex, One of a Kind Products in Factory of the Future.* Commission of the European Communities (ed.), ESPRIT'89, Proc. 8th Annual Esprit Conf., Kluwer Academic Publishers, Dordrecht.

A Manufacturing Business Process Reengineering Method: Design and Redesign of a Production Control Model

J. O. Strandhagen, Dr.ing.
T. Skarlo, Research Engineer
SINTEF Production Engineering, N - 7034 Trondheim, Norway
Telephone: +4773593138•, Telefax: +4773597117
e-mail: ola.strandhagen@protek.unit.no, terje.skarlo@protek.unit.no

Abstract

Design and Redesign of a Production Control Model are methods for business process reengineering of manufacturing companies. It is a practical approach to assist companies that want to make the way they control their manufacturing processes a competitive advantage in meeting future market and customer demands. It is based on application of a mixture of techniques and methods from well known control principles. It has been developed and tested in a dozen Norwegian companies.

Keywords

Business Process Reengineering, Production Planning and Control, Material Flow, Layout, Production Control Model.

1 INTRODUCTION

In order to compete successfully, firms must achieve excellence in managing their manufacturing operations. Companies need to change the role of manufacturing from being "internal neutral" to "externally supportive" based on Hayes and Wheelwrights (1985) stages.

Several studies on performance have been performed. The best known is probably "Made in America" (Dertouzos, et. al, 1989) which studied how the US industry should change in order to regain competitiveness over Europe and Japan. A stronger focus on manufacturing was one of the major findings. In the Norwegian TOPP project a study of the performance of 50 Norwegian industrial companies was performed. This study showed that product development, quality management, production control, procurement and internal material flow all had a higher relative importance than performance, Bredrup et. al (1994). From Bredrup (1994), we also know that traditional accounting and performance measurement systems are not adequate to measure customer satisfaction. Neither has the focus on ISO 9000 seemed to improve the performance of those companies being certified by it. Many ISO certified companies experience less flexibility in adapting to new customer demands. There is also a clear new focus on the total value chain view, introduced by Porter (1985), which requires a new way of thinking when designing manufacturing systems.

The new and current situation has not been coherent with the manufacturing and administrative processes of today's companies. There has been a need to develop methods and tools to redesign or re engineer the business processes of manufacturing companies. The traditional methods of redesigning the production system have been one of three; Either to choose and apply one of the "new" production control principles like Just in Time, OPT, or one of the more traditional ones. Or to directly copy solutions from other companies. Or the most misused one; to introduce (or even to change) an Electronic Data Processing system (most often based on MRPII) to "solve all problems". At NTH/SINTEF Production Engineering we have been working with these issues in several national as well as international projects. Through these projects, we have developed a practical method "Design of a Production Control Model" for manufacturing business process redesign.

2 DESIGN OF A PRODUCTION CONTROL MODEL

The basic idea of designing a Production Control Model is to make the way you organise and control your production and logistic system a competitive advantage for the company. It is a fact that products have shorter and shorter life cycles. The customers ask for specific designs with extremely short delivery times. The competition is global, making the price margins smaller. The first examples of requirements of 100 % recyclable products are seen.

This challenge can be met by a redesign of the manufacturing business processes by developing a Production Control Model (PCM) based on;

- A vision of the future market (pull)
- An analysis of the existing system
    - the production system
    - the product characteristics
- Knowledge and experience of existing theories, methods and techniques

When the scenarios of the future market are defined, the challenge is then in a systematic, but yet creative, way to analyse the current situation, identify potentials and limitations, synthesise the solution into a model, and implement it.

The different topics to consider when analysing the existing system are the following processes and aspects;

- Production planning and control
- Inbound and outbound logistics
- Layout
- Capacity adjustment principles
- Product design aspects related
- Material flows
- Organisation
- Monitoring functions
- Performance measurement

A proposed definition of a Production Control Model is;

> A Production Control Model is a description of how a production and logistic system is organised and controlled. It contains information about layout and material flows. It is used in redesign of a manufacturing system, as an operative control document, and for training purposes. The model defines a set of customer focused measurements to be used in the company.

The term model is used in the meaning as a description of a real world system, used to communicate and explain how the system should be organised and controlled. The following figure illustrates the process for redesign of a Production Control Model. The process is outlined as three streams, one for analysis and synthesis, one for education and training, and one for management discussions and decisions. The following sections will describe the analysis and synthesis in detail.

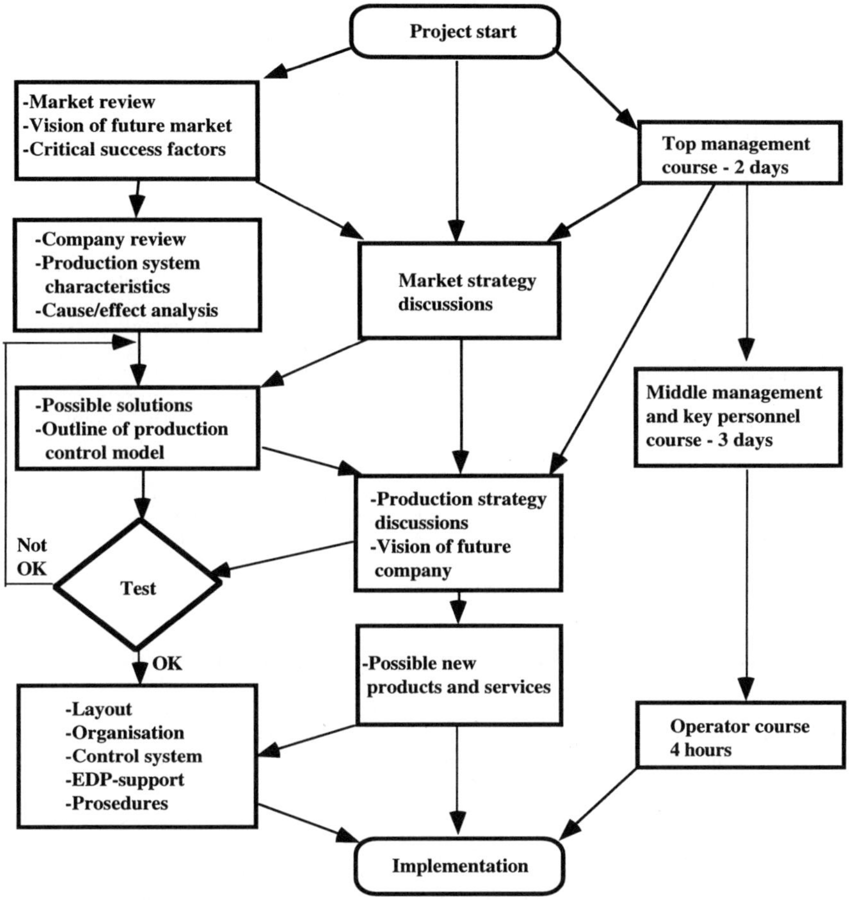

Figure 1 The process of design of a Production Control Model.

3 THE ANALYSIS PHASE

The analysis phase can be characterised by three main activities; data collection & investigation, establish visions for the future company, and define performance measurements. It is of course important to aim for a systematic approach. The objective of the activity is to ensure understanding of the basic features of what is to be controlled. If the collection of data becomes too much oriented towards a specific solution, there is a tendency to concentrate on the first solutions that came up. Other possibilities may thus be left unexplored. Therefore a structure that is independent of the solutions, is applied. In short words, the following stepwise approach is recommended. Each of these steps are explained in the next sections;

- Identify what is important to achieve in the market - the visions of the future
- Collect data and analyse the current situation in the company
- Identify the characteristics of the production system
- Use cause/effect analysis
- Perform training and education of key personnel
- Define performance measurement system

3.1 What is important to achieve in the market?

The whole idea of a business process redesign is to improve competitiveness. It is therefore necessary to investigate how the future income can be increased by better performance on delivery terms. It is not sufficient to ask the market department what is important. One must also look at the development in other countries, within other business sectors and from the customer's point of view. It is important to have both a dynamic and a strategic perspective of the market development. Early achievements give a competitive advantage. However, it is difficult to give credible estimates on how much these improvements will be worth, and which competitors will achieve the same. The gain obtained by possible improvements may be short lived. The companies will always have to endeavour to search for new areas of possible improvements. At least the following future requirements must be identified;

- The expected delivery times
- New functionality and designs
- Future product variety
- The expected accuracy of deliveries
- The degree of customisation
- Price potentials of better performance

Common for all these items is that the development of them will influence directly on the production system.

3.2 Collect and analyse data about the current situation in the company

The most resource demanding activity in this first phase is probably the collection of data and information to describe the current situation in the company, and how the company performs today. This means how the market recognises the company's services, the efficiency of the production system, and how well prepared the company is to meet future challenges. The following areas are examined and evaluated;

- The company's services to the market
- Material flow and layout
- Tied-up capital regarding production
- Reliability of the production system
- Existence of unfavourable policy, culture, etc.
- Product design
- Production flexibility
- Inbound and outbound logistics
- Outline of the current PCM

Concerning the service to market, the following items are examined; delivery times, delivery performance, the ability to customise solutions, the quality of finished products, the market shares and the price level. On the other hand it is also important to find out how the market recognises the company's services, how much the company is paid for their services offered, and how the market recognises the most important competitors.

Another important aspect is the design of the products, as this is one of the two most important factors deciding on which control principles that can be applied (the production resources regarding flexibility and capability being the other). The factors of design to consider are; modularity in design, whether new designs are based on existing components, whether new products fit within existing product groups, design for ease of processing, assembly, automation, repair and production control and whether production people involved in product design. This should establish how production requirements are taken care of in product design, and how designers and production people are co-operating. The ability to increase product variety within existing processes is strongly limited by how the design issues are highlighted.

Burbidge's (1979) material and production flow analysis is well known as a mean to analyse the flow through the production systems. Aspects of particular importance in addition to pure material flow aspects are; throughput time, set-up times, bottleneck resources, transport distances, size of batches, stock levels, changes in area of responsibilities, intervening priorities and unnecessary handling.

Due to increasing product diversity, short product life time and focus on delivery performance, production flexibility is examined. The study concerns both the production system's abilities to handle day to day requirements and happenings, and on ability to handle possible future requirements. This can be analysed by examining the ability to; produce a mix of jobs (short set-up times), to change to other products than today, to change routing when breakdowns occur, to operate profitably at different volumes and the possibility to expand the production capacity.

The capital tied up in machines and equipment, as well as inventories must be identified.

Supply and distribution are predecessor and successor to the production. Delivery performance depends on efficiency and reliability in the whole logistic chain. Items to be examined are; the number of suppliers and transporters, the stability in the number of suppliers and transporters, the need for inspection of goods received, the use of EDI, and limits and regulations to be followed.

A production system requires both reliability and flexibility. Flexibility to handle the unexpected, reliability to avoid the undesirable. Both actual reliability, preventive efforts and possible reliability improvements are important issues. Following items must be considered; the supplier's delivery performance, the machinery breakdowns, the reject rates, the need for safety stocks, and the use of quality assurance and maintenance programmes.

Unfavourable results forced by executive policy, company culture, misunderstood loyalty, etc., can be obstacles to improvements. Major findings have to be identified. Examples are overuse of costly machinery, major belief in large batches, high utilisation rate synonymous to high productivity.

The company is after all in business and has a more or less successful way of controlling its production. Development of a new PCM will benefit from knowledge and understanding of the existing one. This means examining the production control principles, the dispatching rules, the use of pull and push, use of forecasts, which control areas exist, whether scheduling functions are applied, the actual role of planning personnel and the degree of computer support.

All these questions must be asked and answered, and the findings taken down and presented to all people involved. The major findings of this study must be commonly accepted as the description of the company.

3.3 Identification of the key characteristics of the production system

On the basis of this information, the next step is to identify the key characteristics of the production system and its surroundings. Questions to be asked are;

- Does the company sell products or production capacity ?
- Is there an even or uneven workload throughout the year ?
- How many product types and variants of products ?
- What are the typical number of processing stages ?
- What is the degree of use of common components ?
- Is production mainly repetitive or not ?
- What are the typical customer orders size ?
- What are the typical delivery times offered ?
- What are the possibilities of parallel operations ?
- What are the **real** serial effects ?
- Are there any identified bottlenecks ?
- Is there a product dependency regarding production process ?
- What is the stability in the production system ?
- What is the degree of out sourcing ?

The answer to these questions will give the key characteristics of the company. It is these characteristics that will limit, as well as give alternative, solutions.

3.4 Cause/effect analysis

Disclosing possibilities of improvement and outlining of a PCM is just a part of the process. A number of problems has to be dealt with before the production system has become controllable. When solving a problem, one must aim at attacking the root causes and not the symptoms. An important part of the analyses is the use of cause/effect diagrams (Kume, 1985), and the involvement of operators in the problem solving process. The cause/effect diagram shows the connections and dependencies in the problem complex. In addition to giving a survey, it is used to choose hypotheses regarding causes. The hypotheses are checked out by data collection and data analysis. The cause/effect diagram shows both possible effect and limitations, for example in production control improvements.

3.5 Training and education of key personnel

The whole idea of the method is that the PCM is the company's own solution to their challenge. To be able to take part in creating the solution, a substantial amount of training must be performed. The training is a mixture of practical examples and theory, where explanations and solutions are supposed to pop up during the training seminars.

3.6 Define performance measurement system

It is a common finding in the projects we have performed that many of the performance indicators that are applied, are not related to customer satisfaction nor profitability. And they are not indicators that are used to evaluate and improve, but simply to measure. An important aspect in improving performance is to identify indicators that are directly related to customer satisfaction or the **ability** to satisfy customers. Delivery performance is an example of the first, flexibility being an example of the second. To evaluate the production system indicators that cover the production's ability to produce the right products, at the right time, at the right quality and with a minimum use of resources, must be identified. They must cover this ability both in the present time and in the future. The measurements will be different in different companies, and they must include indicators for;

- Delivery performance
- Flexibility
- Claims from customers
- Resource utilisation
- Work in progress
- Planning and scheduling performance

Some companies spend too much resources on uncritical productivity monitoring, other spend too little effort on measurements. Often there is a mismatch between the effort on monitoring and the use of the data from the measurement. It is important that the chosen indicators really give productivity indications. One must also be aware that if indicators of little importance are measured, the fact that they are measured gives them importance.

4 SYNTHESIS

The analysis phase should already have initiated some ideas of possible solutions. But now the highly iterative synthesis phase is where these and constantly new solutions are merged, discussed and revised. The activities are;

- Identify basic principles and solutions that are applicable
- Make rough outline of Production Control Model
- Test of solutions
- Merge and integrate solutions
- Make complete Production Control Model

4.1 Identification of basic principles and solutions that are applicable

The first activity is to identify a set of possible solutions. The list below should be considered as an example of a practical approach of applying some of the basic techniques of all those well known control principles;

- Identify and confirm T-points and customer order decoupling points, CODPs
- Find customisation potentials from standard components
- Make layout flow oriented
- Combine group technology and cyclic production
- Introduce capacity adjusted order point
- Split of production into standard and special products
- Apply simple level based control where applicable
- Identify control areas
- Define information flows

Identify and confirm T-points and customer order decoupling points, CODPs
Identical components used in several products can be stocked at the level where it spreads. This stock point (T-point) allows very short delivery time, depending on sufficient size it will absorb demand variations and it will be suitable for level control. T-point requires frequently use of the product or part. The T-point may coincide with the customer order decoupling point, CODP. But often the CODP is at a later stage in the production process. Both the T-point and the CODP must be identified, as they are the most important factor for deciding on which control principles to be applied where.

Find customisation potentials from standard components

An ideal solution in many contexts is to supply customised products composed by standard components. This means to produce standard components and mount on order. Production control at a T-point will often be order controlled consumption and level controlled supply.

Make layout flow oriented

A flow oriented layout improves strongly the controllability. The methods for analysing and redesigning the material flow are well described in Burbidge (1979).

Combined group technology and cyclic production

Products that can be processed in the same tool set-up or with the same machine adjustments can be processed in sequence. In this way positive effects by producing one-of-a-kind or small series are obtained. It is the product family that constitutes the series. By emphasising the right order, this serial effect might be increased. The solution is a cyclical production.

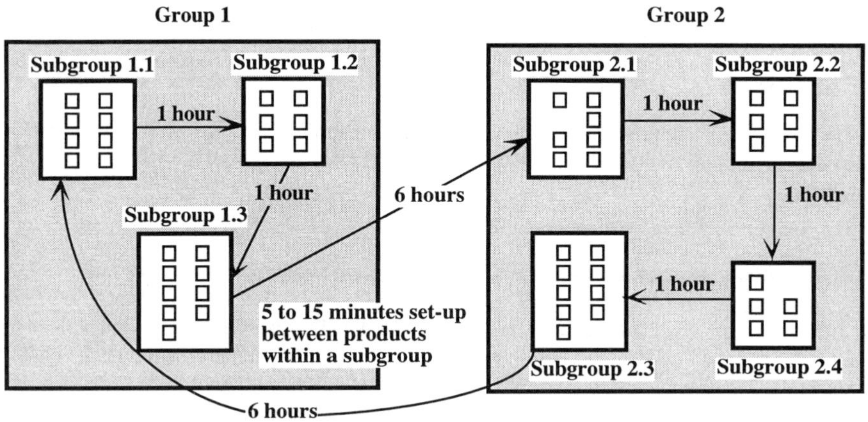

Figure 2 Group technology - Cyclical production.

Introduce capacity adjusted order point

A capacity adjusted order point takes into consideration the actual load on the production system. Possible production orders based on stock level or customer orders are grouped into **must, should** or **may.** A **must order** is critical with regard to delivery time and must be produced. A **should order** will soon become critical, or it is profitable to produce now due to favourable combinations with other products. It is up to the local manager or foreman to decide whether this order should be started or not. A **may order** is possible to start, but it should only be started if there are neither **must** nor **should** orders in the queue.

Split of production into standard and special products

Earlier the efficiency of a machine was solely measured by the number of processed units per hour. Lately more focus has been put on the flexibility of the machine. There still exist many high efficient machines with low flexibility (i.e. they are time-consuming to set up between the various products). In many companies there is an imbalance between the range of machines and the need for machines. A solution which has proven feasible in many companies is to separate the products into standard, processed in the old machine, and special.

Thus the amount of special products will be so small that a flexible machine with a relatively low capacity may be sufficient.

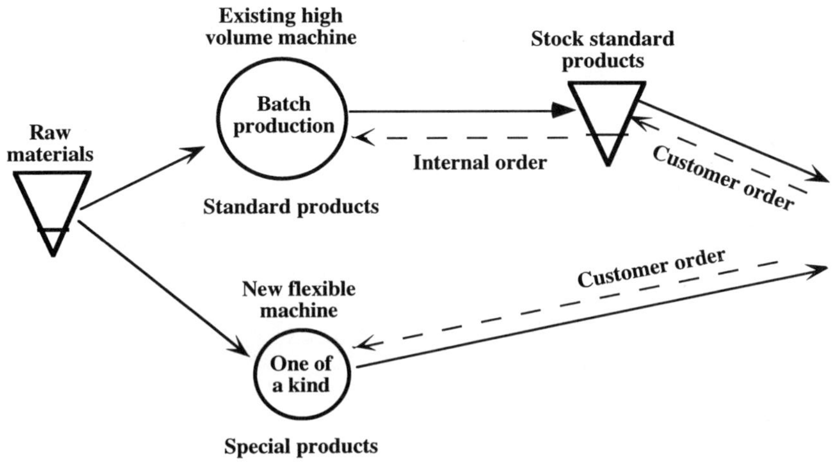

Figure 3 Split of standard and special production.

Apply simple level based control where applicable

As the use of MRPII systems, scheduling systems, bar codes and terminals has increased, the belief in using simple level based control techniques has decreased. Even for simple components advanced functions for calculating order sizes, printing production or purchasing orders, etc., has been implemented by use of computers. This has required a lot of administrative effort. We see now, probably as a result of the need for Just In Time performance, that many companies introduce KANBAN control principles. And that simple level based techniques like the Two-bin principle, have been reintroduced in many control systems. The basic idea is of course to use as little as possible effort and time on controlling standard components, and let these components be self-controlled by for instance KANBAN. And let the people and EDP systems concentrate on the complex, difficult to control customer designed components.

Identify control areas

Following the new layout, the identification of T-points and CODPs, and the control principles, new control areas must be defined. We know from several studies that the more often a production order crosses a border of responsibility, the more unnecessary delays occur. New control areas must be identified based on two major criteria. First, it is important that one manager/foreman has as few control principles as possible within his area. Secondly, the time required to perform the control tasks should be considered. These two criteria are much more important than for instance number of persons or machines, square meter, or number of product types.

Define information flows

The information flows are roughly defined by the control principles that are chosen. But it is important to exactly define the routes information flows, the responsible for these flows, and the routines to be followed.

4.2 Make rough outline of Production Control Model

Based on the chosen principles, the new layout, the information flows and the specific solutions, a rough outline of PCM should be made. The model will indicate solutions, and is normally not more than a few pages of writing. An example of rough PCM is partly shown below. This example is a company producing furniture, and the rough PCM looks like this;

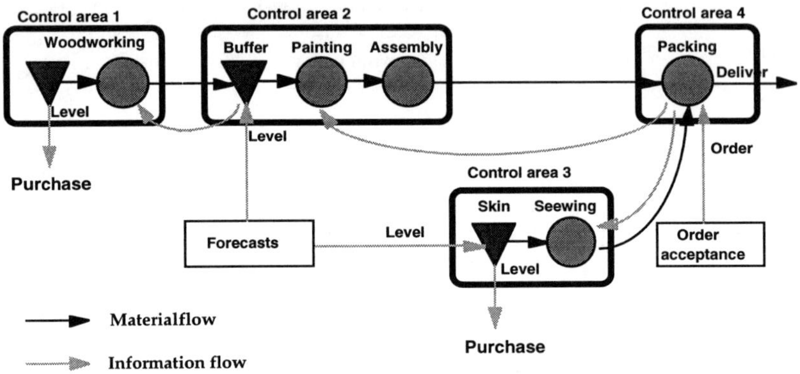

Figure 4 PCM for a Norwegian furniture company.

In addition to this figure, a list description of the control principles applied in each control area is defined. For the painting control area the following description is made;

> *Control area 2; Painting*
> Processes
> Control of buffer of unpainted pieces, polishing, painting, spraying, assembly
>
> Control
> Receive orders on paper from Control area 4. Orders should be painted/sprayed and assembled within 8 working days. This area is free to mix jobs within these 8 days to minimise set-up cost of painting/spraying. The control points of the buffers are decided based on yearly forecasts. Extra capacity should be required from wood working, Control area 1.

Such descriptions are made for all areas, as well as a set of common rules for the whole production system.

4.3 Test, merge and integrate solutions

The next step is to find out if the proposed model will have the anticipated results. In simple cases the test can be carried out by using a spreadsheet and/or running a pilot implementation. When dealing with more complex cases, we recommend stochastic discrete event simulation. The objective of the test is to find out whether the proposed solutions works as expected, gives the expected production volume, whether the flexibility is sufficient for coping with the expected disturbances and what use of resources that have to be estimated.

The next step is then to merge and integrate the solutions, identify discrepancies and correct them. New solution must be found. It is important that this is done with an open mind, as it is obvious that not all the initial solutions are feasible.

4.4 Make complete Production Control Model

The detailed and integrated solutions are included in complete Production Control Model. It is certainly not the final one, as it should be continuously changed as the surroundings and requirements change. The difference between the complete model is three things; It is far more detailed, the solutions are tested and integrated, and the use of EDP support is included. The model, still only existing on paper, must now be implemented.

5 IMPLEMENTATION ADVICE

A few remarks and some advice about the implementation process are needed. The emphasis on training and education is already mentioned. Another aspect is the need for top-management involvement. The whole idea is to make production control a competitive advantage in meeting future customer and market demands, making it a strategic issue.

The change-over process is critical. The change from old to new conditions has its own problems. It is of great importance to identify problem areas and to take preventive actions. These can include extraordinary limitations in delivery programme, full stocks, extra crew on expected bottlenecks, etc. There is a need for a specific plan to cover the change-over preparations and the change-over period. For some people every change is a fear. They struggle against all new. For others the new PCM entails that they lose importance and position. Some of them will struggle against too. It has to be taken into consideration that somebody will use implementation problems to prove they are right. The new PCM is not robust yet; small problems can grow big. Management has to ensure that positive and loyal people cover all positions of importance. The production capacity will possibly suffer during the change-over period. This possibility has to be co-ordinated with the sales department and even with customers. Build up of stocks and/or avoidance of sales campaigns just in this critical period can reduce risk for delivery problems.

6 SUMMARY DISCUSSIONS

This method has been applied with success in more than a dozen Norwegian manufacturing companies. They have all experienced radical improvements in performance. Examples of this will be presented at the conference, as well as in separate papers to be published. A major criticism to the method is that it is still very dependent on consultant's assistance. At NTH/SINTEF we have developed handbooks to assist the method, but up to now no companies have applied it without assistance.

The method should also be developed further with supporting methods for identification of problems and creating solutions. One example is the identification of area for KANBAN introduction, as well as maintenance of KANBAN systems. Another future area is integrating modelling tools, from traditional SADT diagrams to the enterprise modelling techniques, in the method.

7 ACKNOWLEDGEMENTS

This method has been developed over more than 10 years, based on the experience gathered in a series of industrial projects, as well as projects financed by the Norwegian Research Council. We want to express our gratitude to all these companies as well as the Research Council. Special thanks to the companies Håg a.s and Dynoplast A.S Stjørdal who have been extremely co-operative in our joint projects. The term Production Control Model was introduced by our former colleague Tom Quistgaard, and we would also like to mention the important influence from the works and books of Professor Asbjørn Rolstadås and Professor John Burbidge.

8 BIBLIOGRAPHY

Jan Ola Strandhagen and Terje Skarlo are both working at NTH/SINTEF Production Engineering in the area of Production Management. Skarlo has a background from offshore engineering and construction companies, and has been a researcher at SINTEF since 1988. Strandhagen, with i dr.ing. degree from 1994, has been with SINTEF for ten years. He is Research Manager of the section for Production Management since 1992.

9 REFERENCES

Bredrup, H., Bredrup, R. and Rolstadås, A. (1994) PMS contribution to improved enterprise productivity. *Proceedings from IFIP WG5.7 working conference on evaluation of production management methods*, Porto Alegre/Gramado, Brazil.

Bredrup, H. (1994) Measurement systems based on standard costing inhibit JIT implementation. *Proceedings from eighth international working seminar on productions economics*, IGLS/Innsbruck, Austria.

Bredrup, H. (1994) Standard illusions: ISO 9000 as an alibi for quality. *European Quality,* **1**(5).

Browne, J., Harhen, J., Shivnan, J. (1993) *Production Management Systems - A CIM Perspective*. Addison-Wesley Publishing Company.

Burbidge, J.L. (1979) *Group technology in the engineering industry* . Mechanical engineering publications Ltd., London.

Dertouzos, M.L., Lester, R.K., Solow, R.M. and The MIT Commission on Industrial Productivity. (1989) *Made in America: Regaining the productive edge.* The MIT Press, Cambridge, Massachusetts.

Hayes, R.H. and Wheelwright, S. C. (1985) Restoring our Competitive Edge: *Competing through Manufacturing*, John Wiley & Sons, Boston, Mass.

Goldratt, M. and Cox, J. (1984) *The Goal: Excellence in manufacturing.* North River Press, Croton-on-Hudson.

Kume, H. (1985) *Statistical methods for quality improvement*. AOTS Chosakai, Ltd., Tokyo.

Porter, M.E. (1985) *Competitive advantage: Creating and sustaining superior performance.* Free Press, New York, N.J.

Rolstadås, A. (1995) *Performance management - A business process benchmarking approach.* Chapman & Hall, London.

Shingo, S. (1985) *A revolution in manufacturing: The SMED system.* Productivity Press, Massachusetts and Connecticut.

19

Bottleneck-oriented Logistic Analysis as a Basis for Business Re-engineering

T. Penz, P. Nyhuis
Institute for Production Systems, University of Hannover
Callinstraße 36, 30167 Hannover, Germany
Tel: +49-511-762 3494
Fax: +49-511-762 3814
E-mail: pz@thor.ifa.uni-hannover.de

Abstract

To ensure short delivery time and high delivery performance it is necessary to have control over manufacturing procedures and to keep lead times to a minimum. At the same time production profitability must be guaranteed. The interaction between these (and other) objectives and occurring conflicts can be investigated with bottleneck-oriented logistic analysis, which makes it possible, for example, to determine ideal states and realizable values for lead times and inventory for existing or planned production structures. In addition, the system behaviour of the production can be investigated with varied side constraints; also the effects of measures taken to influence the logistics potential can be described qualitatively and quantitatively. This allows an objective-oriented optimization of the manufacturing procedures and structures with regard to business re-engineering.

Keywords

Bottleneck-oriented logistic analysis, logistic operating curves, logistics positioning, simulation.

1 INTRODUCTION

The strategic importance of production logistics will undoubtedly continue to increase in the future. Particulary under conditions where far-reaching changes in progressively shorter periods are required, they have significantly influenced the fulfilment of market demands and company objectives. Regarding the concept of maximizing the efficiency of the production sequence, production logistics should cater for the realisation of such objectives as maximizing schedule performance and loading, as well as minimizing lead time and inventory.

Employing only the currently available logistic systems and instruments in industrial companies is not enough. Rather, it must be guaranteed that the existing interdependence between the logistic objectives in these systems is adequately considered, in order to facilitate a target-oriented structuring and control of material and production flows.To this effect, general logistic models are required to describe the dynamic correlations between the afore-mentioned logistic quality features. Furthermore, the actual and the scheduled production flow of orders should be realistically described.

The funnel model and the throughput diagram represents such a general model (Figure 1). Throughput diagrams enable a qualitative precise description of the dynamic behaviour of work centres. Furthermore, the four primary quality features of production (schedule performance, inventory, lead time and loading) can be depicted and a numerical calculation of the corresponding determinants can be effected (Wiendahl 1987). In this way, the pertinent relationships between the logistic objectives can be demonstrated and a mathematical description derived therefrom.

However, research has shown that evaluation based solely on the graphs and key data obtained from simulated as well as real production sequences is often inadequate to fully describe the dynamic behaviour of production systems, especially the relevant correlations between the logistic objectives. For this reason it has been found useful to condense the results of various stationary states into what are termed "logistic operating curves". These allow an identification of the dependence of diverse objectives on the changes in parameters. In the meantime, it has also been proved that a target-oriented evaluation and optimization of process flows and structures is possible using such logistic operating curves.

Logistic operating curves facilitate the representation of the logistic determinants of performance, lead time and range as a function of inventory (Figure 1 at the bottom). It can be seen that the performance changes only negligibly above a specific value of inventory. At this point, there is always enough work in process so that work interruptions as a result of obstructions in material flow do not arise. Below this inventory value, however, increasing performance losses occur as a result of insufficient work supply.

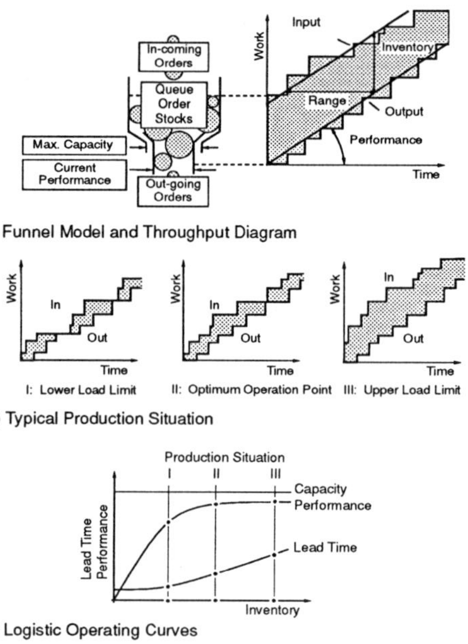

Figure 1 From the Funnel Model to the Logistic Operating Curves.

The lead time respective range in contrast increases proportionally with the inventory above a specific inventory value. A reduction in the inventory leads to a decrease in the lead time. However, the minimum value of the lead time cannot fall below a certain minimum, determined from the operation time of the orders and the transit time, as the case may be. With the aid of the operating curves the interdependencies between the logistic factors can be shown for any work centre. They illustrate the dilemma of the forward scheduling, revealing the mutual enhancement between some of the logistic objectives, as well as a reciprocal interference in others. As a result, there is no single objective whose value could be maximized or minimized; rather, the effects of measures respecting all the elements must be considered simultaneously.

2 POTENTIAL APPLICATION POSSIBILITIES OF LOGISTIC OPERATING CURVES

The logistic operating curves identify the determined system behaviour of a production corresponding to the quality features when the most important adjustable parameter of the production control changes: this parameter is the inventory. Additionally, it is possible to represent different operating curves with varying production and order structures. These can be compared with each other, thereby providing a means of assessing the effects of measures carried out in the production flow under logistic aspects. Figure 2 shows in a summarised overview that, using this technique, new possibilities for quantifying logistic rationalization potentials are presented.

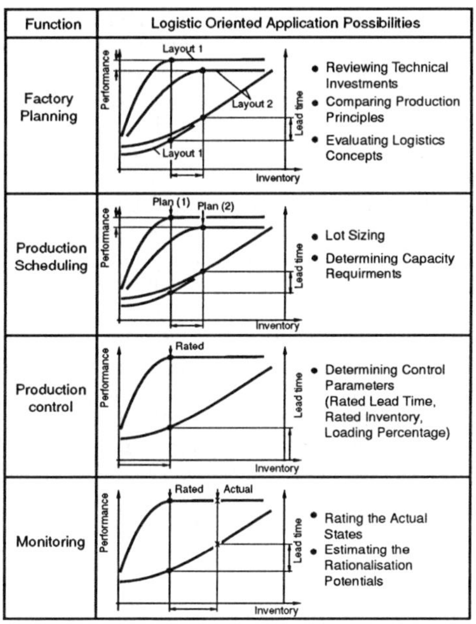

Figure 2 Possible Application for Logistic Operating Curves.

In this way, operating curves can be employed to evaluate the process flows within the scope of a monitoring. It can be shown which lead times and inventories could have been achieved with the existing structural constraints without significant bottlenecks in the material flow and accompanying performance losses.

Besides, a differentiated determination of the rated lead times and other control parameters for the production control is possible. Representing the quality features in a diagram enables a decision to be made as regards the most important feature, depending on the current production and/or market situation, as well as the constraints specific to the work system. The relevant parameter adjustments which should be made to the PPC system can also be decided upon at any time. At the same time, it can be shown which consequences can be expected with regard to the logistic quality features with a change in parameters.

The afore-mentioned areas of application aim at indicating the existing logistic potentials, analysing deficits, or sustaining the logistic state. Should it turn out, within the scope of these applications, that objectives cannot be realised without supporting measures, the operating curves can also be enlisted for a logistic evaluation of planning alternatives. Investment decisions (deployment of new transport systems, introduction of new production technologies,capacity expansions, ...), the choice of an alternative lot-sizing process or measures in the field of business re-engineering (design of manufacturing islands, outsourcing, ...) can be reviewed with regard to their importance for the realisable lead times, inventories and loads.

3 METHODS OF DETERMINING LOGISTIC OPERATING CURVES

Before introducing the different types of operating curves and their forms of determination it should again be pointed out that operating curves are used to describe possible diverse production states. A momentary state in the production always corresponds to a single operating point on a curve. The curve itself shows how the pertinent production or work centre would have behaved if a different inventory level had been set, the other constraints remaining constant.

This clarification indicates that, using analysed data of process chains recorded at different instances, real logistic operating curves can be reconstructed. This assumes a significant change in the inventory within the evaluation period, resulting in different operation points, with other constraints remaining constant. This refers essentially to the structure of the capacities, and the orders to be handled. Since the necessary evaluation time is normally very long, thereby making a fulfilment of the above-mentioned demands very difficult, real operating curves for practical purposes will seldom find application.

A second possibility is the simulated operating curve. Simulation systems provide the means of reproducing a real system using computer programs. In this way, the behaviour of the system with a specific change in the simulation conditions can be described, analysed and optimized. In the past, several researchers have presented simulation methods which describe the relationships between the different production objectives, in the afore-mentioned fields of application (e.g. Rice/Gross, 1990; Kuprat, 1991; Kuhn, 1992; Zäpfel, 1992).

The third type of curve is the ideal logistic operating curve. Based on an analytical consideration of ideal production flows and the derived ideal process key data, ideal production flows were modelled (Nyhuis, 1991). By this means, an order is handled directly on its arrival at the work centre. Hence there is no interoperation time between two operations. Under this restriction, the inventory is determined directly from the work content of the orders to be processed. This depicts the lowest limit of the inventory, which cannot be gone below even under ideal conditions, without causing inevitable loading losses caused by the inventory. Limits and ideal operating curves for the lead time and the range can be established (Nyhuis, 1993).

In the meantime, analyses of several simulation results has made the derivation of an approximation equation possible, enabling a mathematical determination of the operating curves. The approximation equation for these theoretical curves as well as the corresponding parameters and the application constraints are fully described in (Nyhuis, 1993). In order to give a rough sketch of the application possibilities of theoretical curves the parameters of the approximation equation are explained in the following.

The two most important parameters of the equation are the "maximum possible performance" and the "ideal minimum inventory". These two factors, which also determine the ideal operating curve, are derived from work-centre-specific or production-order describing determinants, which are normally present in every company.

Consequently, the 'maximum possible performance' variable is derived from the capacity. The upper performance limit of a work centre is fundamentally determined from the restricted capacity factor (production facilities or personnel). Additionally, it must be observed that, owing to capacity-reducing disruptions in production (e.g. machine breakdowns), the maximum possible loading can be limited further. Furthermore, as long as the capacity is given in loading hours rather than in standard hours, this must be corrected with the performance rate when determining the maximal possible performance.

As illustrated above, the ideal minimum inventory describes the value needed under ideal conditions to make use of the maximal possible performance. It is determined essentially from the work content of the orders: the minimum inventory depends particularly on the variance of the order times besides the mean value. Via the order times, the determination of the operating curve is also affected by the lot size, setup times and the processing times. In addition, it can if necessary be taken into account that an order in stock could be related to the minimum interoperation times because of transportation, or other factors (e.g. cooling times of hot parts).

Furthermore, via structure parameters, the coupling of the relevant work centre to the production flow can be described. In this way, it can be taken into account whether each work centre possesses its own inventory buffer or whether a common buffer exists for several work centres. In the second case, the risk of an interruption in the material flow for a particular work station is reduced, so that the necessary aggregate buffer can be reduced.

Should the parameters in Figure 3 be present, the operating curve can be directly determined with the said approximation function. A change in any of the parameters results inevitably in a different curve. Consequently, the approximation equation can be employed to investigate and assess the effects of different influencing measures on the logistic potential of a work centre, qualitatively and quantitatively, and if need be to establish a basis for comparisons.

4 APPLICATIONS

The application possibilities of determined logistic operating curves are manifold. It was illustrated in Figure 2 that these curves can be employed to predict the qualitative and quantitative effects on the logistic objectives.

The following examples stem from an industrial project which was carried out jointly with a manufacturer of measuring instruments. The primary goal was to investigate the application possibilities of determined operating curves as supplement to a commercial monitoring system.

The particular task here was to examine whether the real process flows in the department of "insertion of printed-board" could be represented well enough with the approximation equation. After confirming this, and based on the determined operating curve, the existing logistic potentials and the measures aimed at by the company for reducing the lead time and inventory were discussed and analysed. In a subsequent investigation, it was examined whether the predicted improvements in the production flow could be implemented.

In the first stage of the project, a comprehensive analysis of the lead time and the inventory was carried out. The results are shown in a highly condensed form in Figure 3. With the aid of this graph, which depicts both the logistic operating curve and the vital logistic data, the most critical work centre as regards order flow (i.e. giving rise to lead time) can be identified. Furthermore, using the logistic operating curves, it is possible to determine the work centres at which a reduction in the lead time could be effected, using control measures alone.

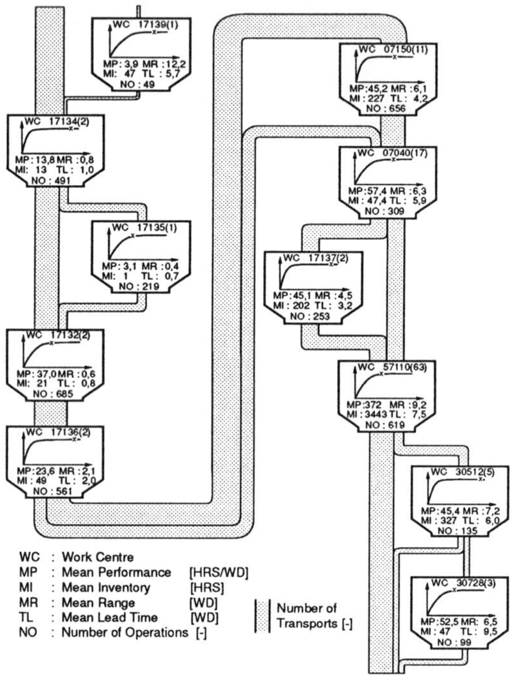

Figure 3 Logistic Score Data in the Material Flow (Extract).

The measured operation point of these systems is located well to the right of the transitional range. In contrast, for work centres where the operation point is located in the transitional region, it is necessary to effect structural changes (orders, capacities, layout,...) in addition to the control measures, if significant reductions in the lead time are to be realised. Two examples from the investigation in which different procedures could be derived will be discussed in the following.

In the first work centre the concern was with 'SMD insertion automats' (WC 17137) (two single work stations) which were fully loaded on a three-shift operation basis.The throughput diagram of this work centre for an investigation period of 6 weeks is shown in Figure 4. The trends of the work input, output and inventory are shown as a function of time. Additionally, the lead time elements have been entered, which describe the individual lead times of the different orders through the system.

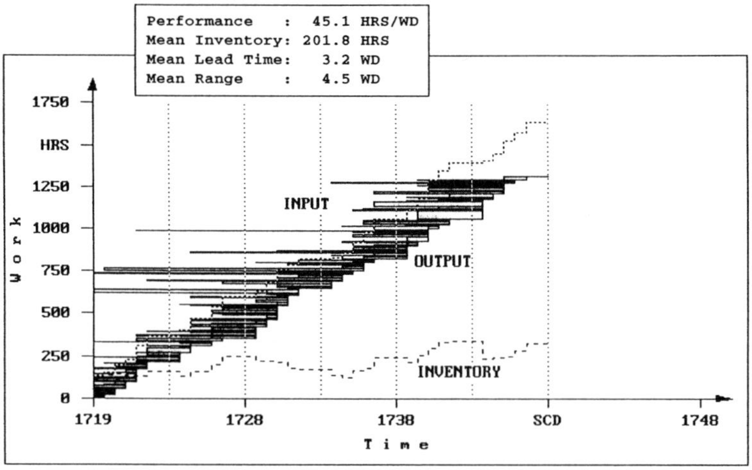

Figure 4 Graphic Results of the first Analyses for the 'SMD Insertion Automats'.

It can be observed from the diagram that fluctuating inventory was present at the work centre during the evaluation period. The work input was obviously not tuned to the output or the capacity of the work centre. Moreover, the lead time elements show that many changes in order sequence were carried out. The existence of changes in order sequence indicates that inventory reductions are possible, for these changes in sequence can only be carried out, if several orders are simultaneously present in the queue.

On the whole, the results show that the existing mean inventory of about 200 hours is clearly higher than the level necessary to maintain performance. In order to quantify this, the operating curves for the subsequent analysis were determined from the minimum inventory (resulting from the order times of the completed orders, see above), the ratio of current inventory to minimum inventory as well as the realised mean performance, with the aid of the approximation function. These curves as well as the actual values of performance, range, lead time, and inventory are shown in Figure 5.

In addition, 'appropriate value range' of the key data is indicated. The limits of this range were set at 1.5 to 2.5 times the ideal inventory in agreement with the company. The graph shows firstly that the measured operating state lies clearly in the over-load region. It can be seen that it is possible to reduce the inventory and the lead time significantly merely by instituting appropriate inventory control. Even presuming that high loading must be ensured at this work centre, it can be assumed that an mean inventory of about 60 hours is sufficient to guarantee the necessary performance. This would result in a reduction of the lead time of 3.5 work days to 1.5 work days. The restructuring measures planned by the company for increasing the insertion rate could hence be shelved, since at the time in question no advantages of any consequence could be expected from them from the logistic standpoint. For this reason an investment of tens of thousands German marks could be saved.

The subsequent analysis carried out after nine months (over a period of six weeks as well) confirmed the previously established assertion with respect to the existing rationalisation potential. The analysis showed that the inventory for this work centre could be reduced, on an average, to 94 hours (by almost 55%). In the case of a higher daily performance (47.6 instead of 45.1 hours per work day), the range and the lead time are reduced from 4.5 to 2.0 work days (WD) and from 3.2 to 1.9 WD respectively.

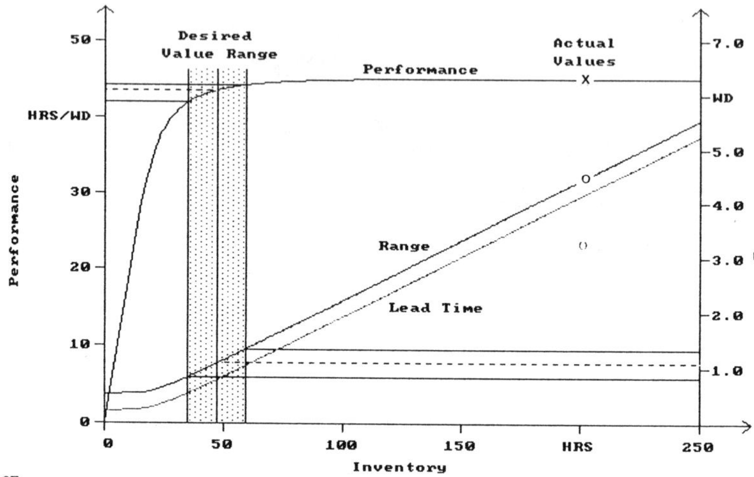

Figure 5 Determined Operating Curves for the 'SMD Insertion Automats'; 1. Analysis.

The example discussed above shows that it is possible to realise considerable reductions in the lead time and the inventory, without altering the structural constraints. Within the scope of the conducted analysis, there were also a number of work centres where it could be proved that, in order to increase the logistic potential, changes in the structure of order time or in the capacity situation were necessary.

An example of such a work centre is the 'Segment Production' (WC 57110) (manual insertion of component with 63 work stations of which an average of 48 were in one-shift operation). Figure 6 illustrates the production flow measured in the first investigation. The diagram shows that a fairly uniform inventory was present at the work centre on the whole. The input is fairly well geared to the output. But at the same time the numerical evaluation of the feedback data shows that the range is 9.2 WD and the mean lead time 7.5 WD.

Since these lead time values were clearly higher than all those obtained from the other analysed work centres, and the 'segment production' represented the most important capacity of the production department, the cardinal importance of this work centre from the logistic viewpoint is accounted for. The operating curves showed that these extremely high values of the lead times can still be considered as reasonable, under the prevailing conditions. Reducing the inventory and the lead time solely through production control measures would not have been possible for this work centre, if the company had not accepted performance losses at the same time.

While discussing the analysis results, it came to light that the required high level of lead times and inventories was attributable in particular to the existing structure of the order time. The analysis of the order times showed that the distribution of the order times was marked by wide mean variation, as well as high mean value. However, since the function for determining the ideal minimum inventory is quadratic with respect to the standard deviation (Nyhuis, 1991), it could be concluded that orders with a high work content should be reduced specifically to achieve a harmonization of the order structure. Accordingly, after the analysis, the company subdivided all orders with order times higher than 50 hours and pooled requirements. Furthermore, it was decided that the orders with order times higher than 30 hours should be handled by two or more workers in parallel. These two measures, affecting roughly 18% of the operations and about 55% of the total work content, led to a reduction of the ideal inventory by 35%.

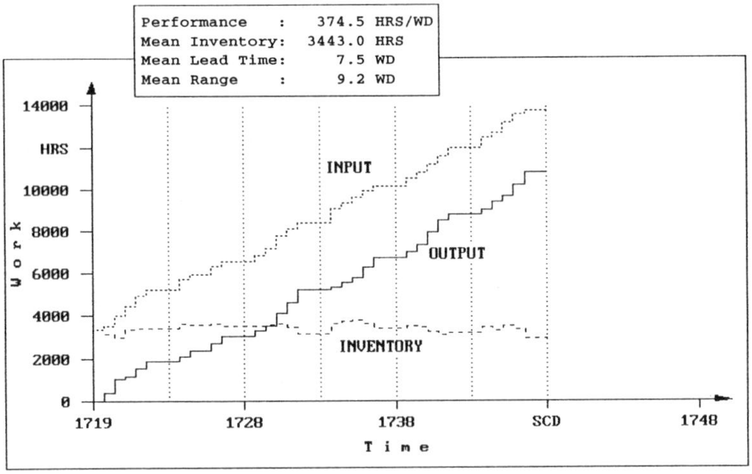

Figure 6 Graphic Results of the first Analyses for the Work Centre 'Segment Production'.

The success of these measures was reflected in the process flow measured during the subsequent analysis and in comparison between the operating curves and the logistic data for the two analyses (Figure 7). Inventories and lead times could be reduced by more than 30% without incurring further performance losses.

However, it must be stressed here that is was not alone the measures aimed at harmonizing the order times that brought this reduction. Rather, the company also succeeded in utilizing the ensuing potentials by reducing the inventory to the targeted level through production control measures

7 SUMMARY

Using the operating curves, the interdependencies between the logistic objectives of lead time, inventory and performance can be described as a function of the company-specific constraints. Hence, they lend themselves as tools for demonstrating the effects of measures aimed at influencing these objectives, and consequently support corporate decisions, particularly in the area of production planning and control (PPC) and factory planning.

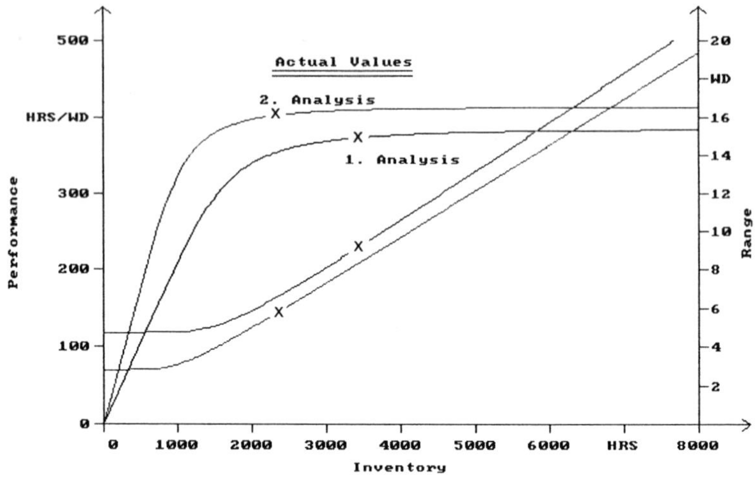

© IFA C2414E

Figure 7 Comparison the Determined Operating Curves for both Analyses; Work Centre 'Segment Production'

Hitherto, the generation of such logistic operating curves was only possible via simulation. The mathematical model for determining the operating curves enabled a satisfactory projection of real process flows in production despite the at first sight considerable simplifications. The pioneer prototype applications of determined curves with monitor systems show that with this model differentiated predictions respecting the attainable lead times and the inventories can be derived.

8 REFERENCES

Bechte, W. (1984) *Steuerung der Durchlaufzeit durch belastungsorientierte Auftragsfreigabe bei Werkstattfertigung*. Fortschritt-Berichte VDI, Row 2, No. 70, Düsseldorf.

Kuhn, A. (1992) Modellgestützte Logistik - Methodik einer permanenten ganzheitlichen Systemgestaltung, in *Rechnergestützte Fabrikplanung '92*. VDI-Berichte 949, Düsseldorf.

Kuprat, Th. (1991) *Simulationsgestützte Beurteilung der logistischen Qualität von Produktionsstrukturen*. Fortschritt-Berichte VDI, Row 2, No. 228, Düsseldorf.

Nyhuis, P. (1991) *Durchlauforientierte Losgrößenbestimmung*. Fortschritt-Berichte VDI, Row 2, No. 225, Düsseldorf.

Nyhuis, P. (1993) Betriebskennlinien zur Entscheidungsunterstützung in der Produktionsplanung und -steuerung, in *Neue Wege der PPS* (ed. H.-P. Wiendahl), gfmt-Verlag, München.

Rice, W. und Gross, R. (1990) Simulation Technique and the WIPAC Curve for Balancing Capacity against Inventory. *APICS*, 33rd International Conference Proceedings, New Orleans.

Wiendahl, H.-P. (1987) *Belastungsorientierte Fertigungssteuerung. Grundlagen, Verfahrensaufbau, Realisierung*. Hanser-Verlag, München, Wien.

Zäpfel, G. and Hödlmoser, P. (1992) Läßt sich das KANBAN-Konzept bei einer Variantenfertigung wirtschaftlich einsetzen? *Zeitschrift für Betriebswirtschaft*, 62nd Annual, No. 4.

20

Measuring and re-engineering logistics chains in the construction industry

T. Wegelius-Lehtonen
Helsinki University of Technology, IIA Research center
Otakaari 4A, FIN-02150 Espoo, Finland
Tel. 358-0-451 3664, Fax. 358-0-451 3665
E-Mail: Tutu.Wegelius@hut.fi

Abstract

The construction industry needs new tools and approaches to develop their business processes. The objective of this paper is to introduce a new framework for re-engineering logistics chains in the construction industry. The re-engineering process is based on co-operation between construction firms and material suppliers. Designers and subcontractors also participate when necessary.

Two tools are introduced as a part of the framework. These tools measure the effectiviness of information and material flows. The objective of the tools is to identify opportunities to improve logistics chains of different construction materials. With these tools logistics chains can be benchmarked to find out best practices and improvement potentials of the firms.

One objective of the paper is to emphasize the importance of a process view of construction logistics. The focus is on material and information flows between designers, material suppliers and construction firms.

Keywords

Logistics chains, measuring tools, re-engineering process, construction industry

1 INTRODUCTION

Construction projects typically involve many actors: an owner, an architect, design offices, a main contractor, subcontractors and material suppliers. In addition, the products (buildings) are usually unique. The construction site organisations and other actors change project by project.

In this situation, many people in construction firms think that one of the main problems is the lack of repetitiveness and this is why the coordination of material and information flows is difficult. That assumption is wrong. What people can´t see is that the construction industry is a typical "one-of the kind" -production industry (Burbidge 1993). Products are different project by project, but work and material delivery processes are almost the same.

Material costs are over 50 % of the building costs (Asplund and Danielson 1991). Therefore, material deliveries play an important role in constuction processes. Materials can be grouped into two categories: a-class materials that are customized materials (design to order) and b-class materials that are standard materials (make to order or make to stock). At the moment, there are big problems in the both material groups. Design changes, timing, shortages, handling and storing, damages, waste and packaging are the main problems.

At the same time, there are many challenges that construction industry should respond to. In competitive industries firms always have cost saving requirements. This is especially so when the economy is in a recession. Cost efficiency is not enough for competitiveness. Profitable construction firms increasingly have to be also time-based competitive. That means shorter delivery times. Re-engineering the logistics processes is needed if the construction industry wants to respond to these challenges.

This article offers some new tools and approaches to develop business processes in the construction industry. In the next chapter, a new framework for the re-engineering process is introduced with two measuring tools. The framework and the tools are applied in chapter three and the key results of a case study is described. More general results are presented in chapter four and conclusions are made in chapter five.

2 SOLUTION METHODOLOGY AND THE MEASURING TOOLS

2.1 The new framework for the re-engineering process

The new framework for the re-engineering process is based on a process view and co-operation of the firms operating in delivery chains. Because of a lack of partnership thinking, re-engineering process typically requires a development project. The project involves a construction firm as well as a material supplier. The reason is that the measuring and re-engineering process is carried out material by material.

The best results can be reached if the whole consortium of material suppliers and construction firms develop their processes at the same time. This way benchmarking can be used and firms are more active and curious to get results.

There are five key steps in the re-engineering process:

1. Identifying delivery processes and measuring the current performance
2. Benchmarking current practices
3. Re-engineering a new ideal model
4. Testing new practices on pilot projects
5. Putting new tested practices into use

The first step in the re-engineering process is to evaluate the current performance and practices. For that purpose, some example construction sites are chosen and the material deliveries of the chosen material groups are analyzed by the measuring tools. The more there are example sites, the better possibilities there are to find out different practices and to benchmark them. The idea is to search for present best practices and improvement potentials.

After analyzing and benchmarking current practices the project continues in the development groups. These groups re-engineer the logistics chains further by creating cost saving solutions to the problems that were evaluated in the first step.

Both material suppliers and construction firms participate in these development groups, as well as a reseacher. The best experts are those who are working with material delivery processes every day. At material suppliers those are staff from sales, production and dispatching and at construction firms the best experts are purchasers and site foremen.

An important stage of the project is to test new ideas on pilot construction sites. The proposals for improvement are tested in practice and the results are evaluated. Perhaps the most difficult task is to put the best, tested practices widely into use. The systematic education and information will help to disseminate the new ideas.

2.2 Tools for measuring costs and time

Two measuring tools are used during the first step of the re-engineering process. The first tool is an activity and cost analysis tool and the second is an accuracy and delivery time analysis tool. The objective of the activity and cost analysis is to find out the costs of the material flow and, at the same time, to show costs of unnecessary work in the delivery chain. The objective of the accuracy and delivery time analysis is to clarify the structure of the delivery time and the accuracy of performance.

Usually the activity and cost analysis and the accuracy and delivery time analysis are implemented at the same time because the results of the analyses support each other. Therefore the material group and the sites that will be analyzed are generally the same in the both analyses.

Activity and cost analysis

The activity and cost analysis is based on the theory of activity-based costing. The principle of activity-based costing is that the operations of the company are divided into activities, which use different resources (Brimson 1991). The same method is used here for analysing activities in the logistics chains and the costs of these activities. The activity and cost analysis focuses on material flows.

The first step of the activity and cost analysis is to identify all activities in the logistics chains. For standard materials (b-class materials), the business process starts with placing an order and ends when materials are assembled. For customized materials (a-class materials), the business process starts when architectural design starts. This stage in the analyzing process is called activity analysis. An example of the grafical representation of the activity analysis can be seen in the figure 1.

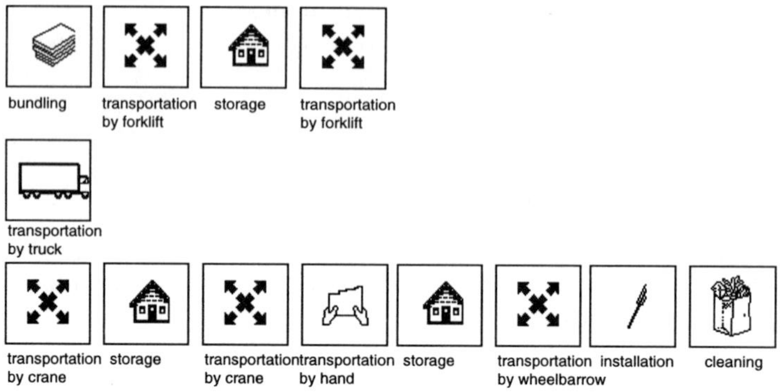

Figure 1 The logistics chain of the plasterboard deliveries on a case site.

After identifying the activities, the second step is to measure the costs of the activities. For that purpose one has to know what are the resources used by each activity. For example workers, warehouses and transport facilities are these kind of resources. It is also necessary to know what are the costs of the resources. For that purpose, a specific study has been made (Bergström J. and Gröning M. 1992). For instance, standard costs of different transportations and storages have been determined.

The costs of each activity can finally be calculated when the use of resources has been found out. After calculating the costs of each activity and the capital costs tied-up to the process, the cost structure of the logistics chain can be presented by graphs. An example is shown in the figure 2.

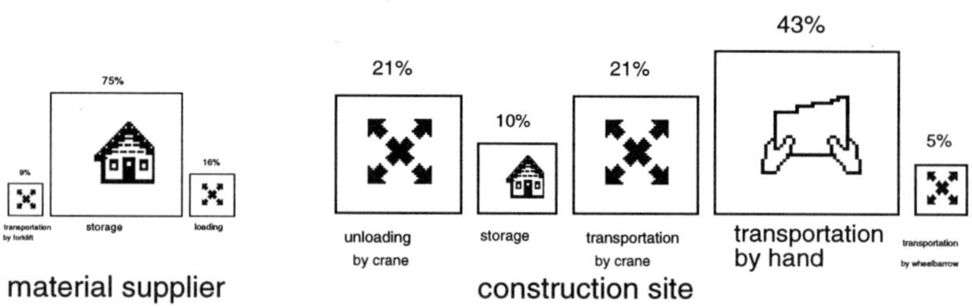

Figure 2 The logistics costs of the supplier and the case construction site. The area is correspond to the costs.

Accuracy and delivery time analysis

The accuracy and delivery time analysis has been developed to find out time lags in material and information flows. With this analysis, planned and actual timing data is gathered and showed graphically. The analysis is based on the theory of controllability engineering developed by Prof. E. Eloranta (e.g. Eloranta and Nikkola 1992). Controllability engineering aims at finding the improvement potential in business processes by using quantitative methods. The accuracy and delivery time analysis focuses especially on information flows.

Using accuracy and delivery time analysis is very simple. First the important milestones in both the material and the information flows are defined and included to the analysis. For customized materials it is necessary to analyze also accuracy and lead time of the design process. It is essential to analyze both the planned and the actual points of time. This way accuracy of the material delivery process can be studied. The ordering day, the planned and the actual day of manufacturing and the planned and the actual delivery day are examples of these milesstones.

The second step is to collect the analysis data from different sources. To get reliable results, it is necessary to use documented data from the planned and the actual time points. Documented data can be found for example in delivery orders, production plans, construction site diaries and installation plans. If documented plans are not available, the processes are probably not planned and controlled well enough.

The most informative way to present the results of accuracy and delivery time analysis is to use graphs. This way time lags and delays in the delivery processes can easily be clarified. For example the analysis shows if a material delivery has arrived to the site many weeks before installation or if the customized material has been manufactured many weeks before the planned delivery day. Figure 3 illustrates the results of accuracy and delivery time analysis on a case site.

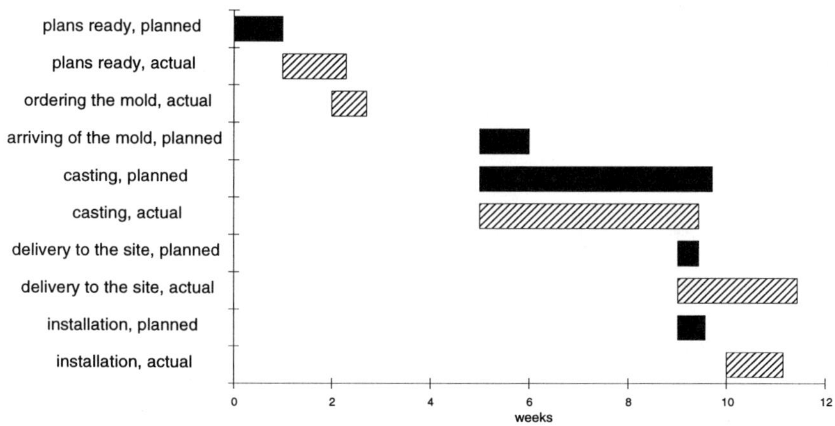

Figure 3 The structure of the delivery time of concrete elements on a case site.

2.3 How to measure and re-engineer plasterboard deliveries - an application example

Background

One of the earliest re-engineering project, which used the framework introduced previously, was co-operation project between three major Finnish construction firms and a large Nordic construction material supplier. The project was carried out between the fall 1992 and the spring 1993.

Plasterboard was the material, the logistics of which was re-engineered. Plasterboard is heavy and bulky construction material which is used for interior walls. Plasterboard is typical standard b-class material which is usually made to stock.

Identifying current performance

The measuring tools were developed during this re-engineering project of plasterboard deliveries. Current performance was identified at five construction sites. One of them was a office building and the rest were residental buildings.

A typical delivery chain of the plasterboard deliveries is illustrated in the figure 1. The logistics chain of the first case site shows that the most of the activities of the delivery chain take place at construction sites.

The figure 2 shows that also the costs of the material flow are highest at the site. In particular, several handlings and movements caused high logistics costs at the site.

Figure 4 illustrates an example of the results of the accuracy and delivery time analysis.

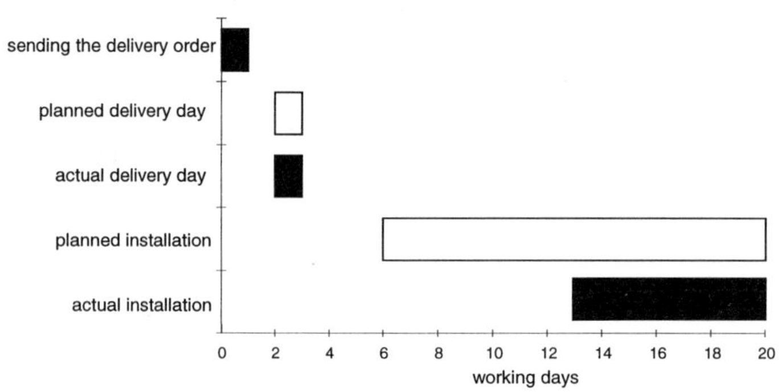

Figure 4 Delivery time and accuracy of the plasterboard delivery at the second case site.

At this second case site the delivery order was sent late and the supplier had only one day time to deliver the material. Still, the material was stored at the site more than two weeks before the installation was started.

Benchmarking current practices

After measuring current performance, the different practices of the case sites were benchmarked using the logistics costs as a yardstick. Logistics costs were divided into three parts: logistics costs of the supplier, logistics costs of the site and the transportation costs. Figure 5 shows the logistics costs of different case sites. The logistics costs were expressed as a percentage of the purchasing price.

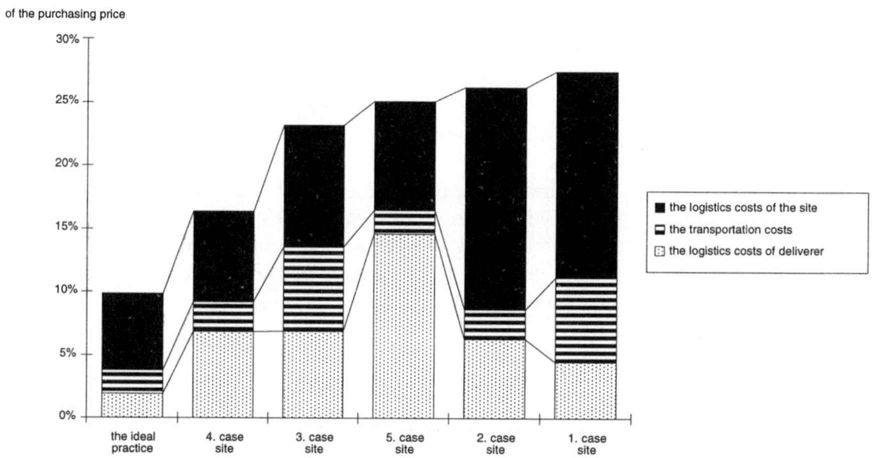

Figure 5 The logistics costs of the plasterboard deliveries on the case sites compared to the ideal practice.

There were relatively great differences in cost between the case sites. The logistics costs varied from 17 % to 27 % of the purchasing price. The assumption that the transportation costs would be largest cost item was proved wrong. The cost saving potential between the worst and the best practice was 10 % of the purchasing price which is much more than purchasing personnel can ever get by price negotiations.

Re-engineering a new ideal practice

Figure 5 represents also the ideal practice which can be reached by combining the current best practices of each stage and by following some guidelines. The ideal practice has the most simple chain of activities. All non-value-adding activities has been removed. The guidelines to the site foreman are:

- Plan and order deliveries 1-2 weeks before need.
- Order deliveries to come to the site just in time.
- Divide purchases into smaller deliveries and plan the time points and the contents of the deliveries accurately.
- Do not order extra pieces to the installation places.
- Move the material direct to the installation place.

If construction firms follow the principles, also the supplier can save costs by making material to order and by combining transportations of different customers.

By these means the logistics costs of ideal practice can be reached. These costs are only 10 % of the purchasing price. Cost savings can be up to 17 %.

3 LARGE COST SAVINGS CAN BE ACHIEVED BY CO-OPERATION

The new framework for measuring and re-engineering logistics chains has been used in several development projects in Finland. So far plasterboard, concrete element, door, window, mortar, timber (e.g. 2x4-inch board), major appliances (refrigerators and stoves) and kitchen cabinet deliveries have been measured and re-engineered. Seven more material groups are just under development.

Activities, costs, accuracy and delivery times as well as the problems of the logistics chains have been identified and explained within these materials. Over thirty case construction sites have been studied to find out different ways to operate. Ideal models have been developed within different material delivery chains by benchmarking performances and by developing best practices further. These ideal models have been tested on several pilot construction sites.

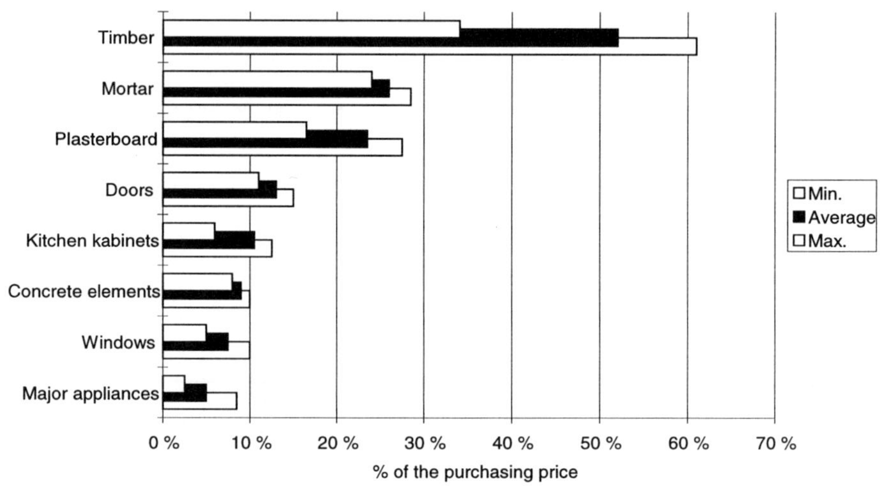

Figure 6 Minimum, average and maximum logistics costs of different material flows.

The summary of the cost analyses is illustrated in the figure 6. The figure shows a large cost saving potential. The highest logistics costs of material flow were more than 60 % of the purchasing price. The differences between the materials were sigfinicant; average costs of material flows varies from 5 % to over 50 % of the purchasing price.

Standard b-class materials usually had higher costs of the material flow than customized a-class materials. Timber, mortar and plasterboard are typical standard material while kitchen cabinets, concrete elements and windows are customized materials.

Also the differences within the material groups were great. For example the logistics costs of timber deliveries varied from 34 % to over 60 % of purchasing price. Expecially these differences offer a good chance to use benchmarking as a tool for development.

Several case study results show that the co-operation of suppliers and construction firms offers many improvement means. The development process is not a zero-sum game but a win-win situation.

4 CONCLUSIONS

The usefulness of the measuring tools depends on the type of material. The standard b-class materials have the biggest logistics costs of material flows. The logistics chain of standard materials consist of many movements and storages. A lot of money can be saved by eliminating unnecessary activities from the chains. Therefore the importance of the activity and cost analysis is emphasized in development of standard material deliveries.

The accuracy and delivery time analysis is most useful in the development of customized a-class material deliveries. Logistics chains of non-standard materials involve many actors in different companies. At least an architect, design offices, a material supplier and a contractor have to interchange a lot of information during the delivery processes of customized materials. With the accuracy and delivery time analysis, the improvement potential of information flows and co-ordination can be identified.

Although the effectiveness of the measuring tools varies, the re-engineering process is always useful by using co-operative methods. The variation of the logistics costs within the material group offers great opportunities to benchmark current practices and to find out best practices. Co-operation increases the chance for cost savings.

The division of the construction materials to a-class materials (customized) and b-class materials (standard) is used, at the moment, only when purchasing materials. However, the division is extremely useful also when managing and re-engineering material delivery processes. Problems are usually similar within these material groups. It is also possible to use similar solutions to improve these delivery chains.

5 REFERENCES

Asplund E. and Danielson U. (1991) *Räta ut byggsvägen.* Svenska byggbranschens utvecklingsfond (SBUF)

Bergström J. and Gröning M. (1992) *Activity-Based cost in the logistics of construction industry* (in Finnish), Helsinki University of Technology

Brimson J.A. (1991) *Activity Accounting; an Activity-Based Costing Approach.* John Wiley & Sons Inc., New York

Burbidge J.L. and Falster P. (1993) Reducing delivery times for OKP products. *Production planning and control*, vol. 4, no 1, pp.77-83

Eloranta E. and Nikkola J. (1992) Economic Aspects in Production Management. *Computers in Industry*, vol.19, pp. 51-64

6 BIOGRAPHY

Tutu Wegelius-Lehtonen is a researcher with the IIA Research center at the Helsinki University of Technology in Finland. Her current research focuses on the logistics of the construction industry. She develops new methods and tools that support co-operative re-engineering projects of business processes. At the moment, she is working with several Finnish construction companies and material suppliers to improve their logistics chains. She received her MSc in industrial economics from the Helsinki University of Technology.

21

Logistics performance measurement - an application to pulp and paper industry

J-M Lehtonen
Helsinki University of Technology, IIA Research center
Otakaari 4 A 20610 Espoo, Finland
Tel: +385-0-451 3694 Fax: +385-0-451 3665
Email: jmlehto@tuta.hut.fi

Abstract

At present, the most common company level inventory performance yardstick is inventory costs in relationship to revenue. This yardstick does not adequately reflect differences in the scope of vertical integration between different companies. The first objective is to develop a suitable yardstick for inventory performance that could enable strategic or company level screening for the need for improvement. This new yardstick is then tested with pulp and paper industry statistics and annual report data.

The second objective is to develop a framework for logistical performance measurement at the operations level. An application of the measurement framework in the distribution logistics of one mill of a large Finnish pulp and paper company is described.

Keywords

Logistics, performance measurement, pulp and paper industry

1 INTRODUCTION

At present, the most common company-level inventory performance yardstick is inventory costs in relationship to revenue. For example, A.T. Kearney's (1993) survey used that one. The inventory costs in A.T. Kearney's report were further divided into capital and operating costs. According to Bowersox et al. (1986) using cost accounting data for logistics performance measurement has two problems. First, the logistics costs are reported differently in different companies and are therefore not comparable. Another problem is that traditional accounting is not capable of specifying true inventory costs. The problems with the reliability of cost accounting data can be overcome by measuring the inventory value. It is also common practice to estimate the true inventory costs as a percentage of inventory value (e.g. Schonberger and Knodd, 1988 or Bowersox et al, 1986).

Another yardstick of inventory performance is inventory turnover, which at the company level is revenue divided by inventory. Both inventory turnover and the inventory costs / revenue relate inventory commitment to revenue. These measurement tools are not valid to screen inventory performance at the company wide level because they do not take into account differences in the scope of vertical integration. This is an especially notable handicap in the pulp and paper industry where degree of vertical integration varies considerably.

The first objective of this paper is to develop a suitable yardstick for inventory performance that could enable strategic level screening for re-engineering needs. This new yardstick is compared to the inventory / revenue yardstick using pulp and paper industry data and is then applied to an annual report sample.

At best, screening tools are only able to highlight problems. Concrete logistics performance re-engineering actions must take place at the operations level. A framework of logistical operations measurement is needed to locate the problem areas, although the right re-engineering actions depend on the particular situation. The second objective of this paper is to develop a framework for logistical performance measurement at the operations level. Traditional cost measurement is unable to take into account the dynamics of logistic chain. The framework should also be able to guard against these amplifying effects of the chain. Then the applicability of the measurement framework is tested in a case study.

2 THEORY

2.1 Strategic level inventory performance measurement

Buzell and Gale (1987) define vertical integration as

Vertical integration = (sales - purchases) / sales.

As somewhat different definitions for value added exists, in this paper value added is defined as revenue less purchases. At any rate, in the light of available statistics the treatment of rents for instance is not a significant issue

Figure 1 shows two logistical chains that perform identical functions. The only difference between the two chains is in ownership structure. The chain on the left is owned by one single, vertically integrated company while the other is composed of three different enterprises. We can now calculate inventory commitment in relationship to revenue for the first chain

Inventory / revenue = 90 : 100 = 90 %.

For the second chain we get three different figures, one for each company

Pulp mill	inventory / revenue = 30 : 40 = 75 %
Paper mill	inventory / revenue = 30 : 70 = 42,9 %
Converting	inventory / revenue = 30 : 100 = 30 %.

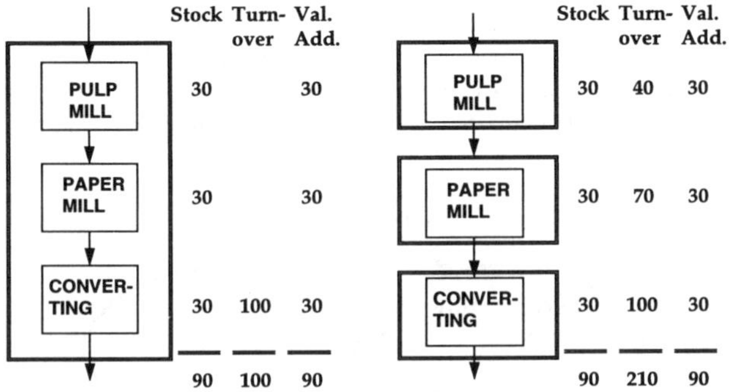

Figure 1 Two functionally identical logistic chains.

If we relate inventory commitment to value added instead of revenue we get for the first chain

Inventory / value added = 90 : 90 = 100 %.

For each the pulp mill, paper mill and converting we get the same results

Inventory / value added = 30 : 30 = 100 %.

The inventory / revenue yardstick gives different values to inventory performance in the two identical example chains depending on the scope of vertical integration. With the inventory / value added yardstick the results are stable. The advantage of measuring inventory performance by inventory / value added is that the different scope of vertical integration is taken into account.

2.2 A framework for logistics performance measurement at the operational level

Theoretical background for the performance measurement framework

The measurement framework at operations level is based on the methodology of controllability engineering (Eloranta and Räisänen, 1986) and the concept of logistical performance cycles (Bowersox et al., 1986).

According to Bowersox et al. (1986) a logistical performance cycle consists of nodes and links. A node is a physical location, for example a factory, a warehouse or a customer. Transportation and communication connections are referred as links. Bowersox et al. (1986) claim that the essential interfaces and decision processes of a logistical system can always be described in terms of individual performance cycles.

The central information acquiring method in Eloranta and Räisänen´s (1986) controllability engineering is zooming and focusing. This means that any aggregated phenomenon is taken into consideration from outlines into details (zooming). At each level the aspect of major relevance is identified (focused) for further analysis (zooming). With the zooming and focusing method the analyst can concentrate only on essential phenomena even at a detailed level.

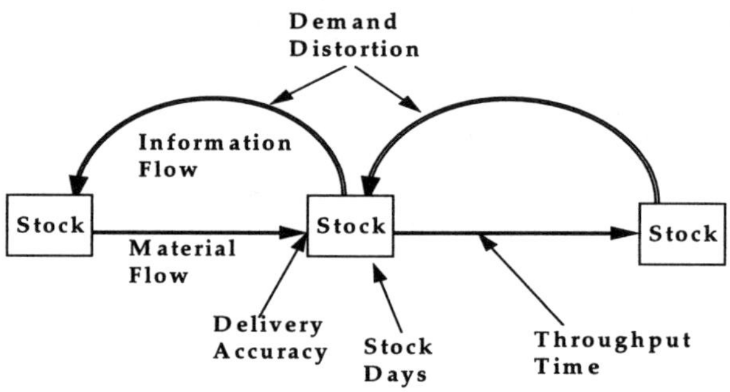

Figure 2 The measurement framework with two performance cycles.

Figure 2 shows the operational level logistics performance measurement framework. The framework consists of two consecutive performance cycles where nodes are referred as stocks and links as information and material flow. Any logistical system can have a number of performance cycles. Controllability engineering methodology should be used to identify the most important ones.

The measurement framework should also be able to find the presence of demand distortion. Towill (1992) names two types of demand distortion or amplification of variability in information flow. The first is Forrester effect, relatively long-term chain dynamics (Forrester, 1961). Houlihan (1988) explains the causes of Forrester's (1961) chain dynamics by a cycle of shortages, over ordering, unreliable deliveries and increased safety stocks. The second one is Burbidge effect which is shorter duration multi-phase noise present in multi-cycle ordering systems (Burbidge, 1984).

The operative level logistical performance measurement framework

The four crucial concepts for operational logistical performance are

- Delivery accuracy
- Inventory commitment
- Manufacturing and transport throughput time
- Demand distortion.

These concepts can be operationalized as follows. Inventory commitment can be measured as stock days which is the inverse of inventory turnover. Throughput time is the time needed from one milestone to another. Delivery accuracy can be measured as the difference between actual and due date. If safety stocks are large, delivery accuracy poor or throughput times long one should be alert for the possible presence of demand distortion. These three yardsticks are already much used and need no further comments.

The demand distortion yardstick needs some explaining. It measures the quality of information flow by calculating the ratio of demand standard deviation in two performance cycles. It is especially designed to find out the presence of chain dynamics. Because it is a ratio of standard deviations of two random variables, a common F-test can be used to test the statistical significance of the amplification effect between two performance cycles.

3 SOLUTION METHODOLOGY AND RESULTS

3.1 Strategic level

Industry statistics

The inventory / value added yardstick was first compared to the traditional inventory / revenue yardstick with pulp and paper industry statistics provided by the Statistical office of the UN Secretariat. The statistics cover years from 1980 to 1991.

Figure 3 shows the development of inventory / revenue in four countries. There seems to be no notable differences in inventory performance between different countries.

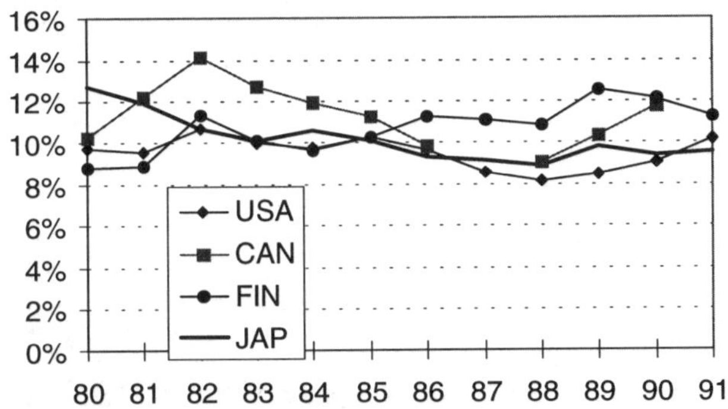

Figure 3 Inventory / revenue.

A completely different picture of reality is shown in Figure 4 where the same countries are analyzed with the inventory / value added yardstick. Two things can be seen from the statistics. Firstly, the Finnish inventory performance was clearly inferior to US and Canadian. Secondly, Japanese were able to increase their inventory performance during the period.

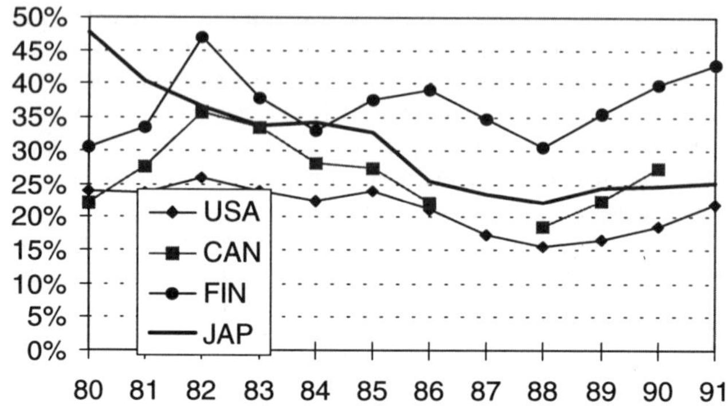

Figure 4 Inventory / value added.

The Canadian vertical integration average during 1980-1991 was 43,9% and American 45,5%. The Finnish vertical integration was only 34,1%. The Japanese vertical integration shows a growth trend. The average Japanese vertical integration from 1980 to 1985 was 29,5%, while from 1986 to 1991 it had climbed to 38,7%. The Finnish vertical integration was lower than US and Canadian and Japanese increased their vertical integration during the period. The different pictures of reality in figures 3 and 4 reflected the differences in vertical integration.

Annual report sample

The inventory / value added yardstick was calculated for 38 pulp & paper company sample to assess the usefulness of the measurement tool in terms of being able to find the improvement potential. The data for this was found in annual reports and covered a five-year-period 1988-1992. The results are shown in Figure 5.

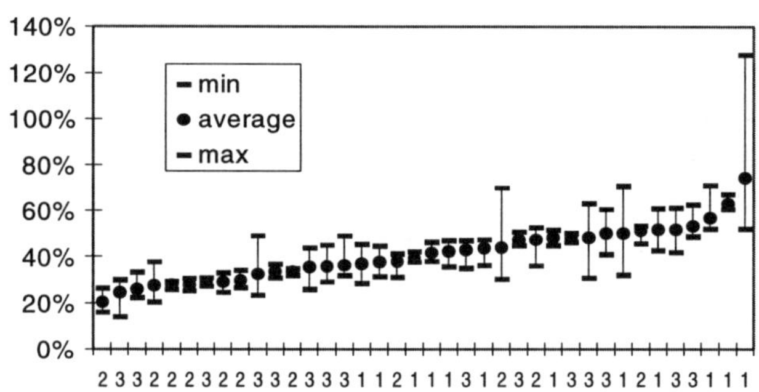

Figure 5 Inventory / value added in 38 pulp & paper companies.
Nordic = 1, North American = 2 and European = 3.

It can be seen from Figure 5 that significant differences exist between companies, which suggests logistics re-engineering potential. When the sample was divided in groups of American, European and Nordic companies, the null hypothesis of no group differences was refuted by Kruskall-Wallis test at $P = 0.05$ level ($H = 8.43$, $\chi^2 = 5{,}99$, $df = 2$).

Interpretation of this result is that in Figure 5 the Nordic companies for some reason need larger inventories to achieve equal value added and are on the right hand side of Figure 5 while the North Americans center on the left hand side. The European companies are distributed rather evenly in the sample. This suggests that especially Nordic pulp & paper companies have room for improvement in inventory performance.

3.2 Operations level case results

Basic information

The applicability of the measurement framework was tested in a case. The company in question was a major Finnish pulp and paper manufacturer and the case was a manufacturing and distribution chain of a large paper mill with 350 000 tons of annual capacity. The study lasted for six months. Some major findings that also demonstrate the applicability of the measurement framework are presented briefly.

Results

Figure 6 shows the delivery accuracy measurement results at one, fairly typical performance cycle. The delivery accuracy is inadequate. Almost a half of the deliveries are late and 25 % are early.

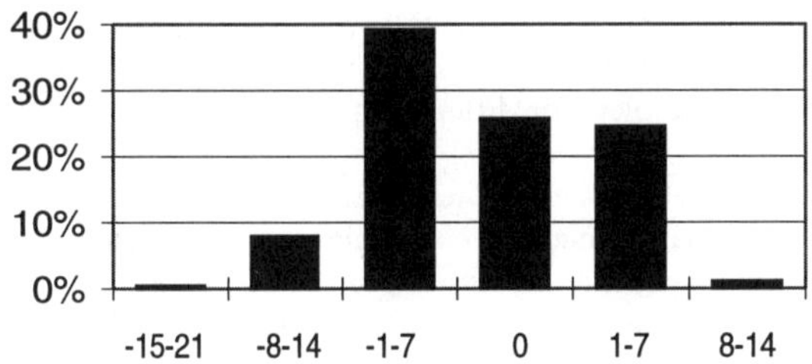

Figure 6 Delivery accuracy in days at one performance cycle.

The demand distortion test revealed weekly demand amplification effect of 1 : 3,19 in the most important single performance cycle, indicating thus problems in the information flow. The data is shown in Figure 7. Based on the F-statistic the hypothesis of equal variances could be refuted with $P = 0{,}01$ ($S_1^2 / S_2^2 = 10{,}16$, $F_{.99,40,40} = 1{,}94$, $df = 50{,}50$)

Figure 7 Weekly demand in two consecutive performance cycles.

Even worse, the delivery accuracy in some other performance cycles could not be measured at all because of lack of any records. The inventory levels were generally found much too large in relationship to stock level variation, indicating excessive safety stocking policy. Regardless of the relatively long sea transport times their effect was a minor concern in comparison to a rather long manufacturing lead time.

Throughout the study, the controllability engineering methodology was found useful in identifying the potential areas for re-engineering actions.

4 DISCUSSION

With pulp and paper industry statistics the strategic level inventory performance yardstick inventory / value added was found superior to inventory / revenue because it takes differences in vertical integration into account. If someone still wants to relate inventory performance to revenue he should be alert for differences in vertical integration both between and within companies.

The reliability of the results in annual report sample was undermined by three factors: small sample size, lack of many American companies to report their employment cost which had to be estimated and the tendency of pulp and paper companies to be diversified in many logistical chains (i.e. business units) sometimes even outside pulp and paper industry.

As such, the operational logistics performance measurement framework turned out to be applicable in this special case. It was able to highlight problem areas but it should be put to further tests.

5 REFERENCES

A.T. Kearney (1993) *Logistics Excellence in Europe*, A Study Report prepared by A.T. Kearney, an International Management Consulting Firm, on Behalf of the European Logistics Association.

Bowersox D., Closs D. and Helferich O. (1986) *Logistical management*, Macmillan Publishing, USA.

Schonberger R. and Knodd E. (1988) *Operations Management*, Business Publications Inc., Texas.

Buzell R. and Gale B. (1987) *The PIMS Principles - Linking Strategy to Performance*, The Free Press, New York.

Eloranta E. and Räisänen J. (1986) *Ohjattavuusanalyysi*, Sitra, Helsinki.

Towill D. (1992) Supply chain dynamics- the change engineering challenge of the mid 1990s, *Proc. Instn. Mech. Engrs*, **206**, 233-45.

Forrester J. (1961) *Industrial Dynamics*, MIT Press, USA.

Houlihan J. (1992) International supply chain management, in *Logistics - the strategic issues* (ed. M. Christopher), Chapman & Hall, Cornwall.

Burbidge J (1984) Automated production control, in *Modelling production management systems* (ed. P. Falster and R. B. Mazumder), IFIP Working Group 5.7, Lyngby.

Statistical office of the UN Secretariat.

7 BIOGRAPHY

Juha-Matti Lehtonen is currently working as a logistics researcher at the IIA research center at Helsinki University of Technology. He received his M.Sc. (Eng.) in industrial management in 1993 from Helsinki University of Technology. At the moment he is 26 years old.

22

A Comprehensive Methodology for Factory Modelling

K. Popplewell and R. Bell
Department of Manufacturing Engineering,
Loughborough University of Technology,
Loughborough, Leicestershire, LE11 3TU, United Kingdom.
Phone: (44) 01509 222925; Fax: (44) 01509 267725
E-mail: K.Popplewell@lut.ac.uk

Abstract

The view on factory modelling presented in this paper is that there is a need for a family of design tools which can be used to support a methodology which will result in the design and improvement of performance of factories, achieved in a satisfactory timescale, at acceptable cost, and yielding effective results.

The methodology embraces issues which range from the initial establishment of business strategy and its interactions with the manufacturing strategy and financial strategy models of a business through the use of approximate modelling methods which are capable of rapid expansion in detail, to the realisation of dynamic outputs as an important viewpoint. The process can be extended down the modelling range by offering interfaces to lower level tools, many of which are already available.

In the authors' view the methodology and the tools it supports form a key business process technology.

Keywords

Factory design, approximate modelling, multi-view modelling, generic models

1 INTRODUCTION

In rapidly changing markets, with the stimulus of multi-national, and multi-business enterprises, it is becoming the norm for a typical factory to have to frequently change its operating practices to meet the dictates of business pressures. This is in part caused by the increasing tendency for enterprises to consist of multi-business groupings: changes in the business status of the enterprise and its markets frequently call for change in the structure of the enterprise, and in particular, its constituent factories. Evidence has been accumulating to support this point, and furthermore shows that the engineering industry increasingly splits the design and manufacturing functions between different sites or indeed different countries.

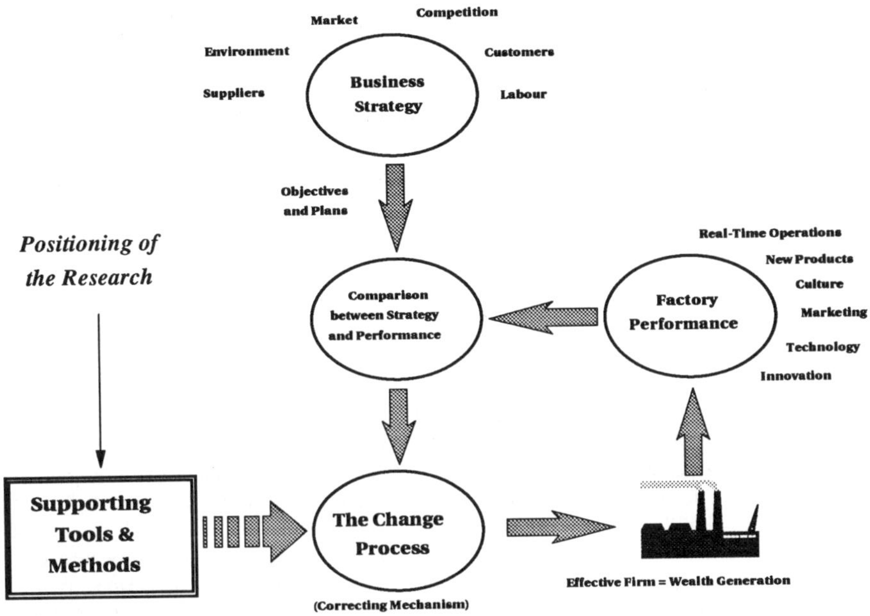

Figure 1 : The business performance enhancement cycle.

There is increasing evidence that these multi-business enterprises depend not only on the information networks controlling and supporting individual businesses but also on the network at the enterprise level linking the activities of the those businesses. It is important therefore for methodologies and tools supporting the alignment of business process with business strategy to be at least capable of modelling the factory activities and information networks concurrently. In some cases the required re-engineering could centre initially either on the reconfiguration of the factory activities or conversely on the information network.

Factory design can either involve the design of a totally new green field site, or be targeted on significant changes in an existing business (ie. business performance enhancement) where the speed and impact of the changes are at a premium. In the contemporary scene reconfiguration of the business is more frequently, though not exclusively, encountered than the task of green field site design.

The challenges posed cannot be met by traditional methods which are typified by potentially inefficient and unreliable decision making, and unsatisfactory information transfer from senior business managers to middle managers who must take responsibility for implementing change. Further, the methodologies and tools currently available are unacceptably slow and limited in their usefulness. We believe there is a strong case for methodologies and software tools to support rapid and competent reconfiguration of factories. Factories can and should be designed as effectively as it is now accepted that products can be designed, and the reported research leads to methods and software tools invaluable in making factory design a systematic and reliable activity. The concept of factory design is to some extent a radical one; many experienced innovators whose techniques are centred on identifying and improving business processes might be surprised by the proposition that factory design is a realisable goal.

As depicted in Figure 1, the industrial change process discussed above can be regarded, in classical control language, as a regulator system where the reference is the business strategy and the output is business performance. If this model is accepted then a change mechanism is needed to complete the closed loop. This change process is a now well established part of industrial life which normally takes the form of a team effort or the work of external consultants to prescribe and implement the necessary change in factory performance. This activity needs proven methodologies supported by software tools: the research reported here is targeted on this area and constitutes a major extension of the facilities which are currently available to the agents of change.

2 STATE OF THE ART

Recent research work on enterprise modelling has made significant contributions on both concept and method. Fox (1993) has developed what he terms a "common-sense" enterprise model, from which he claims that "we lack the software that would enable computers to provide access to information and support decision making across the organisations". Mize's (1992) group explored enterprise rationalisation, concluding that individual management system and production system rationalisation cannot enhance enterprise competitiveness unless both systems are integrated, in line with a vision of improvement at higher levels. Work by Mujtaba (1994) and Malhotra (1992) provides a method by which an enterprise at implementation level can be simulated. The results obtained by Mayer (1991), Cook (1992) and Scheer (1992) are competitive to this research: Mayer uses situation classification frameworks and site specification frameworks to select modelling tools. In contrast Cook uses Taguchi methods to build models, and Scheer shows a preference for information modelling as the core of enterprise design.

There are three apparent competitive lines of thought in contemporary research: the first is based on the use of highly developed, prescriptive, generic enterprise structures. The most dominant examples of this approach are typified by the CIM-OSA architecture currently centred in the AMICE and VOICE projects in the Esprit programme. The concept of the virtual factory model, as typified by the work of VTT in Finland, is a fascinating aid to product design though not a direct competitor to this work. Reference must also be made to the long established and widely used methodology which has been introduced and developed by GRAI group. The approach underpinning this research is significantly different to each of these in that it is assumed always that methodology and business tools associated with business process re-engineering will encourage an uninhibited approach to specifying business strategy objectives and the factories which are designed to realise them. One major competitive feature is the ability to incorporate additional business performance viewpoints relevant to particular aspects of a factory design.

Beside the seminal work of Hammer and Champy (1993), there are a number of publications in the field of business process re-engineering which are relevant, and in particular it is noted that Rigby (1993), Davenport (1990, 1993), and Tunalv (1992) as well as De Toni (1992) have studied the total concept of the business process and how it should be updated; a number of publications dealing with specific methodologies to support the business process engineering activity have been reported by Bennett (1992) and Turnbull (1992).

In considering work done on an international scale, note has been taken of emerging radical views on future trends in manufacturing. The work of Warnecke (1993) on fractal factories presents stimulating challenge to contemporary business process thinking. Japanese views on the post-mass production paradigm present a technologically oriented solution, whilst Hirsch (1994) considers, as in Figure 2 a classification of production management paradigms. Thinking in the USA, typified by work reported by Goldman on agile manufacturing, dominated by the impact of software engineering constitutes a a typical North American approach.

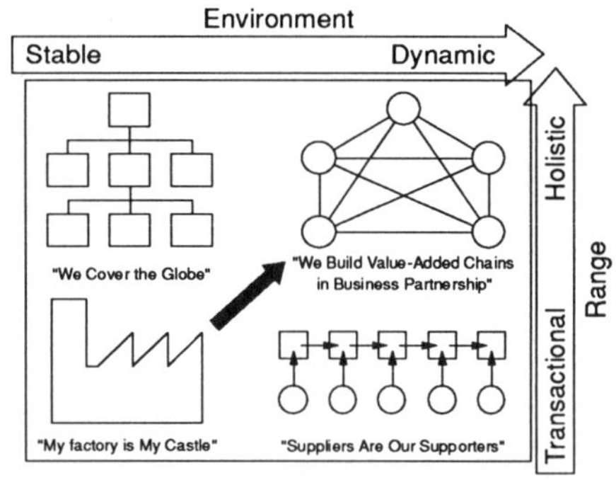

Figure 2 : Classification of paradigms for production management systems (Hirsch 1994).

The biggest international challenges to thought are the astonishing developments which are taking place in industry. Companies drawing upon the support of substantial use of IT are re-grouping into global partnerships. For example the A.B.B. Group consists of a small executive group based in Zurich which controls 5000 profit centres in 130 countries; this re-grouped company is only 5 years old. The authors have gained significant insight in to Japanese practice. However it is considered that this does not represent the plateau of maximum performance as it is noted that the Japanese themselves are anxious to embrace new methodologies and have an increasing interest in software tools. The Japanese experience is not unique: similar evidence is available in Europe and the USA. In the U.K., Lucas Group's approach to manufacturing systems engineering and to factory improvement now offers both a proven track record and distinctive methodology.

3 A TOP-DOWN METHODOLOGY

The structure of this methodology is strongly influenced by the authors' interactions with industry in a number of teaching company and funded research projects. It is essential that the link between top management and middle management - the decision makers and the enablers - should be kept as fluid and as accurate as possible; in turn the middle managers must lead teams working on the updating of particular activity areas who in turn will require software support including ready interfacing to other proprietary lower level tools in order to complete their tasks effectively.

The emphasis should be placed, as in Figure 3, on a top down modelling approach which is adequately fast to be responsive to the rate of change of decisions which can be realistically expected from top management. An important part of this view is the use of approximate modelling methods coupled with dynamic output which assumes that the models are focused on the changing situation rather than extensive historical evidence gained within the business.

A core issue in this research has been the establishment of a consistent link between changes in business strategy and the manufacturing performance of the business. The following four areas of decision making knowledge have been captured by a matrix based modelling approach similar to that of Inamoto (1990 and 1991): business strategy, product design strategy, strategic manufacturing evaluation, and financial constraints. Major characteristics of the Strategic Manufacturing Decision Support (SMDS) tool are the consideration of the time dimension, the flexible integration of multi-disciplinary factors, and easy extension to include other perspectives. This decision aid is being researched in a spreadsheet environment on a PC host so that it can be low cost, simple to use, and capable of tailoring to specific companies' applications.

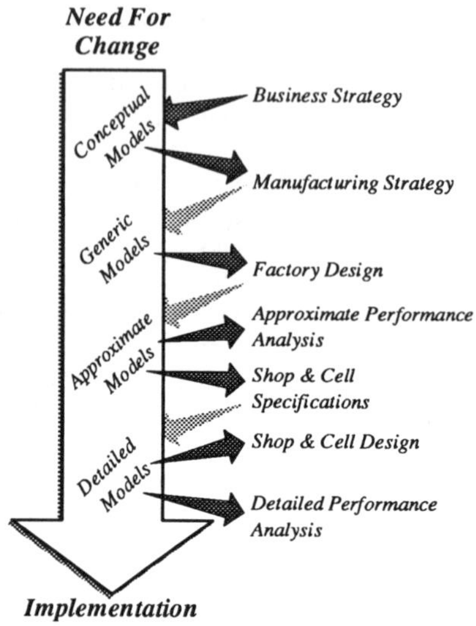

Figure 3 : A top-down methodology for factory modelling.

The complexity of a factory calls for a description approach which can be used to produce a complete but approximate factory model. The model should be complete because all the crucial aspects of the factory should be captured, whilst at the same time it should be approximate because all extraneous details should be ignored.

Whilst the level down to which approximate models are developed will vary from business to business, the effective limit of approximate modelling is the point where a completely satisfactory set of requirements and constraints can be delivered to middle managers, who then may be given the freedom to carry out the redesign within their activity areas. At this level the enablers may readily be able to turn to the use of either proprietary low level modelling tools which they already employ, or other design procedures which allow them to meet the requirements placed upon them.

Recent research has delivered some of the tools needed in applying this methodology, and gives confidence that the remainder can be realised. The ease of modelling is achieved both by the use of a proven user interface coupled with the use of approximate modelling methods, and by a powerful extension to speed of modelling made possible by the introduction of the generic modelling concept. The potentially more important attribute in this concept is the use of multi-view modelling which allows those doing the detailed re-engineering work to make initial access by more than one view.

4 MULTI-VIEW MODELS

The search for appropriate methodologies in business process design has lead to the introduction of a number of paradigms [eg. Hirsch et al., Jovane, Wenzel et al.] to meet a range of related but disparate needs. A major theme in this research is that a paradigm can be integrated into a business modelling software tool which can support a range of model views in a way which harmonises a range of user requirements. Each view allows concentration on a

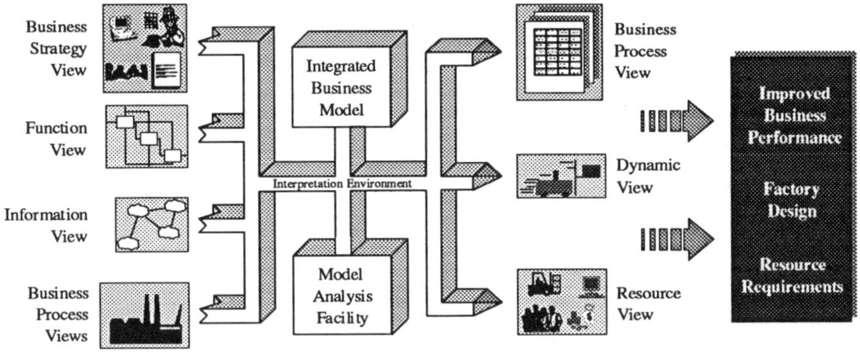

Figure 4 : Integrated multi-view modelling.

particular aspect of modelling, which corresponds to working with a particular conventional modelling paradigm. Thus, for example, in studying the information structures used by the business to specify the business information system, the modeller interacts with an information view of the model, whilst to predict the business' performance he would interact with the dynamic view. However both views share much of the model content, which appears in, and must be entered into, the integrated model only once. In contrast, the sequential use of conventional modelling methods for these purposes requires considerable duplication of model building.

Figure 4 depicts such a multi-view interpretation and identifies two categories of view. Some views are primarily model-building views, in one of which model building would normally commence, and whose output is essentially a presentation of the model as constructed. The function view of the activity structure of the business is such a view. In, contrast, other views, exemplified by the dynamic view, whilst requiring additional information in their construction, would not normally constitute a starting point for modelling, but would provide deeper analysis of the model as their outputs.

The need for integration is highlighted by the degree commonality of information incorporated in the individual models: for example a functional view of the system, identifying the activities which are necessary to achieving business strategy, also requires access to a definition of the structures of information which must flow between the transforming activities. The construction of independent models to meet the needs of each view of the business process necessitates duplication of model definition, potentially to a very considerable degree.

Research has established the feasibility of building integrated multi-view models of this nature, including in demonstrator software a function view, a dynamic view and an initial form of an information view. A taxonomy of further views (eg. a resource view) which may be of value in business process design is being investigated, with a view to constructing a reference model of views identifying applicability and relationship between views.

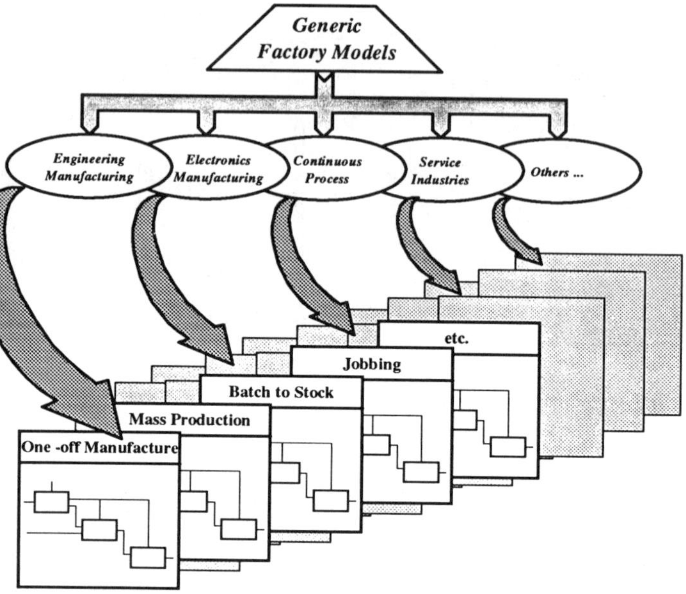

Figure 5 : Generic modelling.

5 GENERIC MODELLING

Generic models provide a starting point for model-building. Each contains, in a generalised form, the common features of a business system operating a particular manufacturing philosophy in a business sector. Initial modelling then takes the form of identifying the generic model which most nearly matches the desired business system; the generic model can then be rapidly modified to meet a particular need.

Generic models can fulfil two distinct roles in supporting business process re-engineering, with its emphasis on the speed of the activity. The task of building a model from a blank sheet is necessarily protracted, and yet it is the authors' experience that common themes run through the structures of successful companies operating in related markets. Thus an understanding of commonality establishes the degree to which generic models can provide a starting point for modelling. This provides accelerated modelling simply because the volume of work involved in specialising a generic model, selected as the closest available to the developing business process design, is much reduced.

To be of value, a range of generic models must be available not only to correspond to different business sectors, but to provide a range of starting points within each sector; they must also allow users to model the global manufacturing interactions which are now coming strongly into view, and must assist users to seek the flatter structures which will be necessary for greater speed of market response. The Warnecke concept of the fractal factory appears to be highly relevant to this work.

Furthermore we see that work reported by Carrie (1993, 1994) et al. has developed ideas on identification of business processes which have a record or probability of performing successfully. Such ideas, incorporated into a set of generic models have the advantage of guiding design to employ best practice. It is however important to emphasise that generic models are not prescriptive: their purposes are to be supportive in rapid model building, and encouraging in adoption of best practice. In building a particular model a generic model will be selected and specialised without restriction.

6 CONCLUDING DISCUSSION

This methodology addresses the need to be able to support design or re-design of factories by modelling in parallel with the natural progression of design decision-making. It is embodies the essential link between the strategic aspirations of the enterprise and initial decisions on the manufacturing philosophies to be adopted and the requirements which must be met by the factory if those strategic aspirations are to be fulfilled. Software demonstrating the feasibility of capturing these links has been successfully researched, and will provide a specification for subsequent stages of factory design.

Approximate models must then be used in design of the structure of the factory, encouraging analysis of the activities the factory must perform, the flows of information and material which these activities generate or require, and the performance requirements of each activity. It is not possible to use more conventional modelling of detailed activities at this stage of factory design, since no detailed decisions are even contemplated at this point in the decision making process. Real decisions on the major functions of the factory (including both manufacturing processes and support functions such as product design, production engineering, production control etc.) are made before consideration of detailed plant and personnel investment starts, and approximate modelling allows support of these decisions at the time they are made, typically in the first month of the factory design process.

Software demonstrators of both approximate and multi-view modelling for three selected views have been successfully implemented. This shows the feasibility of integrating a range of model views, and it is significant that collaborating companies have used these demonstrators and the methodology in factory re-design projects following completion of research activity in the companies.

Both the methodology and commercial developments of the software demonstrators are able to offer enhanced support of the progressively more frequent process of factory design. The top-down approach coupled with use of generic models, allows modelling to provide support on realistic timescales. The use of integrated multiple views allows the model to be examined for a variety of purposes without the need to build several separate models embodying much of the same content.

The feasibility of applying and supporting such a methodology has been established. There is now a need to extend its range of applicability in number of ways. Firstly, developments in both hardware and software technology now make it possible to implement the ideas described above in efficient ways. Much use can be made of established modelling interfaces, and of links from approximate modelling to existing commercially available tools for detailed process modelling. This would provide the modelling software industry with enhanced products, meeting new customer needs, and would provide their customers with supported software to help satisfy their established need to advance manufacturing into competitive future forms.

The range of views considered to date has been sufficient to demonstrate that multi-view integration is possible and valuable, and the views considered have their own merits. There is now need to appraise the range of possible views which might be relevant to factory design, and the feasibility of their integration within the same structure. A reference model of modelling views, based on industry sector, business strategies and manufacturing technology would provide a guide not only to the views which should be integrated, but to those which a particular user of the methodology would need to address.

Related to this is the need to extend the range of generic models to include the range of industry sectors, and business types within sector which can benefit from application of the methodology.

7 REFERENCES

Bennett, D., Forester, P. and Hassard, J. (1992) Market-driven Strategies and the Decision of Flexible Production Systems: Evidence from the Electronics Industry" *International Journal of Operations and Production Management,* **12:2**, 25-37.

Carrie, A.S. and MacIntosh, R. (1993) Improving Integration in Manufacturing Systems, *IFIP Transactions on Advances in Production Management Systems*, (eds. Pappas, I.A. and Tatsiopoulos, I.P.) North Holland, Amsterdam.

Carrie, A.S. and MacIntosh, R. (1994) A structured approach to process redesign, *Proceedings of the BPICS Annual Technical Conference*, Birmingham, 153-167.

Cook, H.E., (1992) Organizing Manufacturing Enterprise for Customer Satisfaction, in *Manufacturing Systems,* National Academy Press (ed. Heim, J.A. and Compton, W.D.), 116-127.

Davenport, T.H. (1990) The New Industrial Engineering Information Technology and Business Process Redesign, *Sloan Management Review* **31:4**, 11-27.

Davenport, T.H. (1993) *Process Innovation:Reengineering Work Through Information Technology.* Harvard Business School Press

De Toni, A., Filippini R. and Forza C. (1992) Manufacturing Strategy in Global Markets: An Operations Management Model, *International Journal of Operations and Production Management,* **12:4**, 7-18.

Fox, M.S. (1993) Issues in Enterprise Modelling, *International Workshop on Intelligent Manufacturing Systems.*

Goldman, S.L., Nagel, R. N., and Preiss, K. (1994) *Agile competitors and virtual organisations.* Van Nostrand Reinhold, New York.

Hirsch, B., Crom, S., Thoben, K.-D. and Kuhlman, T.(1994) New Manufacturing Paradigms - Their Contribution to Improve Customer Satisfaction, Shorten Time-to-Market and Lengthen Time-in-Market, *EITC '94,* Brussels, 26-34.

Hammer, M. and Champy, J. (1993) *Reengineering the Corporation - A Manifesto for Revolution.* Nicholas Brealy Publishing, London.

Inamoto, A. (1990) Conceptual and Architectural Description on a Class of CIM Models, *Proceedings of 1990 Japan-USA Symposium on Flexible Automation,* Kyoto.

Inamoto, A. (1990) The Architecture of Integrated Factory Automation" *Proceedings of 1990 IFAC World Congress.*

Inamoto, A. and Shirakura, T. (1991) Modelling of Factory Automation System by Object-oriented Approach, *Proceedings of 1st IFAC Workshop on Algorithm and Architecture for Real-Time Control,* Bangor.

Malhotra, R. and Jayaraman, S. (1992) An Integrated Framework for Enterprise Modeling, *Journal of Manufacturing Systems,* **11:6**, 426-440.

Mayer, R.J. and Painter, M.K. (1991) Roadmap for Enterprise Integration, *AUTOFACT '91,* 7.1-7.26.

Mize, J.H., Pratt, D.B. and Farrington, P.A. (1992) System Integration: The Forcing Function for Enterprise Rationalization, in *Proceedings of Manufacturing International '92* (ed. D.Durham), 117-126.

Mujtaba, M.S. (1994) Simulation Modelling of a Manufacturing Enterprise with Complex Material, Information and Control Flows, *International Journal of Computer Integrated Manufacturing,* **7:1**, 29-46.

Rigby, D. (1993) The Secret History of Process Reengeneering, *Planning Review,* **21:2**, 24-27.

Scheer, A.W.(1992) *Architecture of Integrated Information Systems: Foundations of Enterprise Modelling*. Springer-Verlag, Berlin.

Tunalv, C. (1992) Manufacturing Strategy- Plans and Business Performance, *International Journal of Operations. and Production Management,* **12:3**, 4-24.

Turnbull, G.K. and Fisher, E.S. (1992) Improving Manufacturing Competitiveness Through Strategic Analysis, in *Manufacturing Systems,* National Academy Press.

Warnecke, H.-J.(1993) *The Fractal Company - A Revolution in Corporate Culture*, Springer-Verlag, Berlin.

8 BIOGRAPHIES

R Bell: Appointed Professor of Manufacturing Technology at LUT 1978. Established research in flexible manufacturing systems with particular emphasis given to modelling methods for cell design. Contemporary research interests concerned with the role of product and manufacturing models in computer integrated engineering, concepts for factory modelling, and research into tool management systems. A major recent interest has been the international activities in IMS, being a member of the European and International Technical Committees throughout the feasibility study. Another major interest has been support for academic development under the aegis of the ODA and UN, with assignments in Brazil, Hong Kong, India, Mexico, Singapore, Sri Lanka and Turkey.

K. Popplewell: Joined LUT in 1985, after sixteen years' experience in manufacturing industry, particularly involving production management, production engineering and computer aided engineering projects. Contemporary research interests include factory modelling concepts, engineering moderation to drive concurrency in simultaneous engineering, and applications of hybrid artificial intelligence methods in support of cooperative project working.

23

BPR, IN THE PRODUCTION SYSTEM, METHODOLOGY AND RESULTS

C. Martín and R. Encina
Production Systems Department, Ikerlan S. Coop.,
2 J.M. Arizmendiarrieta, 20500 Mondragon, Spain

Abstract
This paper presents part of the work carried out in an application case-study. The global project falls within the framework of what is known as Business Processes Re-engineering. To be exact, a strategic analysis of the production system of a toolmanufacturing company has been performed.

0. INTRODUCTION

Traditionally, flexibility has been considered the major advantage of SME's competing with large firms. The application of information technology and modern manufacturing philosophies in large firms has all but reduced this advantage. These technologies and philosophies should be tailored and applied to SME's, enabling them to regain their traditional advantage in the area of flexibility. By improving quality and manufacturing more cost effectively, they will enlarge their market share and boost employment levels. As present, employment levels are under threat from low labour cost locations.

1. METHODOLOGY

1.1. STRATEGIC ANALYSIS OF THE OPERATIONS SYSTEM

We understand Operations System as the group of value added activities: R&D, Quality Assurance, Engineering, Purchasing, Inbound Logistics, Production, Outbound Logistics, Installing, Supporting and Satisfaction Maintenance.

The steps to follow in Strategic Analysis are: Define the Desired Competence Profile, find out wether the Critical Activity in order to reach the desired profile is a transformation activity, if so, the action plan can be established, (BPR Project).

Define the desired competence profile. Developing a model based on some competitiveness criteria (Innovation, Time, Money, Range and Consistency) leads us to define and quantify the competitiveness level of a firm. The specific dimensions for each of the mentioned criteria must be defined from strategies and long term objectives, if they are explicitely defined, if not such objectives and strategies must be identified.

Identify the critical functions. In this step we must find out which functions from the operations system are the key functions in order to fit the desired competence profile, so that they must be redesigned.

Define the action plan. The action plan must be established in order to redesign the critical functions. The result is a range of projects to be carried out in the firm.

2. OUR WAY TO UNDERTAKE BPR IN TRANSFORMATION ACTIVITIES

According to the GRAI Model and self developed research, an operations system can be divided into three subsystems: The Physical System, the Decisional System and the Information System.

The physical system, involves manpower, machinery, materials and procedures, its own objective is to transform parts into finished goods.

The decisional system, it is a hierarchical system of decision making, so that decisions at a higher level (longer term decisions) create a decision context for the lower level decisions (shorter term decisions). Its aim is to develop the decisions which define the management commands to the physical system, it establishes what, how much, who, when, where and how.

On the other hand, the decision system, that is, the group of decision activities taken in a production system can be considered as two-dimensional (figure 1).

A functional dimension which defines the diferent existing functions in the production system.

A hierarchical dimension according to time criteria which establish levels for the decisions featured by the concept of **period** (period of time by the end of which every decision has to be questioned again), and the concept of **horizon** (period of time of influence of the decision).

Despite such concepts, the key for the decisional system is "Synchronism".

The information system, allows the information flow, treatment and saves the required information for decision taking and executing.

	PURCHASING	PRODUCTION	SALES
H=1.5 months **P=1 week**	Materials Planning		Sales Previsions
H=10 days **P=1 day**	Orders to Suppliers	Production Planning	
H=1 day **P=1 day**		Production Programming and Workshop data feedback	Generate Orders

FIGURE 1

3. CASE-STUDY

3.1. INTRODUCTION TO THE FIRM

The firm which is the subject of this case-study is an SME which manufactures tools for wood cutting, it employes about 50 persons, about 35 of them are direct labour.

Their main products are: Reversible blade heads, window groups, cutters, drills and saw blades.

They manufacture both special and standard products.

3.2. OBJECTIVES OF THE FIRM

- Reduce lead-time:

- Guarantee lead-times:

- Be able to process small batches (even one) with client specifications (special products) in a short time.

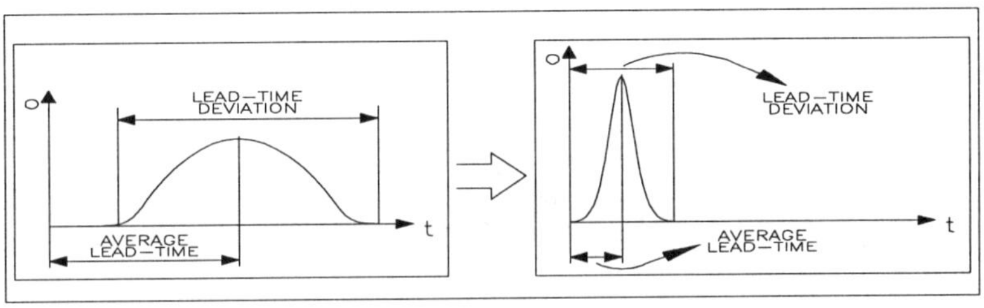

FIGURE 2

Once we get to this point, the firm must translate their objectives (their feeling) into something we can measure, identify, structure and formalize the way the want to compete by means of:

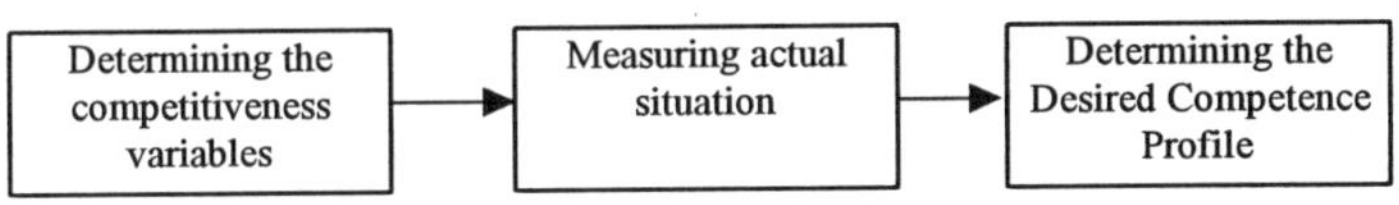

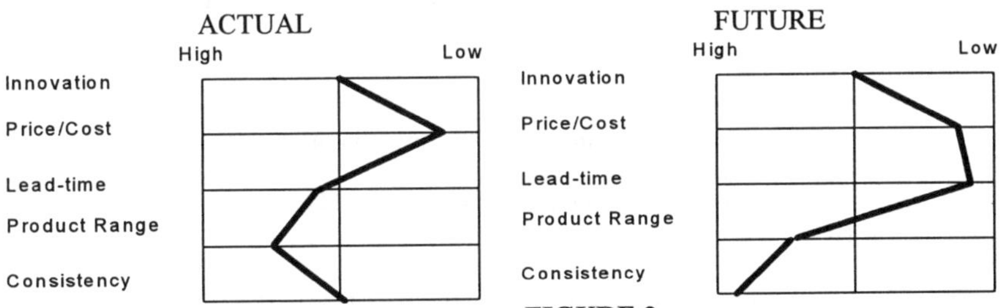

FIGURE 3

3.3. IDENTIFY CRITICAL FUNCTIONS

Analyse fuctions in the Operations System R&D, Quality Assurance, Engineering, Purchasing..., Find out which one is crucial for the proposed objectives.

The Production System had the following features, it was configured in functional groups, placed the way it seemed to be the correct order for product manufacturing, which seemed to be a good solution for simple management and short handling materials time.

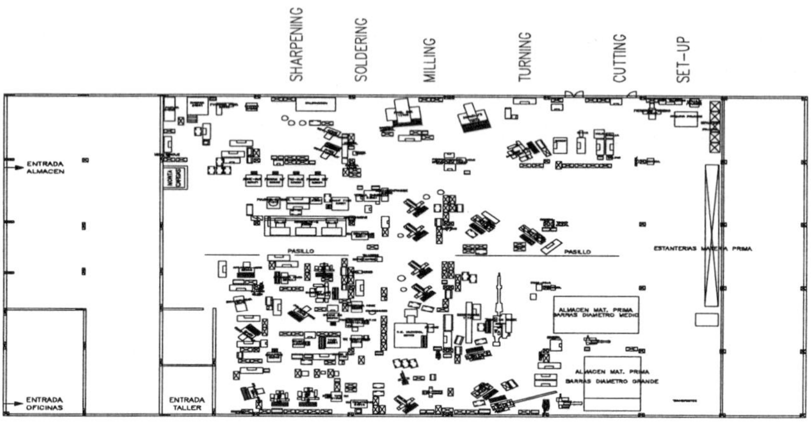

FIGURE 4

- The worker received the orders to manufacture, and he decided the scheduling.

- Proccess not totally determined: there was no established route (process.

That is why we considered the Production System as the critical function.

3.4. ESTABLISH THE ACTION PLAN.

Re-Engineer the Production System and its management in order to get next goals.

Serve: what the market demands, when the market demands it, as much as it demands, with the required price.

Decreasing: risk of obsolescence (products and facilities), cost, risk of investment.

Increasing: quality, Production System transparency.

Therefore we chose Group Technology, which is a philosophy for product-process rationalisation, with the objective of simplifying the Production System.

Simplify the Physical, Decisional and Information Systems, via creating small firms inside the global firm which work with absolute synchronism. And hence making it more lively, decreasing and determining for all of them the same **Programming Period** and **Production Cycle**.

4. PROJECT DEVELOPMENT

4.1. ANALYSIS

4.1.1. Product Analysis (Levels of Product)

Product analisis is carried out in order to measure the flexibility offered by the product (standard parts and components), and required technology for its aseptic transformation.

The variables which determine the flexibility of a product are as follows:

-Variety of finished products which have started from a semi-manufactured product.

-Relationship between operations prior to the semi-manufactured product and after it till it becomes a finished product.

Product analisis (always taking into account the process) has two main objectives: first to find out parts and components which can be manufactured or not by means of actual facilities, and second,to establish a product hierarchy, for which we proceed as follows:

- Put parts and components apart for each product analysing the diversifying effect they have over the final product.

- Study relationships with other products.

These two points aim, first, to simplify the system and its management, and second to transform the commercial catalogue into the production catalogue. To do so, there is something else we must do; Product-Market analisis, which identifies the levels of the process where finished product appears, as different clients order products in different levels of the process.

From the production catalogue analysis we worked out:

a) The degree of flexibility of the product is very low, the product is diversified after very few operations, so it makes no sense thinking of a Production System based on semimanufactured products. It makes more sense to think of a short production cycle, as well as low supplying time.

b) The levels of product were identified as:

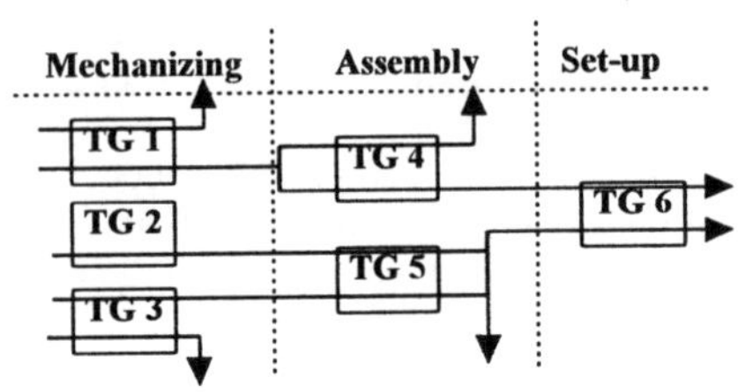

4.1.2. Process Analysis

Process analysis: The process analysis implies a detailed study of existing manufacturing processes in the firm, so, existing technologies, routes and flow diagrams.

Therefore, the first step in process analisis is determining the required facilities, own facilities and subcontractors, where they are and handling systems, (always looking at the materials flow from the beginning to the end of the process of each product).

We determined the manufacturing process of every single product in the production catalogue, these diagrams take into account:

- Products to manufacture
- Machine for each operation
- Operation (manufacturing or subcontracting)
- Technological constraints

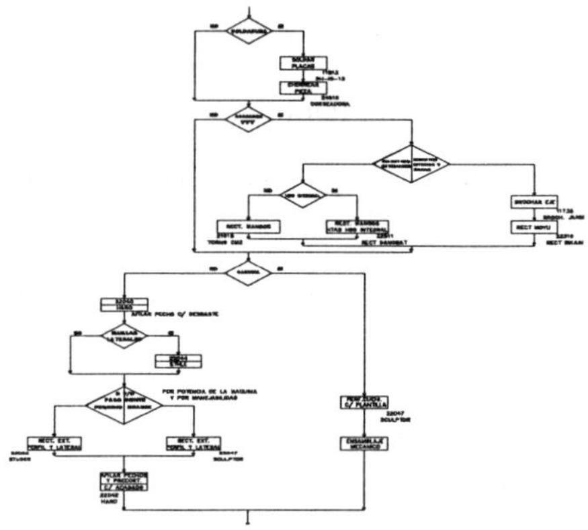

FIGURE 5

4.1.3. Product-Process Analysis

Product/Process analysis: this analysis gives the grouping of products into families according to the different processes, creating the base for the creation of the technological groups.

The aim of this analysis is to determine the product-process hierarchy which allows the market to be served, diversifying the product as late as possible, without excessive levels.

By determining these families, we are determining future technological groups.

4.2. DETERMINE THE FUTURE PRODUCTION SYSTEM

4.2.1. Define the Production Units or Technological Groups.

We have already defined the production families, and the different levels of the process, so we can go on to make our first attempt to define the production units, (we must remark there is always at least one critical facility in any unit).

What we mean by defining the production units is establishing the number and composition of them. Their lay-out, who is exactly going to work in it or where it is going to be placed is something we determine later.

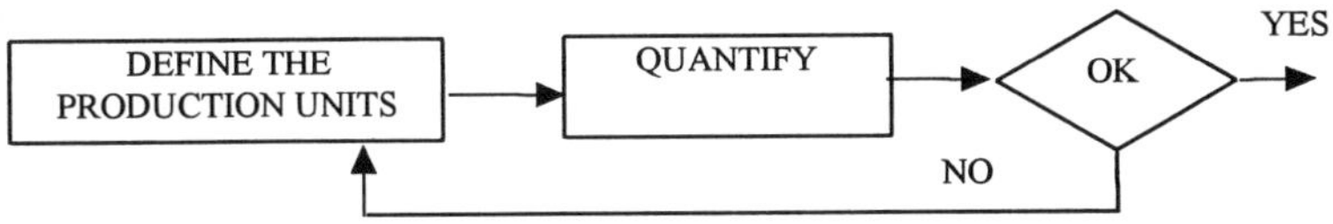

We understand a production unit as an small firm with just one person responsible for manufacturing the finished product or component.

4.2.2. Models Generation

After the first definition, and once quantified the production units, we often find different possibilities.

*a) Investments:*It is quite common to realize that an investment could give our model either more simplicity or shorter lead-time. It is up to the firm to decide what to do, according to their possibilities and objectives.

b) Election: Once the technological groups are defined we must locate them in the factory, according to our defined processes and the physical constraints of the factory, (figure6) shows the four proposed possibilities.

FIGURE 6

*c) Macroprocess:*We have defined each product process, now once the technological groups are defined, we will look at it from a higher level. These macroprocess diagrams represent each technological group as a machine.

4.2.3 Determine the New Lay-out

We must determine the following aspects:

a) Physical lay-out of the technological group, and the product range, which is entirely going to be manufactured inside that group

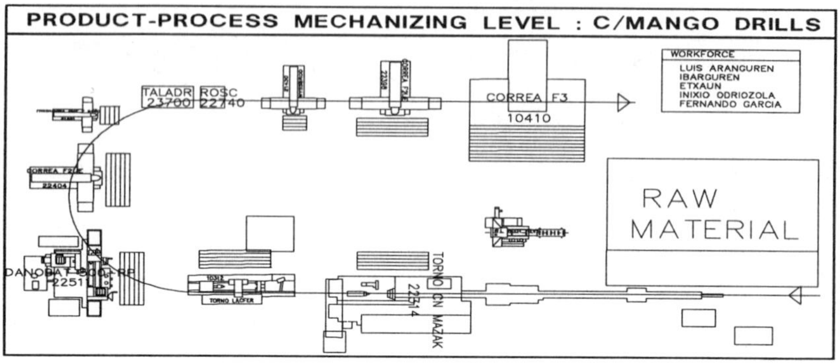

FIGURE 7

c) Required labour, who exactly, the final processes of the products to be manufactured, the person responsible for lead-time, quality, flexibility and stock.

d) Management of the group, that is determining the way labour is going to attend facilities and the scheduling criteria

4.2.4 Select the alternative and move

Once technological groups are totally determined we study the global lay-out.

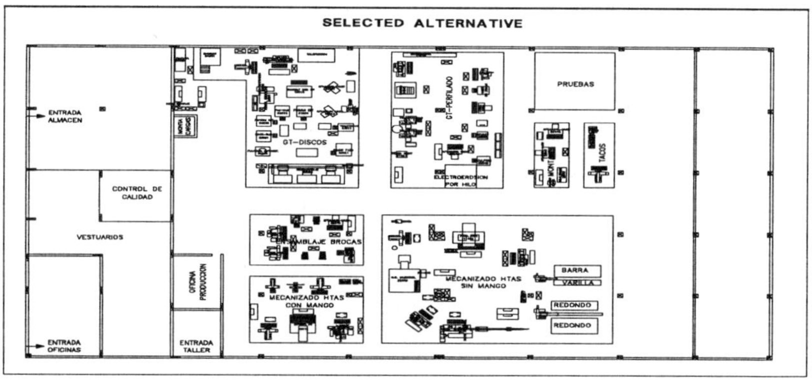

FIGURE 8

Facilities must be moved in order to start up with the new Production System

4.3. . DETERMINE THE DECISIONAL SYSTEM FOR THE PRODUCTION MANAGEMENT

From the value added functions (Purchasing, Sales, Production, etc.) we choose the ones directly connected with production management. We then determine the decision activities for each function, as well as their **period** and **horizon**.

Once the decision activities have been determined, we analyse each one in order to determine the required information. And we represent together the **decisional** and **information systems**.

4.4. PRODUCTION SOFTWARE

Once the production system is simplified according to the mentioned criteria, it is quite easy to manage it automatically, we understand management as workload/capacity planning and workshop programming, (figure 9).

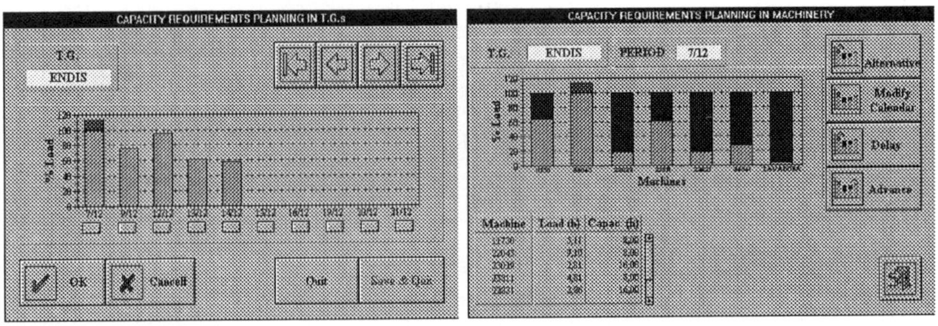

FIGURE 9

4 BIOGRAPHY

Authors: Carlos Martín and Rafael Encina, Ikerlan S.Coop. Mondragón Corporación Cooperativa, Spain.

CARLOS MARTÍN (Industrial Engineer), born in 1968, studied Industrial Engineering (Industrial Organization) at University of Navarra in San Sebastián, and Engineering with Business Studies at Sheffield City Polytechnic (U.K.), Postgraduate in Quality Assurance, he joins Ikerlan S.Coop. in 1993 and works as a researcher in the Production Systems department.

RAFAEL ENCINA (Industrial Engineer), born in 1966, studied Industrial Engineering (Computer Engineering) at Mondragón Eskola Politeknikoa, since his graduation in 1990, he has worked as a researcher at Ikerlan S.Coop. His research interests concern mainly the change of the organizational structure of the production systems

An introduction to a process engineering approach and a case study illustration of its utility

P. Kawalek,
Informatics Process Group, Department of Computer Science,
University of Manchester, Oxford Road, Manchester M13 9PL. England.
Tel: +44 161 275 6183. Fax: +44 161 275 6200. email:pk@cs.man.ac.uk

Abstract.
This paper introduces an emergent process oriented approach to systems engineering. It has come to be known within its domain of development as 'process engineering.' The approach has been developed through collaboration with companies in the services and manufacturing sectors, and through the 'Process Engineering Framework' Project. A case study is used to illustrate the approach which relates to the development of software.

Keywords.
Process engineering, process modelling, process architecture.

1 INTRODUCTION.

1.1 Aims

The subject of this paper is a process engineering approach which is known as the Process Engineering Framework (PEF) after the UK EPSRC project which is developing it. As an introduction to the approach this paper aims to address its nature, position it in relation to related approaches and present two of the most important topic areas within it. These are modelling and architecture.

1.2 The orientation of the approach

At some risk of over generalisation it is suggested that systems engineering is still maturing as a systems discipline. One is reminded of how systems theory was born as biologists in particular faced problems of complexity and order. The sort of problem encountered was summarised by Weiss (1968, p.22) who described how in analysing the cell the enquirer encountered "....rather well-defined and relatively stable complexes of functional and structural properties which are embedded in, and mutually related through, matrices of much less well-defined, more fleeting configurations..." The concept of 'system' is used to understand and describe the properties that parts have when in relationship with other parts. The concept has been defined with differing emphases many times, for example by von Bertalanffy (1968, p.55), Checkland (1981, p.317), and Beishon and Peters (1976 p. 12). Checkland is noted for applying 'soft' systems theory to the study of social problem situations. Leaving to one side the main thrust of his argument, it is useful to note his characterisation of systems through two pairs of ideas; control and communication, and hierarchy and emergence. For the details of this refer to Checkland (1981, pp. 74-92). Although these ideas are not of direct concern in this paper, the reader may note how they recur in the later discussions. More directly relevant to our purpose at this present time is another systems concept, that of systems synthesis. Systems synthesis can be described as concerned with the understanding of the larger system to which a part belongs, appreciation of the properties of this larger system and then recognition of the role played by the part within this larger system (Wardman, 1994). Or, as Weiss puts it "By raising his sights from single objects to their interrelations with others, man reverses his direction from analysis to synthesis" (Weiss, 1968, p.6).

An example

A recent project undertaken in collaboration with an insurance company can be used to describe a practical manifestation of analysis/synthesis. The company have invested heavily in two world-wide databases, one for in progress work and one for active policies. However, currently their business processes are characteristically manual. Most work is done using bulky paper files which contain the policy submissions received from brokers. The files are used by underwriters who scour the details of the submission as part of their risk assessment. They add hand-written notes to the files which describe the reasoning which underpins their decisions about premium quotes and so forth. Administrators use paper forms upon which they hand write summary details of each submission. These are attached to the files. At specified points in the process the information contained in these forms is transferred to a database. Cases which progress to the successful negotiation of a premium will be entered into both databases thereby requiring duplication of effort. Locally, in order to remedy specific information shortages, initiatives have resulted in the development of new databases (e.g. containing brokers names) using systems such as Paradox. Although these do help to remedy the information shortages, they add to the problem of repeated data entries. They require that additional actions are carried out by users so as to keep the databases up to date and in accord with other databases. Therefore, certain items of information are entered in triplicate to databases and at least once to paper forms. This is an example of succeeding with analysis but failing with synthesis. That is, although each of the databases might of themselves be well designed and able to satisfy some requirement, when we consider how the IT system as a whole serves the company's staff it is apparent that the relationship is not satisfactory. The desire to be served by a system which requires data items to be entered once and once only was a something of a leitmotif for the study.

The emphasis in systems engineering today

Returning to systems engineering, and considering it from an organisational perspective, there is evidence that the discipline can be characterised by an increasing concern for problems of synthesis rather than classical problems of analysis (e.g. the bespoke development of a payroll system). The increased emphasis given to systems integration, business led design approaches, holistic process reengineering approaches, workflow, business network redesign (Venkatraman,1991) and to some degree networking more generally, all seem to testify to this increased concern for synthesis. This is reinforced by the aspiration for reuse of components in a lego brick style of development. The approach described in this paper uses ideas from process modelling to explore, define and evolve the relationship between parts of a system. At one level these parts can be seen as the social and technical systems within an organisation, and at another they might be the individual people working for the organisation and their various tools (e.g. databases).

1.3 Process Engineering and Business Process Reengineering.

In anticipation of some terminological difficulties this part of the paper seeks to separate PEF from the vaunted Business Process Reengineering (BPR) which, whilst sharing some of the concerns of PEF, has different origins and emphases. The reader is asked to set aside any preconceptions he or she has about 'process engineering,' 'Business Process Reengineering,' 'process modelling' and the like, and instead think simply of a sociotechnical system. Very simply, the social system can be characterised as being made up of people and their concerns (including culture, politics and structures). An equally simple characterisation of the technical system might describe it as being made up of tools such as the computer systems that are ubiquitous in modern enterprises. These systems can be described as having a relationship in that people in the social system use tools in the technical system to carry out actions. A starting position is then that the relationship between these domains can be understood by analysis and synthesis of actions. In simple terms a process is made up of a number of actions which serve an objective and so we can suggest that the relationship between a social system and a technical system can be understood by exploration of a process which people seek to carry out. Both BPR and PEF use the concept of process dually and simultaneously to see how actions are related to each other and to understand the relationship between social and technical systems.

This said, and although one of the often expressed concerns of BPR is to achieve maximal benefit from IT (Hammer, 1990), characteristically its subject matter is the social domain. Its maxim seems to be that once the capabilities of the technical system are understood, radical but bountiful changes should be made to the social system to exploit these capabilities. In the various writings there seems to be little attention given to the form of the technical system itself (see for example Hammer and Champy, 1993). PEF is different because by origin and nature its primary subject matter is the technical system. More specifically still its domain is informatics (which is made up of information and telecommunication technologies). It draws upon research into the development of support environments which at a minimum level provide integrative mechanisms between defined sets of tools for users. The emphasis today is upon the use of modelling and architecture to develop the relationship between social and technical systems. These topics of modelling and architecture will be considered in more detail later.

The conceptual separation of the social and the technical is very useful and is used for a number of reasons in a number of circumstances. Here it has been used to characterise BPR as concerned with the form of the social system and PEF as concerned with the form of the technical. However, we must appreciate that in a deeper sense the social and technical are not separate at all. The form of the social system is bound up with the form of the technical system and vice versa. In our work we have appreciated this and have extended our methodological concerns to analysis and development of the form of the social system (Wastell et al., 1994). This work, though not the primary focus of PEF, shall continue as we appreciate that the development of the technical system is intimately bound up with the development of the organisation in which it sits.

Finally, although it has been alluded to already it is important to recognise explicitly the characteristic holism which is shared by BPR and PEF. This like many of the points made in this paper relates to the idea of systems synthesis which was introduced earlier. In BPR the emphasis is upon processes which cross functional boundaries, which reach from customer request to customer satisfaction and facilitate change programmes from an enterprise perspective rather than from a organisation function perspective (Hammer and Champy, 1993, p. 35). In PEF the messages are characteristically more technical; that process is a way of integrating roles, and that the various transactions between tools and roles that are undergone in satisfying a customer need can be understood as a 'long transaction.'

2 MODELLING.

2.1 Introduction

Although the focus is upon the form of the technical system, the process engineering approach of PEF starts with models of process in the social system that it seeks to serve. Hierarchical and coordinative models are both used. Hierarchical models are useful because they allow us to develop control structures which are important for process change. They are not dealt with in this paper. Coordinative models are useful because they allow us to explore the composition of components in a process and the communication between these components. These models may represent an existing situation or a desired one. The difference is obviously critical in real world projects and for the purposes of this paper it is helpful if you bear in mind that all the examples presented are a design for an as yet unrealised process.

The case study

The examples in this paper result from an evolutionary, non-radical redesign exercise undertaken by the author. The project was undertaken in collaboration with a British company who operate a number of large chemical plants. The subject of the study was a software development team who are located at one of these plants. There are sixteen software engineers divided into two groups. Most of the work is concerned with the development of software for complex instruments used around the plant. Some of these instruments are classified 'safety related' by the EC and many more are in other ways critical to the operation of the plant. It is therefore imperative that the team are able to develop high quality software and the team have a very good track record in this respect. The project is motivated by three goals in particular. First, it seeks to evolve the social system/technical system relationship in order that certain minor frustrations such as the need to discontinuously maintain a quality tracking system are overcome. Secondly, it seeks to allow a greater degree of flexibility over the way in which certain, non-safety related projects are currently handled. Thirdly, recognising that the organisational and economic circumstances of the team are changing, the project is concerned with making the process easier to evolve in the face of future, as yet unknown circumstances.

2.1 Coordinative Models.

In order to understand the relationships between components of a system, we need to be able to express what these components are and how they interact with each other. The following model is a fragment of the design for a new process. It shows the components of a process and the interactions between these components.

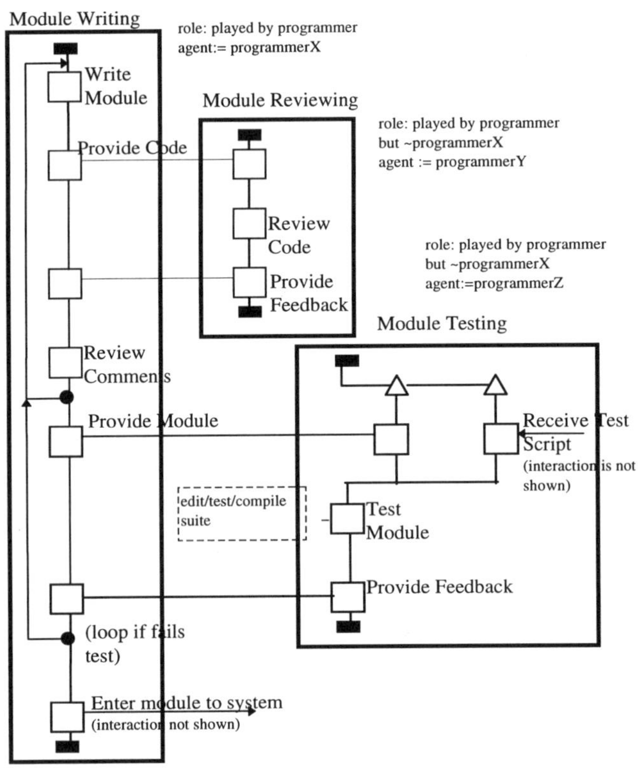

Figure 1 An example of a Role Activity Diagram.

The notation used in this example is the Role Activity Diagram (RAD). This originates from the work of Holt et al. (1983). There is not yet a RAD standard although a complete definition has been given by Ould, (1992). A RAD is a state based diagram in which the vertical lines between boxes represents different states. The boxes are actions. Interactions between roles are a special kind of action and are identified by a horizontal line linking each end. Where one end of an interaction is not shown the horizontal line has an arrow head indicating direction. The triangles represent the commencement of a thread in a part refinement. Each thread in a part refinement can be thought of as a sub-state to the main thread. Therefore as this implies there is no ordering between threads within a part refinement. In the above example loops are allowed. These are lines with arrow heads and a black circle at their commencement. The start and finish of each role is marked by a black rectangle. Finally, the role itself is represented by the rectangle grouping around an action thread (i.e. 'Module Writing,' 'Module Reviewing' and 'Module Testing'). Roles are made up of actions which are related by the internal structure of the role, and thereby the role has the property of emergence. Ould has defined a role as "..a set of activities which, taken together, achieve some particular goal" (Ould, 1992). Roles are played by people and so we might speculate that roles for a lecturer might include 'lecture giving,' 'notes preparing,' 'research direction setting' (gerunds are the convention). Although there are good reasons for wanting a more rigorous definition than this within PEF, this will do for our purposes in this paper. All the roles in this fragment are undertaken by programmers. There are rules which require that the same programmer does not write and review the same module or that he/she does not write and test the same module. These rules are shown as annotations on the diagram. There is one other annotation which shows that the tools which make up the 'edit/test/compile suite' are used in the 'Test Module' action. Annotations which show the technology required by actions are important. Each action shown in the diagram could have been annotated this way. Only one example is given in the example because of the problem of clutter in the restricted confines of the page.

In the RAD, as well as actions, structuring of actions and interactions we see information about role players and technology. Fully annotated it would be a map of a social system/ technical system relationship. We see person to person and person to technology relationships. Technology to technology relationships can also occur although they do not in this fragment. RADs are valuable as representations of current process or as blueprints of process designs. They show co-ordination within and between roles. It is argued that their production and reading can help develop shared understanding amongst individuals involved in a study, that they can help make covert problems manifest and can help rationalise process and procedure (Ould, 1992), (Kawalek, 1994). All of this is important. However, perhaps more interesting is to ask what we could have if the RAD were not just a passive model but somehow enactive. What if instead of just being a representation or blueprint, the RAD represented an encoded prescription of the behaviour we want of the technical system? Thus, for the programmer playing the role of 'Module Testing' the system would manage the interactions with other roles to obtain the module and the test script. It would present the programmer with an icon 'Test Module.' On clicking this icon, the tools necessary for the programmer to carry out the action ('test/edit/compile suite') would be presented. The programmer could finish the test and use the system to send the feedback to the role which created the module.

Process modelling languages
This simple sketch of the concept of an enactive model describes the essence of a process modelling language. The development of PEF is particularly closely associated with an early example of this genus which is fittingly known as the Process Modelling Language (PML) (Bruynooghe et al., 1994). PML is a high level language within which the basic building block is a role. PML roles carry out actions and have interactions with other roles. To this extent there is a clear mapping between PML and RAD primitives. PML was developed as the language of the IPSE 2.5 software engineering environment (Warboys, 1991). It has been further developed as the language of the derivative ICL ProcessWise Integrator (PWI) which is used for support of business processes in general. PWI provides a user interface and a PML application interface so that external applications (e.g. 'test/edit/compile suite') can be integrated into the user's on-line work context. The form of this work context (i.e. which actions and technology are available) will vary according to which roles the user plays and the state of the process of which these roles are a part.

2.2 The relationship between passive and enactive models.

The enactive model encoded in a language like PML serves as a centre of co-ordination and control within a system of many different parts. Thus, Snowdon (1992) has talked of using process modelling to develop IT systems from "..an overall systems level." The attraction of using a high level process modelling language lies with the establishment of constructs which serve both analytical purposes in organisational design activity and as a prescription for the behaviour of the technical system. This suggests that the RAD fragment shown previously could be dually a design for organisational behaviour and an enactable prescription for the way in which the technical system will serve the organisation. Thus the requirement for the designer to invert an organisational view in order to consider the design in the language of the IT system, which is essentially a language of calculation, is much reduced or even eradicated.

Methodological issues
It is important to consider the methodological issues involved in the translation of a model from a passive representation to an enactive, working component in the behaviour of an organisation. Is it really possible to take a simple step between passive and enactive domains? The author has observed development practice for PWI and some similar systems. Currently, for a number of reasons including non-functional considerations, the 'translation' between passive and enactive domains is not normally straightforward. Indeed there is a notable distortion of the role structure of the passive model in developing the enactive model. The temptation is to see the development of such systems as more akin to a conventional process through which requirements are met by an (inverted) programming of the system. One significant issue is that passive models created as part of (social system) redesign exercises are not normally intended to be rigid prescriptions of behaviour. They are simplified interpretations of real world behaviour which are usually used to represent a typical case to users. However the reality of work tends to be characterised by exceptions and the need for creative, extemporising behaviour by users (Fikes, 1982). Even in the very simple RAD fragment which was shown previously we can see how, interpreting the model strictly (as software would do), the 'Module Writing' role will have to undertake and complete the action 'Write Module'

before passing the code for review by 'Module Reviewing.' This may be perfectly reasonable for many cases but it precludes any other behaviour. Perhaps tackling a particularly difficult module the 'Module Writing' programmer wishes to half complete it and pass it over to 'Module Reviewing' for initial comments whilst continuing to complete the rest of the module. This is very reasonable behaviour in the actual domain which places a lot of emphasis upon peer reviews and informal support. However, if we were to simply take this RAD and convert it into software the system would not support this reasonable way of working.

This means that we need to explicitly design the behaviour of the enactive model. However, it is the author's contention that the structural integrity of the model should be consistent across passive and enactive domains. This is achieved in the case study by preserving the roles ('Module Writing,' 'Module Reviewing,' and 'Module Testing') and the pre and post-conditions to these roles across the passive and active domains. All that has been changed is the internal structure of the role so that a pleasant and flexible working environment is created for each user. It no longer strictly enforces typical case behaviour but provides a context for flexible handling of many cases. Incidentally, these design decisions were taken in consultation with users and managers. The following example expresses as a RAD the behaviour of the enactive PML model for the role 'Module Writing.' The other roles have been omitted from this diagram.

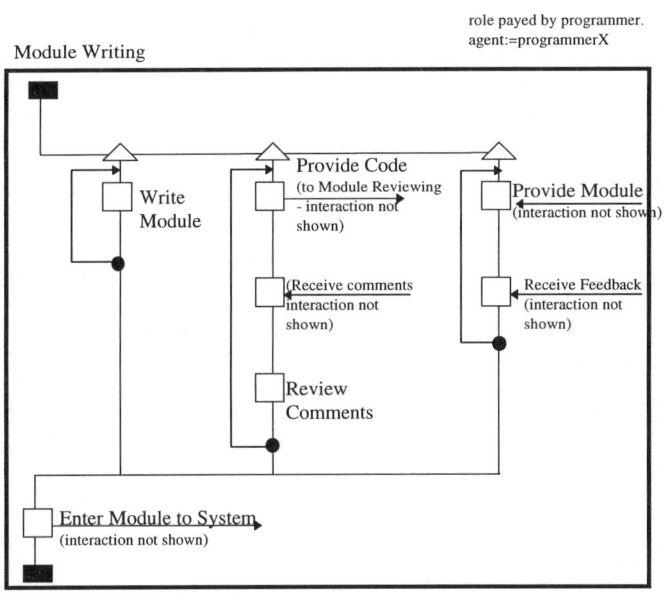

Figure 2 A representation of the behaviour of the enactive model

Thus far we have seen how process modelling can be passive and enactive. It can be described as passive where it is an analytical activity whose subject matter is the form of the social domain. Passive models are used for development of understanding, representation of complexity and the development of designs. Process models can be described as enactive where they are used in real time to define the behaviour of the technical system and hence the real world relationship between social and technical systems. The simple example of a RAD and PML is used to illustrate how the structural integrity of a model can be maintained across the passive and enactive domains.

3 ARCHITECTURE.

3.1 The enactive model as an architectural component.

It is important to consider the implications of the enactive model for the overall system architecture. How does it affect the form of the system to have this model, which has been described as a kind of hub of co-ordination and control, as a component within it? Although research in this area is still in its very early stages we can speculate that the architectural implications of the active process model could be of profound consequence.

We can start by considering the infrastructure of the software development team in the case study.

- They have a structure of process, roles and actions with associated rules.
- They use non-electronic media such as paper for many informal and some formal purposes.
- They have a number of IT applications, databases, networks and platforms. For purposes of illustration applications of note are the edit/test/compile suite which has its own database, a text editor and a quality tracking database.
- The active process model will be used as a coordinative mechanism and will, when activated influence the way work is performed.

We can explore how the enactive model can serve as a coordinative mechanism between some components in this infrastructure. The effect of this is to enable a greater degree and order of synthesis within the system as a whole. The degree of synthesis that is beneficial is a design issue. The reader should not infer that process engineering will motivate a pendulum like swing from a need for integration to a fully (perhaps overly) integrated system. Indeed, the use of the enactive model to integrate components within a system can be thought of as a way of achieving the loosest possible level of integration between components of the system. From this may arise benefits of flexibility.

The relationship between people and tools.

Repeatedly in this paper two important aspects of the enactive process model have been emphasised. These are discussed by Warboys (1991). The first is that it acts as an "upward facing" framework for supporting the interaction of users in a human organisation. The second is that it acts as a "downward facing " integration framework for disparate tools and databases. The following diagram depicts the enactive model in this role as a coordinative component within a system (social and technical) of people, applications and tools. It was first presented by White and Kawalek (1993).

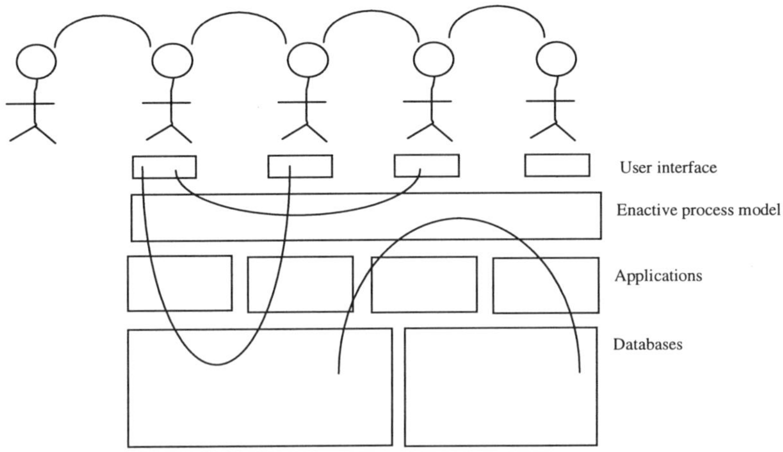

Figure 3 An interpretation of a system architecture incorporating an enactive process model.

The arcs in this diagram represent four different forms of co-ordination. These are as follows;

- Inter-personal (not mediated by IT),
- Inter-personal mediated by the enactive process model,
- Inter-personal mediated by a shared database,
- Between databases (or alternatively between applications) mediated by the enactive model.

It would be valid to denote other forms of co-ordination in this diagram. However those shown are particularly interesting in that we see in points two and four respectively the upward and downwards roles of the enactive process model. The diagram also recognises forms of co-ordination outside of the enactive process model (points one and three).

An example

To illustrate this a small part of the RAD model in Figure 1 is considered. It relates to the interaction between the 'Module Writing' and 'Module Reviewing' roles. Exploring this in more detail shows that the programmer playing the 'Module Writing' role uses the edit/test/compile suite to carry out the action 'Write Module.' The edit/test/compile suite has its own database with version control. It does not share this database with any other application. The 'Provide Code' interaction sends a prompt from the 'Module Writing' role to the 'Module Reviewing' role. On picking up this interaction the 'Module Reviewing' role is connected into the edit/test/compile suite where the module is examined. In carrying out the action 'Review Code' the role has access to a simple text editor for writing a report. All completed reviews must be kept in the Quality Tracking Database. When the review has been completed the programmer playing the role clicks a 'Finished' icon provided by the active model. The enactive model will then enter the review into the Quality Tracking Database and place a 'Review Comments' icon on the screen of the 'Module Writing' role. When this icon is clicked the programmer concerned will have access to the Quality Tracking Database to read the review. This is expressed in Figure 4.

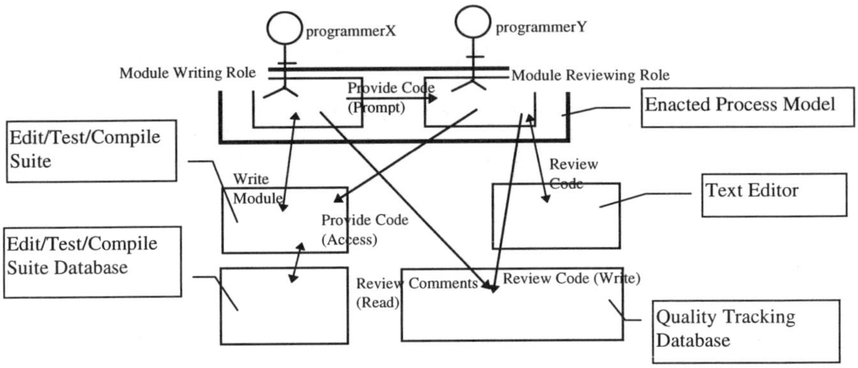

Figure 4 An interpretation of the architectural role of the enacted process model in a small part of the case study example.

Structural holism and functional clarity

This simple example gives us the opportunity to identify two important facets of the PEF approach to architecture. First, it is holistic in a structural sense in that as well as developing models which represent and contribute to the dynamism of human behaviour in organisations, it develops models of the way in which the technical system contributes to the organisation. The approach is concerned simultaneously with interactions supporting the organisation's processes and acting as an integration framework for tools. These are the upwards and downwards roles described by Warboys (1991). Thus we recognise, after Heidegger (1977), that objects such as computer systems become part of a background of 'readiness-to-hand' to their users. The users are not concerned with them as such but with the actions they seek to accomplish. If we wish to consider the actions that people carry out then we have also to consider the tools upon which these actions rely. Warboys (1991) argues that modelling approaches which suggest a separation of the upwards and downwards will have a "...short life."

This holism leaves us with a problem of complexity. One of the ways in which this complexity must be managed is through functional clarity. We have seen earlier how this can be achieved in the RAD models where roles boundaries were defined, their interrelationships mapped and their technology dependencies were annotated. We see it also in the architectural approach whereby we have conceptually separated co-ordination and control from the rest of the system capabilities and classified these other capabilities as applications and databases. Nothing in this architectural approach should be understood to be anything other than an initial response to a very difficult problem.

4 CONCLUSION.

This paper has positioned the PEF approach as concerned with relationships between parts of a system. At its most abstract the relationship of concern is that which exists between social and technical systems. This is worth pondering. This relationship has the potential to become ever more complex. On the one hand a climate of commercial competitiveness and insecurity requires that people and organisations are flexible, creative and innovative. On the other hand the number of technical solutions available in the market-place continues to multiply. Attempts to try and control the import of new tools or the creation of new databases in an organisation are likely to be confounded by the growth of networks and the easy access to hardware. Ultimately the very complexity of the domain may serve to stifle those aspects which are most precious, namely flexibility, creativity and innovation. Any wish to bring structure and simplification to bear will have to start with fundamentals. We need to understand the essential dependencies which bind the work of one person to the work of another. We need to ask what a user wishes to do, what he or she needs from someone else, and what he or she expects to deliver. We can understand these things by understanding process.

5 ACKNOWLEDGEMENTS.

The architecture section "borrows" from a project undertaken jointly with Dick Thomas of Thomas Partnership. Thank you to Mark Greenwood and Martyn Spink for comments.

6 REFERENCES.

Beishon, J. Peters, G., (1976) *Systems Behaviour*, Second Edition, The Open University Press, Harper and Row, London.

Bruynooghe, R.F., Greenwood, R.M., Robertson, I., Sa, J., Warboys, B.C., (1994) PADM: Towards a Total Process Modelling System in *Software Process Modelling and Technology* (ed. Finkelstein, A., Kramer J., Nuseibeh, B.) Research Studies Press Ltd., Taunton.

Checkland, P. (1981) *Systems Thinking Systems Practice*, John Wiley and Sons, Chichester.

Fikes, R.E., (1982) A commitment based framework for describing informal cooperative work, *Cognitive Science,* **6**, 331-347.

Hammer, M., (1990) Reengineering Work: Don't Automate, Obliterate, *Harvard Business Review*, July-August 1990.

Hammer, M., Champy, J., (1993) *Reengineering the corporation, a manifesto for business revolution*, Nicholas Brealey Publishing, London.

Heidegger, M., (1977) *The Question concerning Technology*, Harper and Row, New York,.

Holt, A.W., Ramsey, H.R., Grimes, J.D., (1983) Coordination system technology as the basis for a programming environment, *Electrical Communication* 57(4).

Kawalek, P., (1994) Comments on the use of RADs in case studies, *IOPener*, Newsletter of the IOPT Club, Volume 2, Number 3.

Ould, M, Process Modelling with RADs, Parts 1-3, *IOPener*, Newsletter of the IOPT Club, Volume 1, Number 5 to Volume 2, Number 2.

Snowdon, R.A., (1992) Process modelling is more than sequencing work, *IOPener,* Newsletter of the IOPT Club, Volume 1, Number 3, February 1992.

Venkatraman, N., (1991) IT Induced Business Reconfiguration in The Corporation of the 1990s, *Information Technology and Organizational Transformation* (ed. Scott-Morton, M.S.), Oxford University Press, Oxford.

von Bertalanffy, L. (1968) *General System Theory, Foundations, Development, Applications*, Allen Lane, London.

Warboys, B.C., (1991) The practical application of process modelling, some early reflections, *Proceedings of the First European Workshop on Software Process Technology,* Milan, AICA Press.

Wardman, K.T., (1994) From Mechanistic to Social Systemic Thinking, A digest of a talk by Russell L. Ackoff, *The Systems Thinker*, Volume 5, Number 1, February 1994.

Wastell, D.G., White, P., Kawalek, P., (1994) A Methodology for Business Process Redesign: Experiences and Issues, *Journal of Strategic Information Systems*, Volume 3, Number 1.

Weiss, P.A., (1968) The Living System: Determinism Stratified in *Beyond Determinism: New Perspectives in the Life Sciences* (ed. Koestler, A., Smythies, J.R.) Hutchinson, London.

White, P., Kawalek, P., (1993) A framework for business process management, *IPG report*, Informatics Process Group, Department of Computer Science, University of Manchester.

7 BIOGRAPHY.

Peter Kawalek is a Research Associate of the Informatics Process Group in the Department of Computer Science, University of Manchester. He has undertaken ten collaborative projects with industrial partners. Currently, as well as working on further collaborative projects he is contributing to the EPSRC funded 'Process Engineering Framework.' The results of this project shall be published.

25

Design of activities in shop floor management: A holistic approach to organisation at operational business levels in BPR projects

Eric Scherer
Institute of Industrial Engineering and Management BWI
Swiss Federal Institute of Technology ETH, 8028 Zürich, Switzerland

Martina Zölch
Work and Organisational Psychology Unit IfAP
Swiss Federal Institute of Technology ETH, 8092 Zürich, Switzerland

Abstract

Process orientation in production and business administration is a major means of increasing flexibility and speed while decreasing costs. During the past years the notion of human resources as a potential and means of ensuring flexibility in decentralised organisations gained importance. Even though systematic approaches to modelling and design of business processes and process organisation do exist, they mostly are related either to the derivation of targets on a strategic enterprise level or to specification of information technology and structures. Conversion of strategic targets into design of process activities at an operational level is usually not done in any systematic way and leaves a gap in existing methodology. This paper presents a human oriented approach to analysis and conceptual design of activities within a business process reengineering framework and its application to shop floor management.

Keywords

Business process reengineering, process organisation, task analysis, organisation development, human resources, participation, socio-technical system, shop floor management, manufacturing execution system

1 INTRODUCTION

Customer focus and strategic target orientation is a commonly recognised approach for business process reengineering (BPR). At the operational level of an enterprise it results in a complete change of organisation: tayloristic departments are replaced by target oriented teams. Each team is assigned specific tasks and performs a number of activities required to fulfil as part of processing an overall business process. To ensure a smooth processing, a well suited information and communication system is necessary. An information system provides all information necessary to perform a certain activity, supports the processing by supplying specific tools and controls the overall processing of tasks through a work flow system.

Several state-of-the-art integrated software systems provide these basic features and can be adapted to the demands of various industries and enterprises. This leads to a large variety of possible system layout and configuration based on one standard software. In production management and especially at operational level activities often are automated that previously were performed manually. In the shop floor domain this leads to a high tension between automation of technological activities and automation of business activities, i.e., the intersection of the physical and informational system (Scherer et al. 1994). Therefore system specification during the introduction of a new information system, leads to several problems on how to fit generic business processes into an existing shop floor environment.

2 APPROACHES TO BUSINESS PROCESS DESIGN

Business processes are information processes. They describe an informational or physical transformation and are performed by human resources or technological resources themselves. Time-based and target-oriented structuring of these processes leads to the formation of the overall process organisation (Picot & Maier 1993, 11). Information technology thereby forms the back-bone of process organisation whereas human resources are a source of flexibility within the system, since by his intelligence a human can act independent and creative (Kosiol 1966, 65). A human can provide flexible reactions and adapt his own work methodology if the overall system suffers situational disturbances (Mertins et al. 1992, 205), e.g., in case of urgent orders or a machine breakdown. Human resources therefore are of key value for an enterprise (Warnecke 1993, 58), not only as bearer of an enterprise's know-how and expertise, but as key to organisational flexibility. This aspect of human resource relation to process organisation is often ignored by BPR projects and by most technical-organisational innovation. This situation is enforced by the emphasis on a primarily technical top-down approach as proposed by American BPR exponents (e.g., Hammer & Champy 1993) and leads to an theoretical system engineered on a flip chart rather than an organisation carefully developed (Osterloh & Frost 1994). This presents a thinking similar to approaches common during the CIM euphoria of the 1980s which - by relying on a primarily technical design - mostly led to failure (Mandl 1993). Still newly designed activities within a business process framework form tasks that are to be performed by real people. BPR projects have to consider process organisation, information technology and human resources equally as potentials for design, and unify them in a holistic system.

Formal methods for specification of business processes

For design and specification of business processes several suitable methods and tools do exist (e.g., Scheer 1994, Curtis et al. 1992, Huckvale & Ould 1993) and have been practically applied (e.g., Keller 1994). Usually activities are described and linked by discrete events. This forms the basic *control view* of the system model. Each activity is related to data and information necessary for execution. Additionally each activity is assigned its information resource, e.g., a computer or software system, and the affected organisational unit, e.g., departments or employees. Existing computer-aided tools and methods offer made easy support for formal process specification even at high complexity. No support is offered for the conceptual design of human resources, information technology and organisation itself (see Klein 1994). Accordingly information system design usually is carried out by experts, still human resource issues are often left to participatory approaches. Participation offers a significant contribution to project management techniques (Morris & Brandon 1994, 213ff), leads to a multipersonal approach to problem-solving and helps to transform employees from being affected to being involved (Baitsch 1985). Participation still does not ensure that human potentials really are utilised within a reengineering project (Zölch 1992). Therefore it is necessary to develop and utilise methods that offer guidance and expertise to utilisation of human potentials.

Principles for human-oriented conceptual design

Each employee assumes a number of tasks, each task referring to the processing of a number of consecutive activities, i.e., a section of an overall business process. As part of his responsibilities an employee deals with activities of several business processes of different characteristics at the same time. The perception of a business process by a single employee presents the process as a limited number of activities he has to perform to fulfil the task assigned to him. The summation of all activities of the various processes forms the task environment an employees is assigned as part of his daily work, Figure 1.

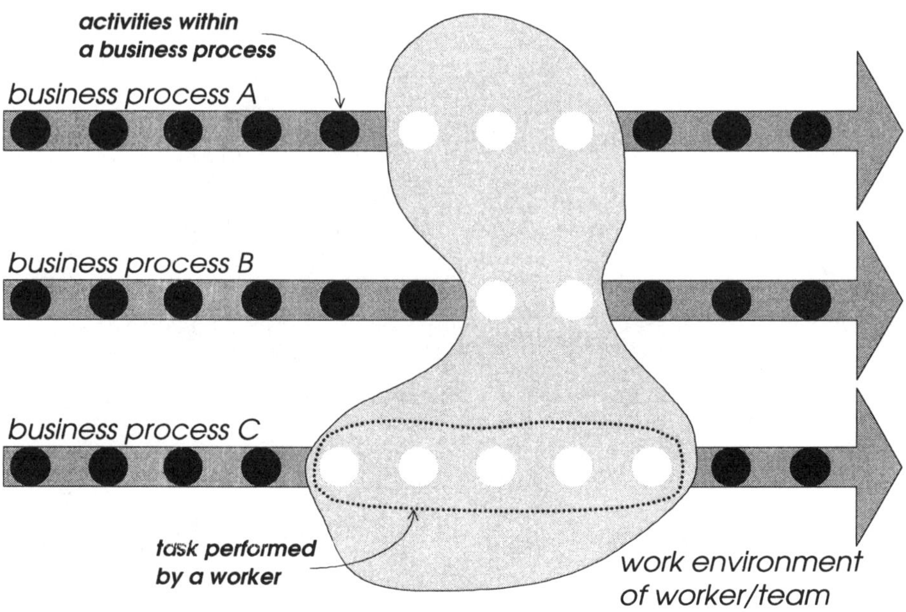

Figure 1 Process view vs. task view of activities.

Ensuring a positive and demanding task environment for each employee makes it necessary to design tasks in a holistic and humane way. Design principles include completeness, variety and richness of meaning of tasks as well as possibilities for social interaction, autonomy, opportunity of learning and personal advancement, time independence and the opportunity of coping with demands without stress and hindrances while tasks are carried out (Ulich 1994, 161). Tasks designed according to these principles lead to an intrinsic motivation of employees and willingness to take responsibility, to gain knowledge through experience, stay mentally alert and develop self-esteem. This leads to a work-oriented perspective trying to utilise and advance human resources. Still complete informational support for all tasks has to be ensured and maintained, i.e., all information necessary has to be provided ergonomically and with easy access. Holistic design of tasks thereby gives access to human resource potentials within process-oriented organisations and helps to foster these potentials. This promotes the overall system performance in respect of flexibility, quality and economy.

Criteria based methods for system design offer a useful addition to BPR methodology and can provide a distinctive description of the informational and social realty of business processes. Work-psychological research has brought up a number of expertise based methodologies for analysis and assessment of tasks within administration and production management (Leitner et al. 1987 & 1993, Volpert & Oesterreich 1991, Dunckel et al. 1993, Weik et al. 1994). Assessment of tasks utilising conceptual criteria can offer advice to business process design at operational level. This leads to a combined top-down/bottom-up procedure within the BPR framework equally utilising human resources and information technology to achieve an ideal process organisation, Figure 2.

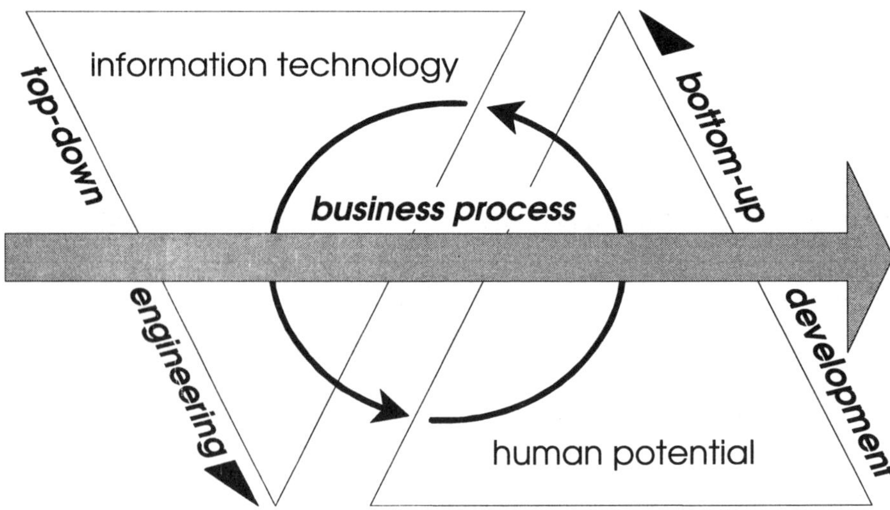

Figure 2 Top-down engineering and bottom-up development of process organisation

3 CRITERIA BASED TASK ANALYSIS AND DESIGN

In the perception of work-psychological research, a task performed by an employee forms the nucleus of each socio-technical system (Ulich 1994, 157ff) and links individual and organisation (Volpert 1987, 14). Hence a task is not primarily perceived as function or duty to be fulfilled as part of a business process or a technical system but as a number of activities to be performed by an individual. The individual becomes an acting character and adopts certain objectives to lead his own action. A task's objective and the activities necessary to fulfil a task can be utilised to assess positive requirements toward the affected employee (Oesterreich & Volpert 19986). This leads to equivalent advises for conceptual design.

The KABA method

The *Contrastive Task Analysis and Design Method (KABA)* is a procedure to analyse and assess work tasks in administration and offices and is originated by work psychologists (Dunckel 1989, Dunckel et al. 1993). It can be utilised to indicate possible consequences of new and existing information technology toward the employees. Originally designed for qualified work in industrial administration, banking, insurance and government, it presents itself useful for administrative tasks in production planning and management and therefore was adopted for use at operational levels within production.

It can be assumed that human acting is goal oriented, object related, social and flexible. Humans can adopt strategic goals as proposed by strategic management for their own acting, object related thinking easily reduces complexity and social behaviour leads to the ability to learn and advance. Based on these assumptions eight key criteria have been derived to assess the human potentials used at a certain system state. Tasks promoting human characteristics and potentials have to offer (1) a large *scope of decision,* (2) *temporal scope*, e.g., time independence, (3) *transparency* and the *possibility to influence the conditions of work*, (4) *a variety of methods of working*, and (5) the *necessity to communicate*. Additionally tasks should be (6) *free of organisational or technical hindrances*, offer (7) *physical activity*, and should be related to (8) *material and social reality*. This allows to compare the abilities of human resources in contrast to information technology.

KABA can be utilised during *redesign of existing process organisation* to identify new human oriented potentials in system design. Therefor existing tasks are recorded and the results of the assessment are used for basic requirement specification, e.g., in case of replacement of an old MRP II system. During *conceptual design of a new process organisation*, KABA offers foresighted assessment to various design variations and stages. This is useful to assess positive and negative effect of proposed software features or computer based communication systems, e.g., electronic mail or video conferencing. Eventually it achieves *assessment of existing systems*.

Application of KABA

The KABA analysis is conditional, i.e., it does not assess the employees themselves and their individual evaluation of the situation but the tasks to be performed and their related work conditions. During analysis, the investigator (1) describes the set of task to be performed by each employee or each characteristical group of employees. The tasks can be derived based on a real system or a prospective one as achieved through a proposed business process layout. Each task is identified and distinguished toward others. Relevant information and control flow is recorded, i.e., how orders are submitted and which work flow information and triggers are used. Each task identified is (2) split up into the activities necessary to perform a task successfully. All required information and related documents, requirements to access information, processing time and utilised information technology resources are recorded. This step of KABA principally leads to a formal description of the control view similar as achieved through process modelling according to Scheer (1994). Based on the knowledge acquired, (3) the impact of information technology is assessed for each activity. Based on the previously listed key criteria it is checked whether the used or proposed information technology has a positive impact on the tasks, e.g., easy access to remote information, or negative, e.g., limitation of the scope of decisions. Based on the knowledge gained, (4) measurements for human oriented system design and utilisation of human potentials can be derived.

Practically KABA is lead by orientational questions allowing the investigator to examine the specific aspects of the work system. The examination leads to a classification according to several criteria and various stages that possibly can be achieved. Figure 3 gives an example of grading.

Level 1	Communication about deviation of determined actions
Level 2	Communication about how to determine a course of action
Level 3	Communication regarding consequences by determining a course of action
Level 4	Communication in regard to making one common decision
Level 5	Communication in regard to making several connected decisions
Level 6	Communication about common decisions in several spheres
Level 7	Communication in regard to development of new procedures

Figure 3 Levels for the criteria „Communication and cooperation requirements“

Based on this classification, KABA indicates necessary improvements and measures to access human potentials and permits benchmarking among different teams, organisation and enterprises. Based on job descriptions or similar documents, each level easily can be interpreted in correspondence to specific activities to be improved. By comparing an employee‘s qualification profile with the profile demanded to fulfil prospective tasks, measurements for personal advancement can be derived. This leads to a perception of tasks by employees not as stress but as challenge and guarantees positive incentives (Alioth 1980, 31).

4 TASK DESIGN FOR SHOP FLOOR MANAGEMENT

The tightening of the logistic supply chain in customer driven production enforces reduction of planning hierarchies and leaves the shop floor level of any production enterprise facing more and more challenges. Reality at the shop floor includes growing external demand, e.g., quality, quantities, variety and speed of products and orders, and internal, e.g., new technologies and complex product structure. These requirements are opposed by heterogeneity and uncertainty within the shop floor domain. Heterogeneity is caused through (1) technologically multifarious machines, (2) different type of production resources, (3) contradicting planning objectives, (4) different production strategies, e.g., KANBAN and make-to-order, (5) non-standardised external and interdepartmental interfaces, i.e., due date allowance, order/lot size, delivery procedures. Uncertainty is created through frequent changes of orders and unpredictable behaviour of the production system, e.g., machine breakdown or illness. Hence all resources for shop floor management, human resources, information technology and organisational structure have to be carefully utilised and designed toward a holistic system.

Case study in electromechanical manufacturing

To provide deeper insight, a short case study is given. At a Swiss manufacturer of optoelectronical instruments a concept for a new production order processing framework was developed and was to be implemented. This included establishing of a number of product oriented, market driven segments (cost centres) for assembly of final products and several technology-oriented, make-to-order segments for internal supply of parts and pre-assemblies. The processing of all orders on an aggregated level was supported by the introduction of a new MRP II software package. This was to release aggregated production orders, issue routings and control workflow between the segments. Within the electromechanical manufacturing segment this led to the decision to introduce a new computer aided manufacturing execution system (MES), or „Leitstand", to replace the previously manual system. Through the MES overall performance should be increased and requirements met as set within the new production order processing framework. Project objectives included among others (1) creation of simple and flexible business processes, (2) creation of more autonomy for operators, (3) increase of reactivity upon work load changes, (4) increase of situational transparency, and (5) humanisation of the work environment. Table 1 gives characteristic numbers for the concerned shop floor segment.

Table 1 Characteristic numbers

Staff	44 person
Number of active product types	110
Number of active assembly types	790
Number of used parts per year	11 million
Number of orders per year (MTO & KANBAN)	6.000
Percentage of KANBAN orders	35%
Average lot size	60
Average lead time (production)	8 days
Average lead time (supply)	50 days
Queuing time/production time	60/40

Analysis included (1) description of order related business processes at the operational level and within the overall order processing framework, (2) task analysis and definition of human resource requirements, and (3) analysis of information system capabilities (also see Bullinger & Hirsch 1994, 20ff). For analysis and design the ARIS methodology (Scheer 1994) was chosen as formal modelling technique, but the KABA methods were utilised for conceptual design of processes and to contrast human potentials and MES functionality. During analysis four classes of employees were investigated at their work place: floor supervisor, foreperson, skilled and unskilled workers. Figure 4 gives an example of the results of the task analysis in case of skilled and unskilled workers. The reference indicated is based on work psychological research and gives a benchmark for a realistic workplace classified humane.

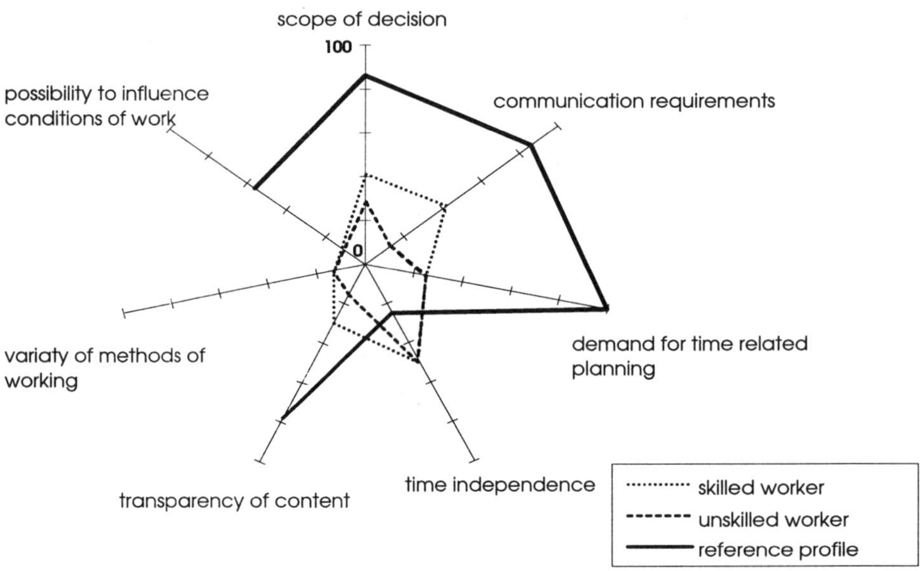

Figure 4 Analysis results for skilled and unskilled workers

Results and measures

Analysis led to following diagnosis: (1) several information is not easy to access or is missing, (2) division of planning and execution of activities causes loss of information and unnecessary interfaces, (3) lack of work-related communication causes loss of knowledge, i.e., only communication on what to do and not how to do it. This resulted in redesign of order processing and related tasks in many respects. Prime measure was a reduction of hierarchies to two levels. Supervisor and foreperson were combined into the role „floor controller", all workers formed teams of 10-12 people. Key task within each group was the distribution and assignement of jobs. In the old procedure each job was split up according to its routing and assigned the relevant machines. In contrast, with the new procedure jobs are assigned to the workers themselves. This was possible since there was no specific domination of machines as restricting resource. Distribution of jobs through staff assignment is done in daily meetings and encourages a more object related perception by the workers. Figure 5 indicates the minimisation of hierarchies.

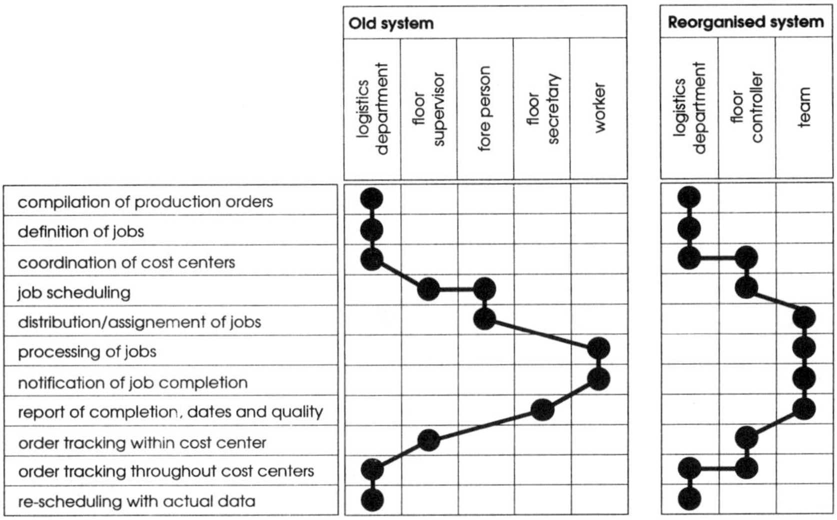

Figure 5 Allocation of shop floor management related tasks

To support staff assignment as key task, MES functionality was slightly adapted to treat human resources as key capacities in scheduling and present them in the GANTT chart. Further contrastive analysis of human and information technology potentials led to the definition of MES functions to be utilised, Figure 6. Through the MES and the allocation of job assignements within the teams it is possible to reduce proactive planning and scheduling tasks for the shop floor controller. This offeres more spare time for further management tasks and coordination with related segments. Reactive scheduling was improved through additional information provided by the system.

5 REFLECTION

Several approaches to restructuring organisation describe an enterprise as organism and propose the formation of decentralised and partially autonomous segments such as teams as a major means to increase productivity (e.g., Warnecke 1993). Thinking in processes and customer focus establishes a strategic guideline for straight forward reengineering of organisational units at an operational level like the shop floor. Modern information technology thereby offers an extensive potential to support business processes (Frese 1994, 133). Information technology provides information necessary to perform activities successfully and supports and controls the overall work flow. Distributed systems enable co-operation and communication among different

	leitstand functions: / **tasks:**	*scheduling*						*report.*	*operating*		*information/ documentation*			
		automatic scheduling	manual scheduling	GANTT chart (monitoring)	checking ressource availability	checking plausibility of schedule	simulation of schedules	reporting work progress	definition of scheduling priorities / objectives	relating dependent orders (order network)	information on schedules of other domains	automatic reporting on finish of jobs	printing out of bill-of-materials etc.	listing of orders on a terminal
	compilation of production orders									●				
standard process (make-to-order)	scheduling of standard production orders	●												
standard process (make-to-order)	re-scheduling / optimization of schedule		●	●			●							
standard process (make-to-order)	surveillance of due date compliance					●								
standard process (make-to-order)	release of order to supply raw material				●									
standard process (make-to-order)	supplying material												●	
standard process (make-to-order)	allocation of responsible team													●
standard process (make-to-order)	controlling work progress of specific jobs			●										
standard process (make-to-order)	recording of job completion, dates and quality							●						
standard process (make-to-order)	updating schedule represantation (GANTT)							●						
standard process (make-to-order)	reporting to MRP II system											●		
exceptions	scheduling of KANBAN orders		●	●			●							
exceptions	scheduling of urgent orders		●	●			●		●					
coordination	coordination with central business units													●
coordination	communication with related production units										●			

Figure 6 MES functions and supported tasks

business segments, teams and distributed production sites without the necessity of direct and personal contact. It is possible to obtain information previously difficult to access and utilise it to make proactive decisions and decisions based on broader knowledge. Information technology seemingly offers a broad variety of design possibilities. Simultaneously it leads to the danger of unnecessary and inefficient automation. At decision-making and execution of single process activities it therefore still necessary to consider individual intelligence only provided by human resources within the overall system.

Strategic decisions to reengineer an enterprise's organisation, e.g., flattening of hierarchies, establishing customer-oriented business segments or introducing a distributed production management software, always have to consider potentials and limitations at the operational level. In case of shop floor management BPR has to be supported by adequate and realistic task design considering individual qualification. It also has to consider the basic conflict between physical and informational tasks as performed by most employees at a operational business level such as the shop floor level. Therefore human resources and information technology equally have to be treated with expertise and conceptual methodology to form a holistic system. This does imply that a technologically feasible solution is not necessarily the solution required for positive system performance (Kötter & Volpert 1993, 131). Contrastive analysis and design of tasks like KABA makes it is possible to design tasks and related activities within a distributed business process framework with a human resource perspective, and thereby utilises their potentials. To ensure lasting efficiency of the newly realised organisation permanent organisation development is necessary. Therefore employees have to have the opportunity to reflect their behaviour and performance as teams and individual and adopt personal goals for their work. This includes the possibility to question tasks and regulations set and change them adequately (see Marks 1991). The ability to reflect processes and organisation from the local perspective of the shop floor level ensures permanent improvement and flexibility within the process framework for production management.

6 REFERENCES

Alioth, A. (1980): *Entwicklung und Einführung alternativer Arbeitsformen*. Huber, Bern.

Baitsch, C. (1985): *Kompetenzentwicklung und partizipative Arbeitsgestaltung*. Frankfurt a. M.

Bullinger, H.-J., Hirsch, B. E. (eds.) (1994): *Leitstände für die Werkstattsteuerung*. Schriftenreihe der Bundesanstalt für Arbeitsschutz, Fb 705. Wirtschaftsverlag NW, Bremerhaven.

Curtis, B., Kellner, M. I., Over, J. (1992): Process Modelling. *Communications of the ACM* 35 (1992) 9, 75-90.

Dunckel, H. (1989): Contrastive task analysis, in: Landau, K., Rohmert, W. (eds.): *Recent developments in job analysis*. Taylor & Francis, London, 125-136

Dunckel, H., Volpert, W., Zölch, M. et al. (1993): *Kontrastive Aufgabenanalyse im Büro - Der KABA Leitfaden*, Grundlagen, Manual, Arbeitsblätter. vdf/Teubner, Zürich/Stuttgart.

Frese, E. (1994): Aktuelle Organisationskonzepte und Informationstechnologie. *Management & Computer* 2 (1994) 2, 129-134.

Hammer, M., Champy, J. (1994): Reengineering the corporation. paperback edition, HarperBusiness, New York.

Huckvale, T., Ould, M. (1993): Process Modelling: Why, What and How, in: Spurr, K., Layzell, P., Jennison, L., Richards, N. (eds.): *Software Assistance for Business Re-engineering*. John Wiley, Chichester et al., 81-97.

Keller, G. (1994): SAP Business Workflow Management - Durchgängige Geschäftsprozessoptimierung mit Hilfe des R/3-Systems, in: Scheer, A.-W. (eds.): *15. Saarbrücker Arbeitstagung 1994 - Rechnungswesen und EDV.* Physica, Heidelberg, 471-495.

Klein, M. M. (1994): Reengineering Methodologies and tools. *Information Systems Management* 11 (1994) 1, 30-35.

Kosiol, E. (1966): *Die Unternenhmung als wirtschaftliches Aktionszentrum.* Rowolth, Reinbek bei Hamburg.

Kötter, W., Volpert, W. (1993): Arbeitsgestaltung als Arbeitsaufgabe - ein arbeitspsychologischer Beitrag zu einer Theorie der Gestaltung von Arbeit und Technik. *Zeitschrift für Arbeitswisenschaft* 47 (1993) 3, 129 -140.

Leitner, K., Volpert, W., Greiner, B., Weber, W., Hennes, K. (1987): *Analyse psychischer Belastung in der Arbeit: Das RHIA-Verfahren,* Handbuch und Manual. Verlag TüV Rheinland, Köln.

Leitner, K., Lüders, E., Greiner, B. Ducki, A., Niedermeier, R., Volpert, W. (1993): *Analyse psychischer Anforderungen und Belastungen in der Büroarbeit: Das RHIA/VERA-Büro-Verfahren,* Handbuch und Manual. Hogrefe, Göttingen.

Mandl, C. (1993): Computerintegrationsprojekte - Die Logik des Misslinges. *IO Management* 63 (1993) 12, 54-58.

Marks, S. (1991): *Gemeinsame Gestaltung von Technik und Organisation in soziotechnischen kybernetischen Systemen.* VDI Verlag, Düsseldorf.

Mertins, K., Albrecht, R., Steinberger, V. (1992): *Werkstattsteuerung - Werkstattmanagement.* Hanser, München.

Morris, D. C., Brandon, J. (1994): *Re-engineering your business.* paperback edition, McGraw-Hill, New York et. al.

Oesterreich, R., Volpert, W. (1986): Task analysis for work design on the basis of action regulation theory. *Economic and Industrial Democracy* 7 (1986) 4, 503-527.

Osterloh, M., Frost, J. (1994): Business Reengineering - Modeerscheinung oder Business Revolution? *ZfO* 63 (1994) 6, 356-363.

Picot, A., Maier, M. (1993): Interpendenzen zwischen betriebswirtschaftlichen Organisationsmodellen und Informationsmodellen. *Information Management* 8 (1993) 3, 6-15.

Scheer, A.-W. (1994): *Business process engineering: Reference models for industrial enterprises.* Springer, Berlin et al.

Scherer, E., Rössle, R., M. Zölch (1994): Ganzheitliche Gestaltung der Werkstattsteuerung. *IO Management* 63 (1994) 3, 59-63.

Ulich, E. (1994): *Arbeitspsychologie.* 3. Aufl., vdf/Poeschel, Zürich/Stuttgart 1994.

Volpert, W. (1987): *Psychische Regulation von Arbeitstätigkeiten.* in: Kleinbeck, U., Rutenfranz, R. (eds.): Arbeitspsychologie. Horgrefe, Göttingen, 1-42.

Volpert, W., Oesterreich, R. (eds.) (1991): *VERA Version 2: Arbeitsanalyseverfahren zur Ermittlung von Planungs- und Denkanforderungen.* Handbuch und Manual. Technische Universität Berlin 1991.

Warnecke, H.-J. (1993): *Revolution der Unternehmenskultur - Das Fraktale Unternehmen.* 2. Aufl., Springer, Berlin et al. 1993.

Weik, S., Grote, G., Zölch, M. (1994): KOMPASS - Complementary analysis and design of production tasks in sociotechnical systems, in: Kidd, P.T., Karwowski, W. (eds.): *Advances in agile manufacturing.* IOS Press, Amsterdam.

Zölch, M. (1992): Partizipation contra kriteriengeleitete Augabenbewertung - eine Scheinalternative, in: Langenheder, W., Müller, G., Schinzel, B. (eds.): *Informatik cui bono?* Springer, Berlin et al., 243-247.

26

The Use of the GRAI Method in Re-engineering

R MacIntosh[1] *and A S Carrie*[2]

1. Department of Management Studies, Glasgow University Business School, 53-59 Southpark Avenue, Glasgow, G12 8LF, UK

2. DMEM, University of Strathclyde, 75 Montrose Street, Glasgow, G1 1XJ, UK

Abstract

One view of Business Process Re-engineering is that it seeks to do for clerical activities what cellular production systems do for the organisation of manufacturing processes, bringing together all the elements of a process into a single workgroup to provide substantial performance improvements. This paper presents an overview of the GRAI method and its use in a structured approach to re-engineering called the Strathclyde Integration Method. An example of this approach is presented, based on a real re-structuring exercise which was successfully carried out in a manufacturing company. The Strathclyde Integration Method, which involves a combination of Data Flow Diagrams and GRAI Grids, has been used in over ten companies. Based on this experience, the advantages and limitations of GRAI Grids as a re-engineering tool are discussed.

Keywords

GRAI Method, Re-engineering, Integration

1. INTRODUCTION

The advocates of Business Process Re-engineering suggest that, as organisations grow or change they become more bureaucratic, their systems become more formalised and many of their business processes are subdivided and distributed among several people in different functional departments and locations within the company. By bringing together these elements of a process into a single work group improvements can be achieved. Industrial engineers may take the view that BPR seeks to do for clerical activities what cellular production systems do for the organisation of manufacturing processes. By reorganising business processes, delays and errors in communicating between different groups may be eliminated. The personnel concerned gain a more complete view of the processes and are able to interact with customers in a more knowledgeable way. This creates an environment in which personnel can be made more aware of the organisation's objectives.

Hammer and Champy (1993) state that re-engineering is "the fundamental rethinking and radical redesign of business processes to achieve dramatic improvements in critical, contemporary measures of performance, such as cost, quality, service and speed." The underlying philosophy is to undertake a fundamental review of the objectives of an organisation and the way in which those objectives are currently fulfilled. These objectives are then translated into key processes or functions which must be performed and radical re-structuring takes place as the business is re-organised around these key processes. Some have claimed that a universal set of processes can be identified which are common to all manufacturing businesses, but consensus has not been reached on this issue. This reflects a lack of any universally accepted methodology. Some advocate an IT-based approach and seek to develop new software engineering tools. Others adopt a more human work design approach and suggest a methodology similar to TQM. Some argue that BPR has a fundamentally different perspective than TQM in that whereas TQM encourages a regime of incremental continuous improvement, BPR seeks to achieve step-change breakthroughs in performance.

2. THE GRAI METHOD

Two fundamental constructs within the GRAI method (Doumeingts, 1989) are the GRAI Grid and the GRAI Net. GRAI Grids provide a diagrammatic overview of the decision making procedures which drive the business. They show the functions of the business and the time scales on which decisions are taken. The major information flows used in decision making processes and the inter-dependency of various decisions can be clearly illustrated. Every decision in a system is mapped onto a matrix according to the function which it performs and the time period for which the decision is relevant. Each location on the matrix is called a decision centre. Each of the decision centres on the grid is then described in greater detail using a GRAI Net. These nets give a more detailed view of how the decision is executed. In the methodology described below GRAI Nets were not used. However GRAI Grids, which are a key part of the methodology, were used.

Functions in the GRAI Grid are not departmental functions in the organisational sense. The GRAI Grid focuses on decision making processes as being the key element of an organisation and defines functions in terms of the decisions concerned. If, for example, decisions and processes from one function on the grid are actually performed across several different departments in the current organisational structure, there would appear to be scope for a re-engineering exercise to be conducted.

3.0 A STRUCTURED APPROACH TO RE-ENGINEERING USING THE GRAI METHOD

The Strathclyde Integration Method (Carrie and MacIntosh, 1993 and 1994) uses a combination of Data Flow Diagrams and GRAI Grids to study business processes. The method has been described in a workbook (Carrie et al, 1993). DFDs are used to provide a detailed model of the processes concerned, showing the relationships between individual processes and identifying the flows of material and information from one process to another. From these detailed DFDs, an overview DFD is synthesised which groups the detailed processes into a limited number of "core" processes which form the basis for a GRAI Grid.

The first step in the method, described in Figure 1, is a project planning activity. It involves defining the boundary for the study and identifying those personnel who will have to be interviewed. From this list of interviewees it is possible to estimate the overall duration of the study and the key review dates.

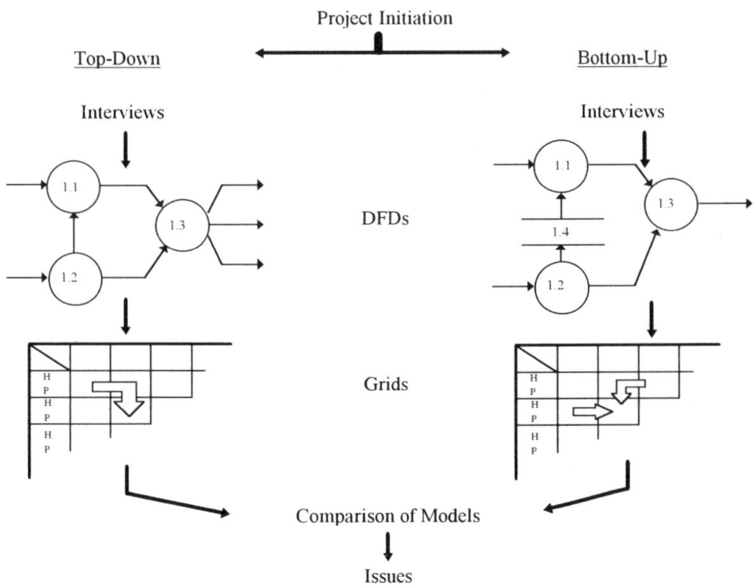

Figure 1 : The steps in the Strathclyde Integration Method.

Having identified the interviewees, two parallel interview sessions begin. This allows a comparison of the relationship between material and information flow from two distinct perspectives to be made. The first set of models, from a top-down perspective, reflect the management's understanding of the organisation regarding material flow, information flow and the decision-making procedures currently in use. The second set of models, from a bottom-up perspective, represent the views of those directly involved in daily transfers of material and information and those responsible for taking operational decisions. Typically these are supervisors, operators and clerical staff. This captures both a formal systems viewpoint of the processes and an informal user's viewpoint. These models are constructed independently to ensure that a valid comparison can be made later. Interviews are held with staff from the various departments or functions concerned in each of the business processes. Each interview lasts approximately 30 - 40 minutes covering the activities and the information flows through the area as well as any decision making procedures performed. Based on these interviews the DFD models are constructed and then validated by the interviewees.

The method focuses on the decisional processes within the DFD models, since it is these processes that have the greatest impact on the overall performance of the organisation. GRAI Grids are used in order to give an overview of these decision making procedures. All the processes within the DFD models are classified as being either decisional or mechanistic. To identify the decisional processes a list is created of all the processes within the DFD model. Processes which are exploded into greater detail are disregarded, since decisional processes always occur at the lowest level within the model. Those processes which are not exploded any further are classified. The mechanistic processes do not have any decisional content. The decisional processes involve some form of decision and for these it is important to note the time scales involved. Table 1 shows a brief example of the approach used to classify processes. Three possible cases may be encountered, an example is given of each. Process 1.0 is a decisional process but is exploded to a lower level in the DFD model. All processes must only be considered at the lowest level at which they appear in the model. Therefore whether the process is decisional is irrelevant, the time scales of any decisions involved are not applicable until the lowest level of detail is arrived at. Process 1.1 has been exploded to its lowest level of detail but is mechanistic not decisional, time scales are only recorded for the decisional processes. Process 1.2 is a decisional process which is not exploded to another level and therefore the time scales of the decision, in terms of periods and horizons, are noted.

Process Number	Description	Exploded Yes / No	Decisional Yes / No	Horizon Period
1.0	Schedule Production	Yes	Not Applicable	Not Applicable
1.1	Check Technical Details	No	No	Not Applicable
1.2	Set Priorities	No	Yes	H = 1 Week P = 1 Day

Table 1 - Example of Process Classification

To create the grid template the columns and rows must be defined. The columns are defined as the functions of the area being studied, e.g. to purchase material, to manage resources etc. These functions are the processes in the overview DFD and the labels from the processes in the overview DFD become the labels for the GRAI Grid columns. The rows of the grid, which represent the time scales of the decisions being made, can be taken from the list of decisional processes, as illustrated by the example given in Table 1.

With the columns and rows defined, the grid template is complete and individual decisions can be mapped onto the grid. For each decisional process, the function which it relates to can be identified and therefore the column it should be placed under. Also the row on the grid is defined by the periods and horizons found in completing the list of all decisional processes. Once all the decisional processes have been mapped onto the grid, the linkages between them may be added in the form of information flows and decision frames. Having prepared GRAI Grids for each set of DFD models, i.e. top-down and bottom-up it is possible to compare the top-down and bottom-up perspectives. Each grid acts as a summary for the validated DFD model from which it is generated. The DFD models and the grids are explicitly linked by a referencing system so that the analyst can easily move from the summary level grid to the detailed and verified DFD model.

4. CASE STUDY

This case study concerns the purchasing function in a company designing and manufacturing complex engineering products, characterised by both deep and wide Bills of Material. The company had suffered a significant reduction in size, caused by a combination of the recession, major cutbacks by their primary domestic customer and the fact that the latest machines were much more powerful than earlier models reducing the number of units required in a market of a given size. The company had existed on the same site throughout its long history but now operated with a substantially reduced workforce. A very hierarchical organisation structure was in place, many of the key processes had been fragmented and spread over a number of departments. This operating environment had remained largely unchanged for some time and the management were ready to consider implementing radical changes. During the process of downsizing the company had reduced its manufacturing capabilities in certain non-core areas and was increasingly reliant on sub-contracting work and purchasing bought-in-finished items. The combination of lower staffing levels and increased activity in the purchasing function meant that this department was becoming a problem area. It was decided to use the Strathclyde Integration Method to re-engineer the company's operating practices concerning material supplies.

From initial discussions with the company it was clear that the study would involve three main areas, production scheduling, purchasing and expediting. The purchasing section consisted of buyers and administrators. Customers, suppliers and all other departments within the company and would be considered as external to the study.

Having identified the departments which would be considered in the study, the next task was to obtain a rudimentary understanding of their operation and interaction. Following a brief tour of the departments concerned, a conceptual model was drawn. This simple model did not adhere to any particular modelling conventions and merely provided a sound basis for the more detailed modelling which was to follow. With the conceptual model complete, the top-down and bottom-up studies were started as parallel but independent activities.

From these interviews a set of DFDs was constructed for both the top-down and bottom-up perspectives. These initial diagrams were then verified by the interviewees. The top-down model contained 21 diagrams and was 6 levels deep, the bottom-up model contained 16 diagrams and was also 6 levels deep. Both models contained a context diagram showing all of the external entities to the study. The context diagram is then exploded into an overview diagram which contained the main processes involved in the study and their interrelationships in terms of material and information flow. Each of these overview processes is then exploded to show exactly how each activity takes place.

In order to identify the decisional processes within these models, a conversion list was used. This listed all of the processes within each of the DFD models, stating whether each individual process was decisional or not and identifying time scales for those processes which were decisional. A GRAI Grid template was constructed using the processes from the overview diagrams of the DFD models as the functions of the area being studied. These functions were to manage production requirements, to purchase material and to expedite. The time scales identified for individual decisional processes in the conversion lists were used to define the horizons and periods for the grid template.

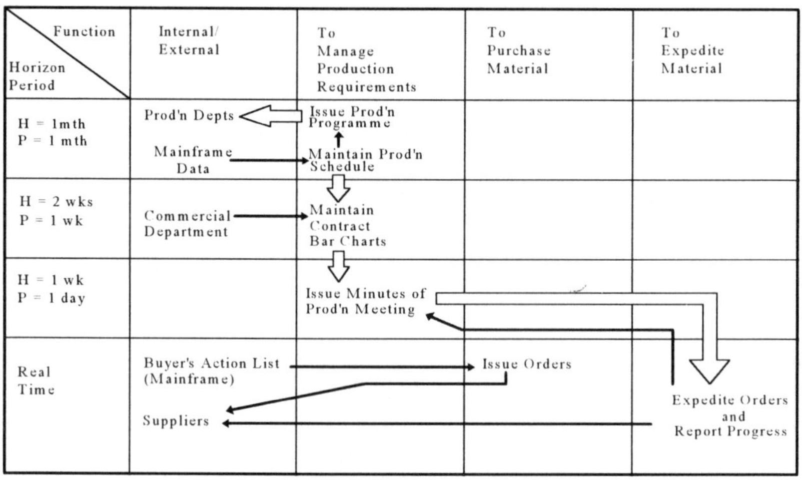

Figure 2 Simplified top-down GRAI Grid of original situation.

Decisions were then mapped onto the grid template according to the function to which they relate and the frequency with which they occurred. Once all the decisional processes had been mapped onto the grid it was possible to link them up using information flows and decision frames. Figures 2 and 3 show simplified GRAI Grids of the original purchasing processes from top-down and bottom-up perspectives, respectively. Both grids share the same template, with the same three major functions identified.

When an order was placed for a machine it was registered on the company's mainframe computer, with appropriate details of the Bill of Materials to be used etc. The computer system generated a list of all the items to be purchased for each machine called the Buyer's Action List. The top-down grid, Figure 2, shows mainframe data being used on a monthly basis to maintain the Production Schedule. Another form of the Production Schedule, called the Production Programme, was used on the shop floor by the production departments. A third scheduling device, Contract Bar Charts, were maintained by the Commercial Department and used as the basis for any discussions with customers. Finally, the minutes of the weekly production meeting were issued as the most up-to-date set of prioritised orders. The bottom-up grid, Figure 3, omits some of the decision making procedures contained in the top-down view. This is normal, since the bottom-up grid is based on interviews with the operational staff who do not concern themselves with many of the higher level, inter-functional activities described in the top-down grid. The bottom-up grid does, however, show that the expediters receive direct enquiries from production departments looking for specific items which are urgently needed.

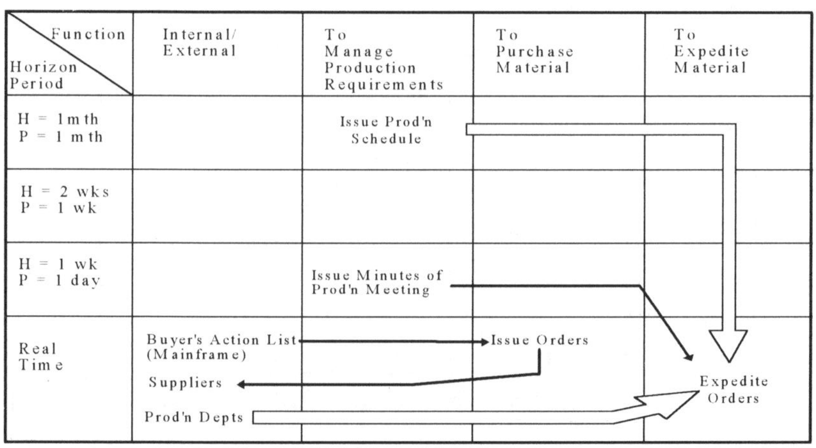

Figure 3 Simplified bottom-up GRAI grid of original situation.

Following the application of SIM, a number of issues were identified which were to be addressed in the eventual re-engineering of the purchasing function. Firstly there were several scheduling devices in use. Quite apart from the duplication of effort that this involved, discrepancies inevitably occurred between these documents with the result that departments were working with different priorities. Secondly, the buyers who actually placed the orders had no prioritising mechanism.

Thirdly, the grids showed a lack of communication between the buyers and the expediter, despite the obvious relationship between the two tasks. Finally, the bottom-up grid showed that the expediters were trying to satisfy conflicting demands from the production departments and the priorities identified in the production schedule (shown by the convergence of two decision frames at one decision centre).

Three major changes were implemented during the re-engineering exercise. Firstly a single production schedule was introduced and all departments within the company worked to this schedule. Secondly, the Buyer's Action List was prioritised using information which was already stored on the mainframe regarding Bill of Material structures and build sequences. Finally the expediting and the purchasing departments were merged. The new purchasing department involved buyers and expediters working in teams to provide a complete purchasing service. These changes are summarised in Figure 4 which shows a simplified GRAI Grid for the re-engineered organisation.

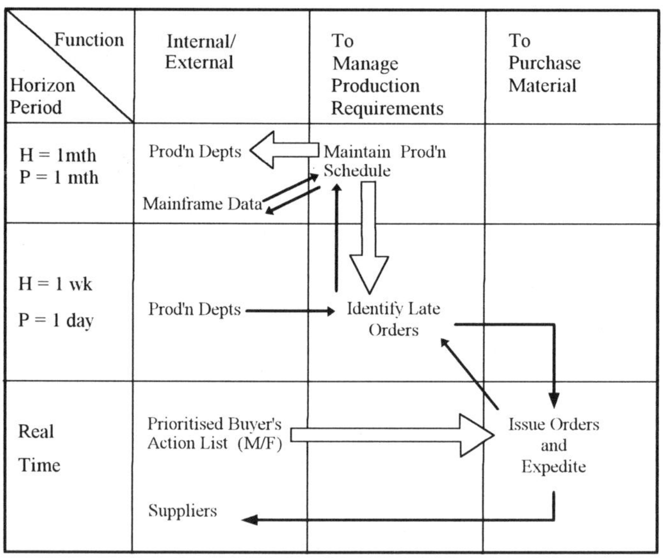

Figure 4 Simplified GRAI grid of situation after re-engineering exercise.

5. CONCLUSIONS

An approach to Business Process Re-engineering has been presented based upon the combined use of GRAI grids and Data Flow Diagrams. The Strathclyde Integration Methodology has been applied successfully in over ten organisations. It combines the strengths of structured techniques with the overview, insight and flexibility provided by GRAI Grids. The paper demonstrates that SIM can be used as the starting point for a radical re-think of an organisation.

As a result of this experience we offer some comments on the advantages and disadvantages of using GRAI Grids in this BPR methodology.

5.1 Advantages of GRAI grids

1 Overview: The GRAI grid however gives a one-page summary of the architecture of a decision-making system, providing an overview in a way no other technique matches.

2 Decision Making Level: By classifying decisions according to period and horizon, the grid sorts decisions according to their place in the decision-making hierarchy. This makes it very easy to distinguish strategic decisions, tactical decisions and operational decisions.

3 Mis-matches With the Organisational Structure: Once constructed, GRAI Grids show the relationship between the organisational structure and the functions which must be performed. If a single organisational department is involved across several functional boundaries then a re-engineering exercise may prove extremely valuable.

5.2 Limitations of GRAI Grids

1 Validation of the Existing System: It is necessary to establish and clearly document the existing system. DFDs may be presented to company staff and validated by them as representing a correct understanding of the current system. GRAI Grids on there own do not give sufficient detail for this purpose. Although GRAI Nets could perhaps have been used, they were considered unnecessarily complex for this application.

2 Definitions of Functions: The GRAI method suggests that functions can be classified according to whether they are concerned with managing three key entities, namely products/materials, resources and time, and combinations of them. We found this difficult to apply in practice. Despite this criticism, the underlying concept proved very useful. Flexibility of column definition has proved to be a positive advantage in SIM.

3 The Absence of a Rigorous Methodology: GRAI Grids lack a rigorous support methodology by means of which every analyst would develop the same grid for a given system, or which would allow other analysts to interpret a grid developed by another analyst in the same way.

4 Modelling of Event Driven Situations: GRAI Grids model repetitive, period driven decision making extremely well. However, difficulties can be encountered when using them to model event driven organisations, such as specialist one-of-a-kind manufacturers who tend to use project management type approaches. Many decisions only occur in reaction to a particular situation or problem and will not be repeated using a regular period and horizon.

5 Dynamic Modelling of Grids: Exploratory research has been conducted into the dynamic modelling of information systems as represented in GRAI grids. This revealed that several additional related constructs, concerned with time phasing, would be necessary to enable the grid to be unable as a dynamic modelling tool:

- Time required to make a decision
- Delays in communicating between decision centres
- Synchronisation of information flows
- Time phasing of decisions
- Stability.

6. REFERENCES

Hammer, M. and Champy, J. (1993) Reengineering the Corporation, Nicholas Breasley Publishing, London.

Doumeingts, G. (1989) GRAI Approach to Designing and Controlling Advanced Manufacturing Systems in a CIM Environment, in *Advanced Information Technology for Industrial Material Flow Systems*, (eds S Nof and C L Moodie), NATO ASI Series, Volume 53, Springer-Verlag, Berlin, 461-529.

Carrie, A. S., Goldie, A. H., MacIntosh, R., Peoples, G. A. and Warrington, S. W. (1993) The Strathclyde Integration Methodology Workbook, Internal Report, Department of Design, Manufacture and Engineering Management, University of Strathclyde, Glasgow.

Carrie, A. S. and MacIntosh, R. (1993) Improving Integration in Manufacturing Systems, in *IFIP Transactions on Advances in Production Management Systems*, (eds I A Pappas and I P Tatsiopoulos), North Holland, Amsterdam.

Carrie, A. S. and MacIntosh, R. (1994) A Structured Approach to Process Redesign, Proceedings of the BPICS Annual Technical Conference, Birmingham, 153-167.

7. BIOGRAPHY

Robert MacIntosh

Robert MacIntosh is a lecturer in Operations Management with Glasgow University Business School. Previously he was a lecturer in the Department of Design, Manufacture and Engineering Management at the University of Strathclyde where he also worked as a Research Assistant on a three year project investigating Manufacturing Integration with funding from the ACME Directorate of SERC and industrial collaborators. He was involved in developing the Strathclyde Integration Method and has recently completed his PhD on this subject.

Allan Carrie

Allan Carrie is the Hoover Professor of Advanced Manufacturing Technology at the University of Strathclyde. From 1990 to 1992 he served as Head of the Department of Design, Manufacture and Engineering Management. He is Chartered Engineer, Fellow of the IEE and Member of BPICS. He has industrial experience as an Industrial Engineer with Babcock and Wilcox and with Northern Electric in Montreal, Canada. He is the author of a book on simulation and author of over 80 papers. For the past year he has been a member of the panel on Manufacturing, Production and Business Processes of the Office of Science and Technology's Technology Foresight project.

27

GRAI Approach : A Methodology for Re-Engineering the Manufacturing Enterprise

G. DOUMEINGTS (GRAI / LAP - University Bordeaux 1 - F)
F. MARCOTTE (INTEGRAI - F)
H. ROJAS (GIAT / GITECH - F)

Abstract

For many companies, Business Process Re-engineering (BPR) is becoming a mean to attaining a competitive advantage. But BPR is too often related to specific processes while it should ensure a global performance by covering multi-processes. For that purpose, methodologies are required to perform an integrated Business Multi-Processes Reengineering.
GRAI approach has been developped for more than 10 years and gradually improved through a lot of industrial experiments, each of them followed by a research step. This approach allows to perform such an integrated Business Multi-Processes Reengineering.
In this paper we describe first the new requirements for BPR and then the GRAI approach. Finally, we present a recent application of the GRAI approach for the re-engineering of a workshop in the most important french company in the domain of defense industry.

Key words

Business Process Re-engineering, Methodology, GRAI approach.

- 1 - Introduction

This paper presents a methodology to perform re-engineering in manufacturing enterprises. We will describe a recent application of the GRAI approach for the re-engineering of a workshop in the most important french company in the domain of defense industry.
The methodology used to perform this study allows to evaluate the weak points and the strength points according to the manufacturing requirements and the enterprise strategy. From this evaluation, a set of actions is defined to ensure the implementation of the designed solutions.

After a general introduction into the general area of Business Process Re-engineering (BPR), this paper presents the GRAI approach which is used to perform effective BPR. The third part is a presentation of a recent application in a french defense company.

In the conclusion, future development for the company and future improvements for the methodology will be discussed.

- 2 - Business Process Re-engineering (BPR)

Companies are under increased presure today to improve performance while being customer oriented. The hight competition in a global international market and the decrease of product life cycles have led manufacturing enterprises to increase their capability, to react rapidly and with coherence, and to focuss on their core-business.

Business Process Re-engineering (BPR) is an approach for achieving improvements in a company's performance in a short period of time. BPR has helped these companies to understand how the different functions or processes are acting. Approaches exist for improving these business Processes, but many of these techniques are focussing on the improvement of individual processes (figure 1).

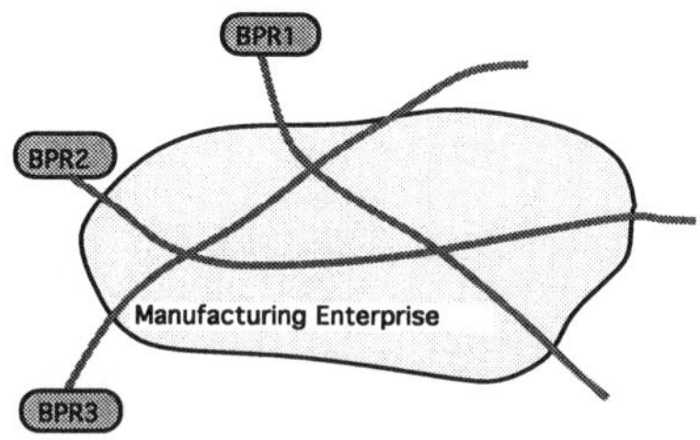

Figure 1 : Business Process Re-engineering

Now, global performance is not a set of local business performances. It is absolutly necessary to improve each business process according to the overal manufacturing enterprise performances, and not only according to the optimisation of each process individually. An efficient BPR technique must ensure Business Process Integration through a multi-process improvement (figure 2).

Company improvements are possible with manufacturing enterprise re-engineering tools covering the whole re-engineering process (from evaluation to implementation) and taking into account global perfomance criteria. The GRAI approach, presented below, is an available methodology to perform such re-engineering.

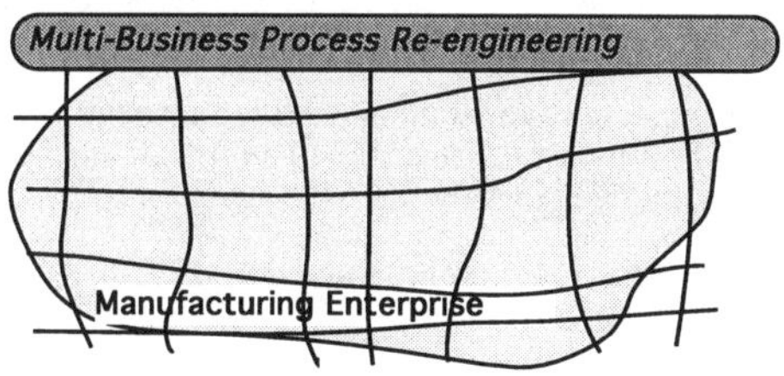

Figure 2 : Integrated Multi-Business Process Re-engineering

- 3 - The GRAI approach

GRAI approach has been developped for more than 10 years and gradually improved through a lot of industrial experiments, each of them followed by a research step. Several ESPRIT projects allowed these improvements (OCS Open CAM System - ESPRIT Project n° 418), IMPACS (Integrated Manufacturing and Planning And Control System - ESPRIT project n° 2338), FOF (Factory Of the Future - ESPRIT project n° 3143), FLEXQUAR (Adaptative System for Flexible high Quality and Reliable Production - ESPRIT project n° 6408). Recently, through the GLOBEMAN 21 project (Enterprise Integration for GLOBal MANufacturing towards the 21 century), an IMS (Intelligent Manufacturing System) project, an interesting study has demonstrated how GRAI approach could be used for the evaluation of manufacturing performances and for global Benchmarking.

The GRAI approach is based on the following elements

- 3 - 1 - The GRAI conceptual model of the manufacturing system

The GRAI conceptual model aims at giving a generic description of what a manufacturing system is, on a control point of view. The control of a manufacturing system will be seen at first with a global point of view, and at the level of the decision center afterwards. The interest of GRAI conceptual model is to be generic enough in order to be usable for any kind of manufacturing system (discrete, process, and services). The GRAI conceptual model is mainly based on the system theory, control theory and hierarchical control theory.
The GRAI conceptual model identifies three systems in the manufacturing system : the physical system, the decision system and the information system. (Figure 3)
The physical system has to transform raw materials, components, into the output Products through the resources.
The Decision System (DS), split up into decision making levels, according to several criteria, each level composed of one or several Decision Centers (DC). In the decisional system we determined two parts : the upper part which is periodic driven, and the lower part which is event driven. We call this second part the operating system : it is the interface with the physical system : it contents the numerical control system, the programmable controllers... : this part of the DS is more event driven : we call it the operating sub-system.

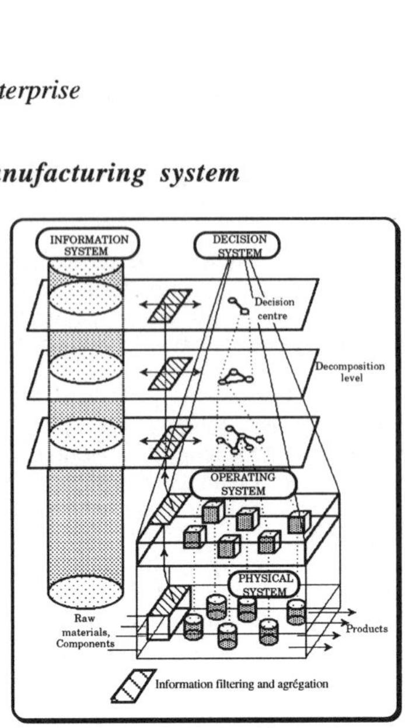

Figure 3: The GRAI conceptual model of the manufacturing system

The information system contains all information needed by the DS : it must be structured in an hierarchical way according to the DCs structure (Decision Centers).

- 3 - 2 - The GRAI conceptual model of a Decision Center

A decisional center at any level of the structure can be decomposed using the same criteria of decomposition: a physical, a decisional and an information (figure 4).

The controlled system is the view of the physical system at the level of decision.

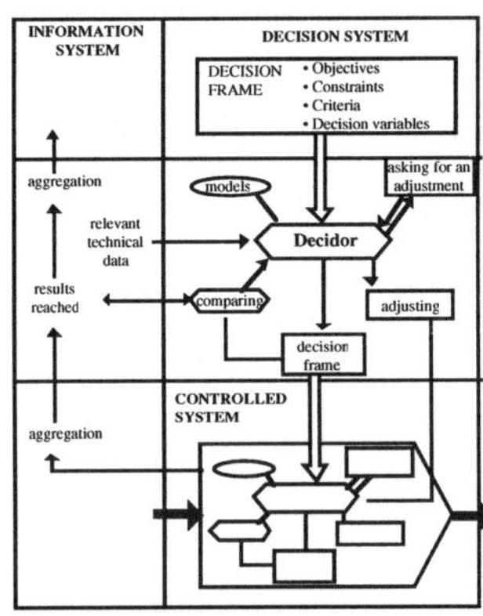

Figure 4: The GRAI conceptual model of a Decision Center

- 3 - 3 - The GRAI concept in Production Management

The GRAI approach is focussing on the decisional aspects of the production system. It allows to model and to analyse the decisional structure of the Production Management System (PMS). In order to identify decision centres in real manufacturing system, according to the concepts defined in the GRAI conceptual model, it is necessary to use criteria, to specify the nature of the decisions and their structuration. In this study, we consider only the periodic driven part of the DS. To structure it, two axis of decomposition are defined : the vertical one is the coordination axis and the horizontal one is the synchronisation axis.

The coordination is required when considering the decomposition of the physical system. The global objective of the factory must be decomposed, step by step, in detailed objectives for the

production cells. The temporal aspect of a production cell behaviour is different than the temporal aspect of the overall production system, and the control decisions are depending on this temporal aspect.
The criteria for the coordination is the temporal aspects of the decision : "Horizon" (The interval of time over which the decision extends (i.e. remains valid)) and "Period" (The interval of time after which we reconsider the set of decisions. In such a structure, the horizon is a sliding horizon).

The criteria for the synchronisation is a functional one. There are 3 basic types of functional activities in PMS : the product management activities (supplying decisions, stock management decisions, ...), the resource management activities (capacity management, resources allocation,...) and the planning activities. The manufacturing activity planning function defines synchronisation requirements which are objectives for the two other functions : to perform a manufacturing activity in an efficient way, we need to synchronised the availability of the products with those of the resources.
Other functions may be introduced when they are influencing the Production Management activities like : the maintenance function, the quality function, the engineering function.... Then it is possible to analyse their inter-relations which caracterize the integration of the manufacturing system.

- 3 - 4 - The GRAI approach Modelling formalisms

Based on the concepts and the criteria previously presented, the GRAI approach provides formalisms to identify the concepts in real manufacturing systems.
Three basic formalisms are used to model the manufacturing system: the GRAI grid and net, the Idefø formalism and the Entity/Relationship model.

The GRAI grid is used to model the decisionnal structure of the PMS, according to the two criteria defined previously.
The GRAI net allows to detail the decision centre.

The Idefø formalism is used to perform the functional view and to model the physical system, controlled by the PMS.

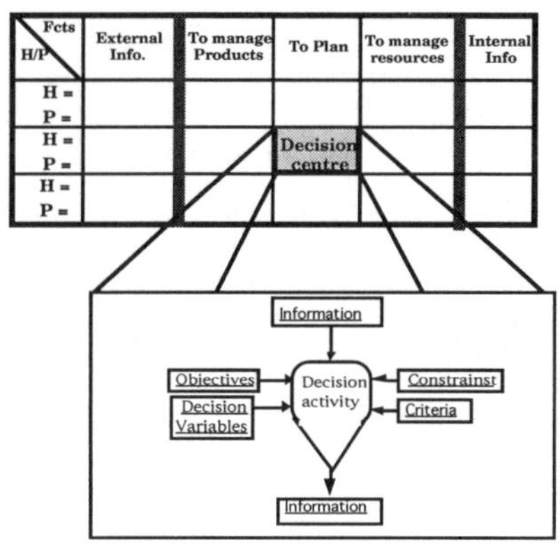

Figure 5 : The GRAI Grid and the GRAI net

Also, the E/R formalism can be used when the information system modelling is required, especially when software developments are decided.

These two last formalisms are presented in the figure 6.

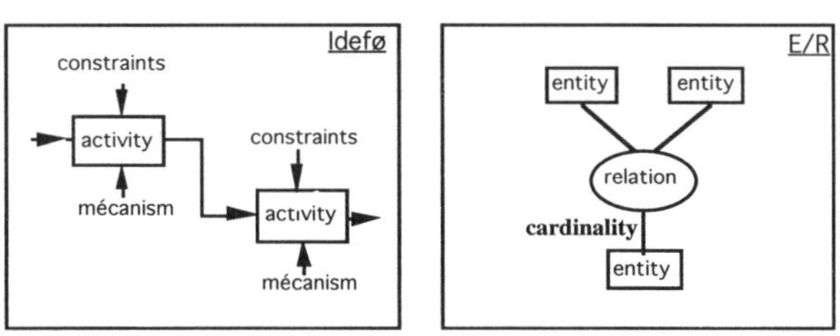

Figure 6 : the Idefø and the E/R formalisms

- 3 - 5 - The GRAI Structured approach

The various parts of a manufacturing enterprise must be designed in a coherent way and efficiently. For that efficiency, it is necessary to use a structured approach which provides a step by step procedure. Four main phases are identified in the GRAI approach : initialization, modelling, design and implementation. The picture 7 summarizes the overall re-engineering process.

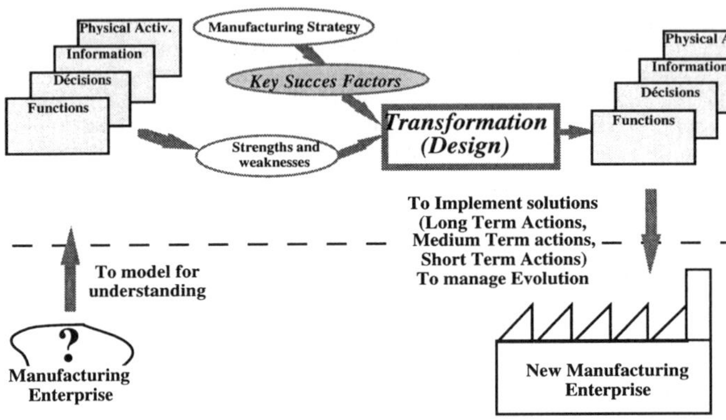

Figure 7 : Overall re-engineering process in the GRAI approach

The application of GRAI is also structured in terms of participants. It requires:
• a project board composed by the top management of the manufacturing enterprise. The goals of this group are to give the objectives of the study, and to orient the study according to the results of the main phases.
• a synthesis group composed of the main future users and main responsible people of the manufacturing system functions. This group has to follow the development of the study, and to check the results of the various stages. Their technical abilities and their suggestions will be used to guide the design of the new system.
• an analyst (several if necessary), whose job is, in particular, to collect all the data needed for the various phases.
• the users group which has to provide more detailled on specific parts of the manufacturing system.

After the modelling phase, specific groups may be defined, according to the specific solutions to be developped.

The main advantages of this approach are the following.

First, there is a set of formalisms, which allow the user to perform an integrated model of the manufacturing enterprise, according to the three views identified in the conceptual model.
The GRAI grid gives an overview of the decisions made in the various functions of the manufacturing enterprise. Through the links between these decisional activities, it is possible to study the effective integration of these various functions.
Further more, for all decision centres the activities to be performed are analysed and designed according to the structure and the organisation of the physical system (the controlled system). This is to ensure the coherence between the decisional system and the physical one.
Finally, the information system is structured according to the requirements of all the decision centres.

Second, the structured approach provides an efficient guideline to manage the project (step by step approach) and it also ensures the validity of the model which are performed (top-down analysis followed by a bottom up one).

These two main advantages give to the GRAI approach a real power to perform BPR in an integrated way.

- 4 - Re-engineering of a workshop : a case study

After the short overview of the GRAI approach, which is the basis of the study performed in a workshop of a French Defense Company, the following paragraph presents the main steps and results of the GRAI application.

- 4 - 1 - Presentation of the industrial contexte

This workshop belongs to an important french defense company. It works in the domain of mechanical industry and in the past was acting as a subcontractor for the various divisions of the company.

Due to the tremendous changes in this industrial sector, the re-structuration of the workshop and the re-orientation of the activities of this workshop was absolutely necessary in order to find new markets. Further more, important financial losses was questionning its future, and a hard social context was bringing pressure on the top management.

The study performed with the GRAI approach was organised in three steps :

- existing situation analysis : the modelling phase,
- specification of the new workshop : the design phase,
- implementation process and planning : the action planning.

The various steps of each phase are presented below.

- 4 - 2 - The modelling phase

The first phase was the modelling phase in which we have elaborated the model of the workshop using the knowledge of the actors. To perform this modelling phase two groups were working in parallel : the Synthesis Group and the Foreman Group.
The Synthesis Group was composed of the responsible persons of the main functions of the workshop : Purchasing, Engineering, Manufacturing, Supplying, Quallity and Maintenance functions (8 persons). This group was in charge of analysing and modelling the overall organisation of the workshop : main decision centres, relations between functions, hierarchical structure and coherence of the PMS.
Based on this global view, the Foreman Group (equivalent to the users group) was in charge of analysing in detail specific complex decision centres at the short term level, or specific manufacturing cell organisation. This group was composed of the main Foreman or team leaders, working in the workshop (7 persons).

The global analysis with the Synthesis Group allows to identify the points to be analysed in detail with the Foreman Group. Such an approach allows to focuss on the real problems, and it avoids to be swamped by a mass of irrelevant detail, which may cost time and money.

To support this phase, 2 persons from INTEGRAI (the research transfert company of GRAI) was guiding the sudy and managing the meetings, and 1 person from the company had been allocated to this study to learn from the GRAI approach and to perform interviews and complementary data collection.

The objectives of the modelling phase are to determine the improvement tracks to be studied during the design phase. For that purpose, the study was organised in three steps :

- training to modelling techniques,
- application to the workshop,
- analysis of the models to evaluate the workshop situation.

The modelling phase provides a set of models :

- functional model of the workshop (Idefø formalisms),
- physical model of the workshop (Idefø formalisms),
- decisional model of the management structure (GRAI grid and nets formalisms),
- information system (Entity/Relation formalisms).

From these models, strengths and weaknesses are identified. The evaluation is performed on the following aspects :

- production management structure (decisional level and functional decomposition),
- coordination principles (decision frames, information flows),
- production flow analysis (physical system organisation).

This evaluation has also been performed according to the business planning of the company. However, this business planning was not precise enough, and the information was not enough disseminated.

From this evaluation, seven tracks have been identified, as being important to improve the performance of the workshop :

- strategic orientations and commercial actions,
- structuration of the physical system,
- organisation of the "Scheduling" department,
- implementation of a hirarchical planning system,
- product quality control activity definition,
- tools to support customer order engineering,
- training on benchmarking procedures.

- 4 - 3 - The design phase

Before to start the design phase, the business planning of this workshop has been collected. This business planning was deduced from a market analysis and also from the evaluation of the strong points of the physical system (products and processes).

- Design phase : organisation

The design phase objectives was to re-inforce the training on the weak points, to look for practical solutions and to define the planning to implement the selected solutions.
The organisation of this phase has been made according to the 7 improvements tracks. For each track, a working group has been created, with participants from the both groups : the Synthesis Group, and the Foreman Group.

For each group, the following approach has been used :

- detailled analysis of the identified problems,
- analysis of existing studies and solution proposal,
- feasibility study,
- selected solution definition,
- proposal for an implementation planning.

Also, to ensure the overall coherence of the results, various meeting of the synthesis group and of the foreman group were organised.

- Design phase : results

The results of the design phase is a global model describing the organisation of the workshop, based on the conclusions of each working group. This model specifies for the three sub-systems (Physical, Informational and Decisional) the physical system organisation, the relations between the various functions of the workshop, and the responsabilities to be allocated.

Physical system :

Based on the evaluation of the current situation, and based on the business planning, a product oriented organisation has been selected. A detailled analysis has shown that the current available machine-tools were sufficient to ensure the constitution of the production cells (figure 8). A geographical implementation has been proposed, taking into account the technical constrainsts related to specific machines. One constrainst has led us to identify one common cell for the last activities of some manufacturing processes (especially quality checking activities). Training requirements and the new team organisation was also defined.

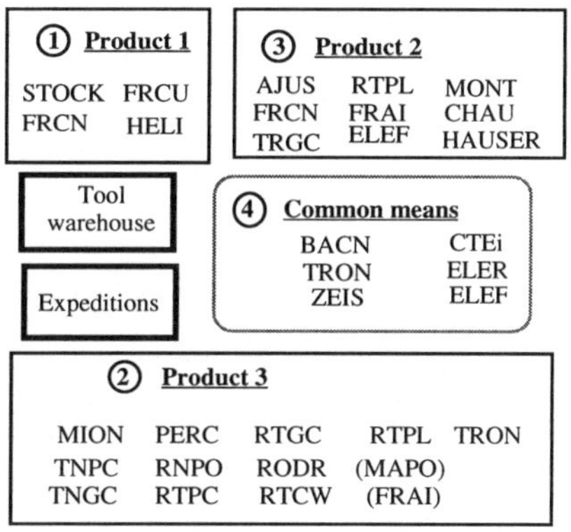

Figure 8 : Production cell constitution

Decisional system :

The interest of the GRAI approach is to allow an integrated modelling. So based on the structure of the physical system and based on the hirarchical planning principles, we have specified the decisional structure of the PMS (see figure 13). In this structure, the three basic planning levels are defined, and the relations between the various functions to.
The detailled description of each planning level is based on the physical system, as it is seen by this level.
From this decisional structure, the various responsabilities in each function were defined precisely (figure 9).

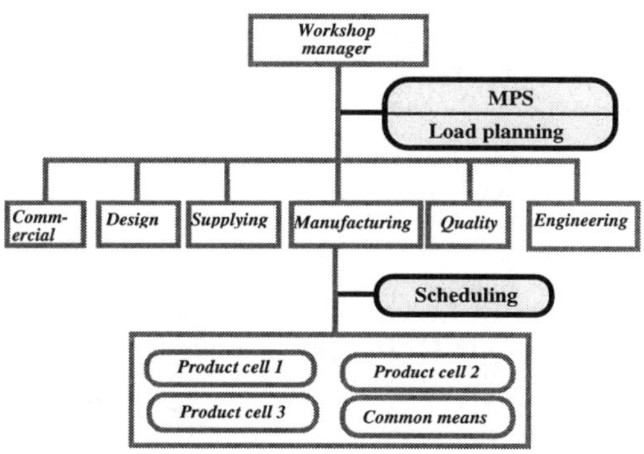

Figure 9 : Function responsabilities regarding planning levels

Information system :

In this specific case, there was no requirement for an information system anaysis. However, after the modelling phase, it appears that it was necessary to implement a scheduling system, first because of the new environment (customer order driven activity, important amount of work-orders, needs for simulation to evaluate lead-times), and also because the existing system was suppose to stop at the end of the next year.
So a specification book has been writen, three suppliers were selected, and one day has been organised with the end users to have demonstration of each selected tools.
The final choice is now on going.

- 4 - 4 - *Action planning*

The overall result of this design phase is a set of organisation principles and technical solutions. Such a result is not sufficient for a manufacturing manager, it is important to define how the solutions will be implemented.
For that purpose, we have specified a list of actions, with the related required investment, the responsible person, and the scheduling of the various steps.

These actions were defined in three groups :
- short term actions (1 month horizon)
- medium term actions (2 to 18 months horizon)
- long term actions (more than 18 month horizon).

- 5 - Conclusion

During this study, a BPR has been performed on the whole manufacturing enterprise. Most of the individual business processes have been studied : customer order processing, shopfloor control, quality checking, customer order engineering processes, long terme management process, product manufacturing processes,.... All these processes have been designed in a coherent way, based on the GRAI approach, which allows to relate all these processes.

Also, based on this integrated design, the manager of the workshop succeed in discussing about the CAPM to be implemented. At the beginning, the company top management was thinking about one software for all the workshops of the company, a MRP based software. Now, in this workshop, they are looking for a simple scheduling package (specification book already mentionned) which is more relevant for their type of production (specific tooling in three main areas).

Three months after the end of the application, the workshop has increased its turn over (≈10%) and many improvements are in progress (level of customer service rate, number of firm orders / Number of quotations).

This study which is examplary, will support next year the application of the TIME method which is now under development in the frame of the EUREKA programme (TIME : Tools and Methods for Integration and for Management of the Evolution of Industrial Enterprises EUREKA project n°824). This TIME methodology aims at providing the European industrial companies with the means of supporting, guiding their long term evolution process, through modelling techniques (GIM and Game View), self audit and benchmarking. The following picture (figure 11) summarizes this conceptual elements.

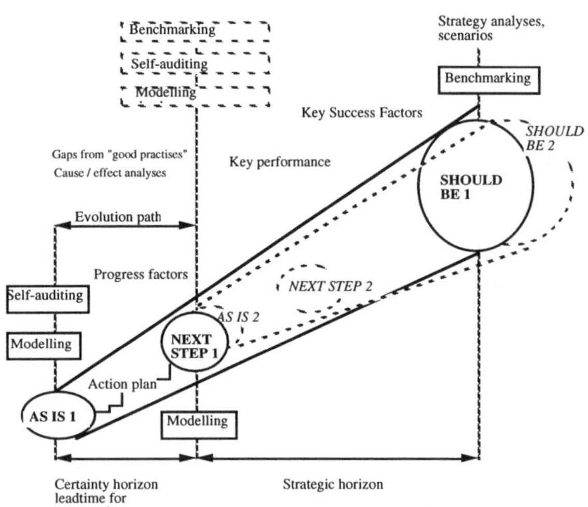

Figure 12 : The TIME methodology for evolution management

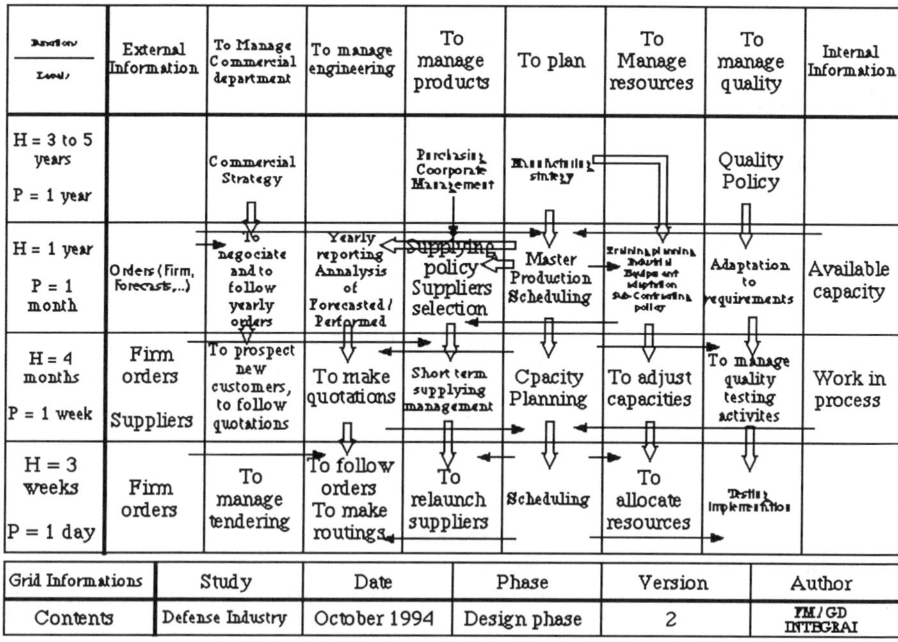

Figure 13 : the GRAI grid : the decisional structure of the PMS

28

Process monitoring for continuous improvement

V.J. Thomson
Werner Graupe Chair in Manufacturing Automation
Department of Mechanical Engineering
McGill University
817 Sherbrooke Street West, Montreal, Quebec, Canada, H3A 2K6
tel: 1-514-398-2597, fax: 1-514-398-7365, me60@musica.mcgill.ca

Abstract

The combined trends of increasing levels of knowledge work, but declining productivity of knowledge workers have hampered efforts to improve overall corporate productivity in the industrialized world. The active management of knowledge work processes promises to provide a significant improvement in organization wide productivity. Process monitoring, a tool which measures realtime process performance, permits proactive management of processes and enables their continuous improvement. The design of a process monitoring tool is described along with industrial projects which demonstrate its usage.

Keywords

Process monitoring, business process reengineering, productivity

1 INTRODUCTION

For more than a decade, the productivity of knowledge workers has been going down rather than up. U.S.A. labour statistics from 1980 to 1990 reveal that blue collar productivity rose 28%, while at the same time white collar productivity declined 3% (see Table 1). In addition, the relative contribution of knowledge work to product value is increasing continuously, while material content and the cost of production have decreased. According to Miller and Vollmann (1985), "The indirect work embodied in logistics, balancing, quality and change transactions now accounts for the lion's share of value added in most production based industries." In the electronics industry, overhead is as high as 80% (Miller, 1985). Increased overhead or knowledge worker levels are also a fact in service organizations in addition to product oriented industries. Overhead distribution in the electronics industry is given in Table 2; the majority of these costs are salaries and support for knowledge work.

Table 1 Worker productivity comparison (Thurow, 1992)
Real output grew 30% in the U.S.A. (1980-1990).

Growth	*Blue collar (%)*	*White collar (%)*
Employment	2	33
Productivity	28	-3

Table 2 Manufacturing overhead costs in the electronics industry (Miller, 1985)

Materials overhead	33%	(procurement, off floor handling, coordination)
General and administrative	20%	(personnel, accounting, management)
Facilities and equipment	20%	(capital, insurance, energy (4% of total cost))
Engineering	15%	(design, planning)
Indirect labour	12%	(maintenance, quality control)

The problem of white collar (knowledge) worker productivity is present in all economies of the world. Shintaro Hori (1993) in an article on Japan's white collar economy states that: "Companies must reorder their functional priorities, shifting their focus from production and product development to white collar work." The article points out that the disproportionate growth of overhead in the sales, general and administration category in Japanese companies from 29.2% (1984) to 33.5% (1992) is symptomatic of productivity problems. Hori proposes that industry must focus on process oriented functions to improve white collar productivity. These combined trends of increased levels of knowledge work, but declining productivity of knowledge workers have hampered efforts to improve overall corporate productivity in the industrialized world.

As well, the future will see more and more emphasis on process. With ready access to all forms of knowledge, what is starting to differentiate companies, and what will continue to differentiate them, is process knowledge. This is reinforced by Lester Thurow's (1993) statement in *Head to Head* that, "in the future sustainable competitive advantage will depend more on new process technologies and less on new product technologies." Process is a competitive advantage; it needs to be managed as a corporate asset.

Manufacturing has been highly successful in increasing productivity through commonly designed and executed processes. The same approach needs to be used for 'soft' or knowledge work processes, such as, design, engineering, procurement, distribution, etc. There is a need for a new culture based on process management which will enable the design, planning, and monitoring of all processes in an enterprise. This will lead to a significant improvement in organization wide productivity.

2 PROCESS MONITORING

Production systems in discrete and continuous manufacturing have operated for many years with highly defined processes and monitoring systems to deliver feedback for continuous improvement. Underlying principles and methods have been developed for these systems to provide the necessary techniques to allow appropriate process definition and process monitoring. The development of these principles and methods has allowed manufacturing industries to have continuous productivity improvement in their production systems. More importantly, continuous process improvement has also led to ever increasing levels of process complexity, which has meant better products and services, even while reducing overall cost structures.

Process monitoring needs to be developed for knowledge work processes. This will allow processes to be planned, measured and continuously improved. In addition, process monitoring promises to allow knowledge work to progress to higher levels of sophistication, and thus, produce better products and services, by assisting the active management of processes, as has been done on the shop floor. In fact, for 'soft' processes, process monitoring systems are even more imperative. For, in knowledge work much more initiative is allowed, and indeed, expected on the part of workers. This means that knowledge work processes are less rigidly defined, and are expected to be highly adaptive to circumstances. Thus, the monitoring of processes must provide significant capabilities in terms of comparison to planned activities, continuous measurement of performance, and prediction of trends in order to manage processes better.

The use of statistical process control techniques, which are aimed at measuring variations in the characteristics of outputs or products of processes, can be used in some cases for monitoring of knowledge work processes. However, specific, statistically measurable outputs are not available for many processes, or in a lot of cases, the output which is easily measurable does not reflect very well the goals of the process, and is, therefore, not very useful in judging the quality of the process. Also, good metrics have not been developed for measuring the quality of the process itself. For example, good engineering process is measured by the output of an engineering group. If this has relatively high volume of somewhat comparable products, such as electronic circuit designs, process performance is implied from 'good' output; but, if it has low volume of not very comparable products, measurement of the performance of process is nearly impossible. A new basis for measurement is needed, and must focus on the effectiveness of the process itself, not on its output. The concept of process monitoring proposed here will provide this capability.

Process monitoring starts with the basic notion of comparing a real process to a model of the desired process in order to provide performance comparisons and feedback. The key idea is the measurement of process parameters, not necessarily to determine any characteristics of the product being made, but to measure the process itself. These parameters need to be easily measured to reflect how the process is doing, and to allow feedback for process control. Here, the principle of 'You can't manage what you don't measure' applies. Thus, process monitoring captures realtime measurements of processes, performs a comparison of these measurements with models of the processes, and then, provides feedback to workers for performance analysis and process redesign. The design of a process monitoring system presently being built is shown in Figure 1.

The major components of the process monitoring system are: a process modeller which facilitates the design and planning of processes; realtime process measurement (not shown) which gathers selected data from actual processes; a process comparison engine which compares the performance of real world processes to process models in realtime; a feedback generator which delivers reports on process performance; a tool kit which contains algorithms and procedures for measuring performance, analyzing process maps, and enabling some changes to processes for low level automation of process monitoring; and a simulator which can 'replay' real processes for process review as part of feedback.

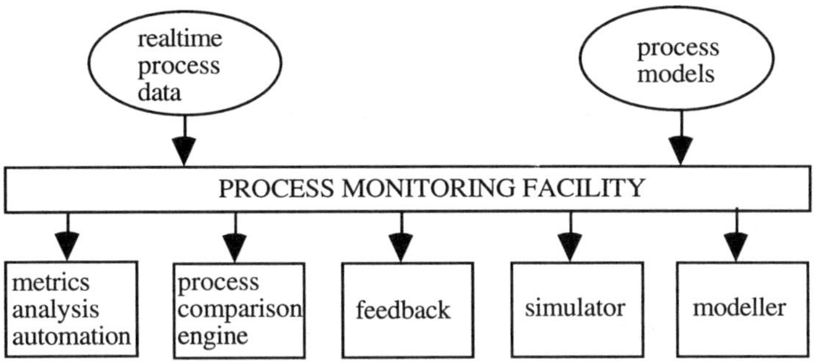

Figure 1 Process monitoring system.

For such a system, data acquisition techniques as well as suitable process modelling and process model simulation methods exist so that in creating process monitoring tools, the main item to be developed is a process comparison engine, i.e., the comparison of data and measurements from real processes with corresponding items in process models. In order to do process model to real process data comparisons, some of the major issues are: the determination of cause and effect in a many step process; the handling of comparison when data collected from real processes and data from process models are at different levels of granularity, i.e., where the levels of detail between the real process and its model are different; and the automatic testing of process rules to determine if they are followed and what are allowable exceptions. For process monitoring systems to work effectively, the systems must be able to traverse quickly large process maps with a large number of interconnections, and to link causes with effect in order to qualify and quantify feedback for process workers about process performance, assignment of process error, and the following or violation of process rules. A particular issue is that due to cost, data cannot be collected for all steps in every process; therefore, techniques are being developed for determining cause and effect relationships in processes for the case of sparse measurements.

As mentioned earlier, process modellers do exist. For process modelling and process simulation, FirstSTEP, a commercial product by Interfacing Technologies Inc., is being used. It is built in object technology (ParcPlace: VisualWorks, based on Smalltalk). For use in process monitoring, FirstSTEP has the advantage that it considers tasks from the process point of view rather than the functional point of view (Pardasani, 1992). This allows simple definition of processes and a simple basis for comparison of processes. FirstSTEP can also simulate models of processes.

However, the design of a simulator for process monitoring needs to be different than a simulator for process models. It is impractical to measure all the data for real processes; therefore, a simulator needs to be able to 'replay' real data while simultaneously considering models of the processes in order to interpolate between measured points and fill in missing information. Intelligent agent technology has been developed which can operate in realtime. Methods are being designed to be added to these agents in order to do the necessary interpretation of models and interpolation of data.

For easy analysis of monitored processes, it is imperative that a good tool kit be developed which contains process metrics and analysis tools. Several types of metrics are needed which can indicate process performance, flexibility, cost, complexity, coordination, etc. Many concepts exist such as, activity based costing, value added, on time delivery, and throughput; algorithms for these are being developed for incorporation into a process monitoring system. Other metrics, such as, an entropic measure for flexibility which has been considered for flexible manufacturing systems (Kumar, 1986) are being developed for use with all processes.

One of the important considerations for managing processes are interfaces, i.e., points at which items are handed off between departments and/or organizations. Process analysis tools are important, particularly, ones such as process connectivity, interface identification, as well as data volumes and timing at interfaces; all of which can assist in interface management.

Presenting feedback information is problematic. The mechanism needs to be simple for easy interpretation, yet sophisticated enough to allow viewing of large systems of processes. It needs to show which processes are not operating effectively or are in trouble, as well as indicating the upstream and downstream processes which can affect or be impacted by problem processes.

Since process monitoring is information intensive, methods for automating data gathering, analysis and reporting of feedback need to be part of a process tool kit. The methods are not for automating the process, but for simplifying process monitoring.

3 METHODOLOGY

At present, process monitoring concepts are being developed in a three step process:
1) design the major system components and the process tool kit using object technology;
2) develop and test the concepts in the laboratory using a process modeller/simulator; and
3) examine simple prototypes inside companies to test concepts further.

Simulations of process scenarios using data acquired from actual processes will also allow the determination of performance characteristics for processes, and thus, desirable features for process analysis tools and metrics. As stated above, the process modeller/simulator used is FirstSTEP. Process monitoring prototypes will be built using VisualWorks and Smalltalk. Object oriented technology has proven to be very effective for building applications which can handle great complexity, and at the same time provide interfaces which are very simple to use.

Intelligent agent technology is being studied to judge its usefulness in tracking the execution of processes and assignment of cause and effect relationships on the basis of realtime data collection, process metrics, and the parent-child relationships of process steps and lower level activities.

Realtime operation of a process monitoring system is deemed to be essential; the appropriate level of realtime response and analysis versus likely usage scenarios are being studied. The challenge of overall system design is to make it simple to use. The total system of process planning in addition to measuring and updating of process data must be transparent to those who operate processes; otherwise, process monitoring will not be adopted.

4 TEST CASE EXAMPLES

There are many activities inside corporations which have a significant impact upon overall performance, and which are difficult to manage with existing systems. Examples are given below which demonstrate how process management and, in particular, process monitoring can provide a good means to better both performance and control in corporations.

4.1 Managing corporate processes in a project environment

The operation of many corporations can be characterized as the performance of projects, e.g., design/engineering companies, product/service development organizations, etc. Projects have different life times and are generally not related to each another, i.e., different deliverables, different customers or both. However, all of the projects make use of a small set of common processes inside the corporation. These processes are used at different times, and possibly, in different ways. It is presently being realized by many corporations that competitive advantage comes from support processes, not the ability to manage projects. In most cases, these processes contain the corporation's core competencies.

To manage projects corporations use management information systems to track time and cost by activity centre for projects. Process information is rarely captured; but, it is the performance of processes common to all projects which greatly affects project performance. These critical processes need to be actively managed. Process monitoring is the solution, i.e., measurement of key parameters which indicate good process/project performance, and systematic review so that both processes and projects are carried out effectively. In addition, feedback must be given to all process performers so that they can continuously improve the processes for the better execution of projects.

The benefits of process monitoring in this scenario are twofold. First, project mangers are able to influence the processes which have a direct bearing on project execution, and thus, improve project effectiveness. Secondly, process owners are able to improve processes and core competencies for overall corporate productivity improvement.

4.2 Engineering specification approval

Many companies bid on contracts singly or with a group of partners. For custom designed and engineered products, once the contract is awarded, there is a requirement to exchange engineering specifications about the to-be-delivered product with the customer, prime contractor and/or contract partners. This is quite common for the delivery of final products or subsystems for large items, such as bridges, ships, aircraft, hydro electric projects, buildings, etc.

The process for the exchange of engineering specifications has many problems: highly variable length of time, highly variable cost, changing circumstances for different projects, and changing specifications which can change production costs. The inability to estimate the length of time and cost to arrive at agreement on engineering specifications greatly impacts the ability to finalize design and to schedule production. Constant changes lead to rework for engineering calculations and product design. Also, one of the most significant difficulties is the changing nature of the problem from project to project; this leads to paralysis in trying to find a solution, since it appears that there needs to be a different solution for each project.

Process management can assist in reducing the uncertainty associated with the process for engineering specification approval by designing a suitable process (Thomson, 1995). However, the major difficulty is not so much poor design of process, but the large number of interfaces both inside and outside the corporation across which flow exchanges of information. It is these interfaces for information exchange which determine the efficacy of the process. A process monitoring system is needed to track key parameters in order to measure process performance, especially in regards to the management of these interfaces, and to provide feedback to both sides of an interface to allow for knowledge about the state of the process and for continuous improvement of process.

An important concept in this case is that the design of the process framework for improved engineering specification approval must contain process monitoring, or at least provide for its later inclusion. For, process redesign alone with not guarantee good process performance; only the active management of interfaces through process monitoring will do this.

5 CONCLUSION

Knowledge work is making up a greater and greater share of work inside corporations in the industrialized world. To improve knowledge worker productivity, knowledge processes need to be actively managed. Process monitoring, i.e., the capture of realtime measurements of processes, the comparison of these measurements with data from process models, and the provision of feedback, promises to provide a tool which will permit proactive management of processes and also enable their continuous improvement.

There are many types of activities in corporations which can best be managed by the gathering and analysis of data about process, not product or service. Two such scenarios were discussed showing how a process monitoring tool can assist in this management. An important characteristic of such activities is that the processes contain key competencies which the corporation needs to manage well in order to operate effectively.

6 REFERENCES

Miller, J.G. and Vollmann, T.E. (1985) The hidden factory. *Harvard Business Review,* Sep-Oct. 142-50.

Hori, S. (1993) Fixing Japan's white collar economy: a personal view. *Harvard Business Review*, Nov-Dec, 157-72.

Kumar, V. (1986) On measurement of flexibility in flexible manufacturing systems: an information-theoretic approach. *Proceedings of the Second International Conference on Flexible Manufacturing Systems: Operations Research Models and Applications* (ed. K.E. Stecke and R. Suri), 131-43.

Pardasani, A. and Chan, A. (1992) Enterprise model: a decision support tool for computer integrated manufacturing, *Proceedings of the International Conference on Object Oriented Manufacturing Systems*, Calgary, May, 182-7.

Thomson, V. (1995) Process management in manufacturing. *Control Engineering Practice*, (to be published Feb 1995).

Thurow, L. (1992) *Head to Head.* W. Morrow & Co., New York.

7 BIOGRAPHY

The author has spent 15 years at the National Research Council (Ottawa, Canada) working with industry on integrated manufacturing problems. He has recently been appointed to the Werner Graupe Chair in Manufacturing Automation at McGill University (Montreal) where he is doing research on process management.

Process assessment and simulation games - Methods and software supported tools in Business Process Re-engineering

Terho Laakso
Researcher, *, e-mail: Terho.Laakso@hut.fi
Jussi Hakamäki,
Researcher, *, e-mail: Jussi.Hakamaki@hut.fi
Kimmo Forsberg
Researcher, *, e-mail: Kimmo.Forsberg@hut.fi
Riitta Smeds
Associate Professor, *, e-mail: Riitta.Smeds@hut.fi
* Addresses: Helsinki University of Technology, Faculty of Industrial Management, Otakaari 4 A, 02150 ESPOO, FINLAND, Phone 358 0 4511, Fax 358 0 451 2107

Abstract

The objective of this paper is to introduce two methods and tools, process assessment and simulation games, which can be used to support business process re-engineering. Process assessment is a software supported analysis procedure to evaluate the present status of the processes. Simulation games are method for process redesign that is supported by the GAMEVIEW software. Tools will be developed as part of EUREKA TIME GUIDE -project (EU 1157). The paper is based on the research conducted at Helsinki University of Technology and in the eight Finnish pilot companies.

Keywords

Business process re-engineering, Process modeling, Self assessment, Self audit, Simulation games

1 INTRODUCTION

Enterprises are becoming more and more aware of business processes, the performance of which is the basis for the overall capability of the enterprise. The evolution of the enterprise proceeds trough incremental process improvement or radical process innovations. Process improvement in general has three objectives:

1. Improve performance in key success factors.
2. Fulfill customer requirements and create customer satisfaction.
3. Accomplish the strategic needs of the company.

However, the methods and tools to support this new process management are still insufficient. TIME GUIDE aims to develop methods and supporting software to evaluate the performance of processes and to design new process solutions. In this paper, two of the TIME GUIDE tools are presented: process assessment and simulation games. Through process assessment, an enterprise gets aware of the weak points in its processes and can define a new set of performance measures that supports process structures and process thinking. Process assessment gives a starting point to the improvement work and to process re-engineering. By participative simulation games new process structures can thereafter be designed and tested.

2 MANAGEMENT OF PROCESS IMPROVEMENT

The improvement arises from individuals active perception of problems or a recognition of opportunities. The needed creativity and innovation can be either encouraged or discouraged by the environment. Creativity and innovation begin with a clear mental picture of an imagined future state (Barrett, 1994). This clear picture creates us a space for possible actions where the prevailing operations and procedures are allowed to be challenged.

Participation is one main principle in successful change management. When all the relevant individuals, who are engaged in an operational process, can participate in its redesign from the beginning, the jointly developed solutions are likely to be better in productivity and work satisfaction, and resistance to change can be avoided to great extent (Smeds, 1994). Participation is thus supported in the TIME GUIDE tools.

Process improvement can be seen as a system of nested cycles 'Figure 1'. In the outer slow cycle, the strengths and weaknesses of the company derived from the market needs and expectations are at first clarified. After this phase the company identifies the key process that has to be developed and how much it needs improvement. Improvement can be evolutionary (incremental) or revolutionary (radical).

In the fast inner cycle, process improvement is based on small-step-development (incremental). The process structure is usually in this case sharpened and/or the performance measurement system is developed to support better the process objectives. From our experience in the Finnish pilot companies, substantial improvements in an enterprise can be achieved by streamlining present processes and by introducing a new process oriented measurement system. This re-engineering approach will be supported by the new tools.

Process assessment and simulation games are tools in the nested business process improvement cycles. The process assessment tool helps companies to analyze processes independently. The simulation games are used for process redesign. They visualize the problems of the present process and can be used to test new process structure.

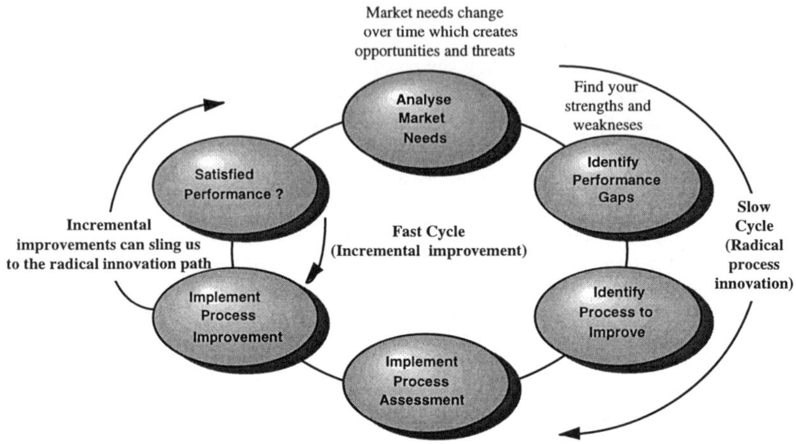

Figure 1 The nested business process improvement cycles.

3 THE PROCESS ASSESSMENT METHOD

The process assessment (PA) project proceeds through a sequence of seven steps 'Figure 2'. The steps are described in detail in the subsequent sections. PA focuses on the key success factors driven from customer requirements, market needs, competitors' actions and objectives of the company.

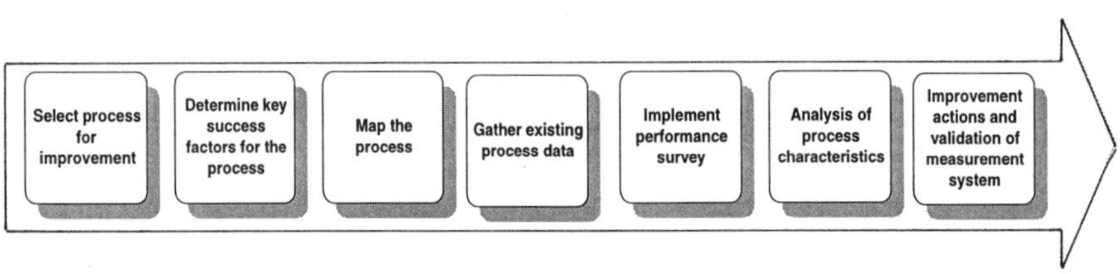

Figure 2 Framework of the process assessment procedure.

Procedures to select process for improvement

There are many methods and procedures available to identify process improvement needs of the company. The main stress should be in the requirements that come from the external environment of the company. However, internal requirements of the company should not be ignored, because there are activities that are important for the company, but not for customers, and vice versa.

J. Harrington (1991) listed following reasons for selecting a process for improvement:

- external customer problems and/or complaints;
- internal customer problems and/or complaints;
- high cost processes;
- long cycle time processes;
- there is a better-known way (benchmarking, etc.);
- new technologies are available;
- management direction based on an individual manager's interest.

Determine Key Success Factors for Processes

Key success factors (KSF) are important characteristics of the outputs of the processes. They are factors that have the most influence on the success of the company. Usually, KSFs are external customer-driven, but in cases where a process does not have an external customer interface, the factors must be defined by an internal customer.

One of the essential precondition for realizing key success factors in a company are an appropriate target setting system. We have to convert KSF's to measurable units. Thus, we can create a link between the overall objectives of the company and the objectives of any single activity in a process.

Process mapping

From our experience in the Finnish pilots the process mapping sessions have had a strong impact on understanding the inner activities of, and relationships between, studied processes. The process map also facilitates communication between people and it helps to share a common view about what is going on in the process.

We would like to point out some general aspects, which should be taken into account in process mapping notations. The notation should have a visual power to depict clearly different processes, but at the same time the notation should be easy to learn. Ease of learning may be viewed differently by the people who draw process maps and by those who read them (Kontio 1994). While the map is a tool for communication, the main emphasis of the notation should be on readability and on straightforwardness.

While mapping the process, we should aim at a sufficient level of description. Otherwise we can easily end up to describe the process in more and more detailed parts. This results in a complex model that is hard to understand, and most of our resources have been spent to the description alone. There is no more strength and motivation left for the improvement and redesign. People do require quick visible responses and actions in order to maintain their enthusiasm for the change project.

Process data gathering

Actual process data gathering begins after the mapping sessions. It can turn out to be a quite arduous job. The information system of the company does not normally support the process point of view, but is usually build up around traditional functional structures.

Our approach to data gathering begins by becoming familiar with the present reporting and measurement system, because it offers a good entry point to reality of the process. We discover at once to which extent present reports and measures are supporting the objectives of the process.

If a wide variety of products is produced in one process, the usage of time and other resources varies heavily depending on product type. In this situation we should focus on a few case products that help us to restrict the scope of the work and at the same time give us reliable information about the process. Overall averages do not give much information about real process performance. The strength of case products are also that we get concrete clues to follow.

Performance survey

The performance survey gathers the employees' subjective believes and opinions about process and company performance. The survey as such commits people to improvement work by involving them into the project at an early stage. Most important, however, is to find out the possible differences in the individual performance estimates. They might reveal a great need for process improvement.

The performance survey in the process assessment project is tailored for each process and each company individually. The survey is partly based on the process map and partly on questions classified according the general process models. The main parts of the questionnaire are:

1. general information gathering;
2. consciousness of key success factors of the process;
3. evaluation of activities;
4. analysis of few selected activities that are most important for the respondent;
5. general system evaluation.

Analysis of process characteristics

In the previous section, the qualitative and subjective performance survey and its objectives were introduced. Here we are concentrating on figure-based analyses. They are performed to find out the status of the process, using measurement system information or other means of documentation. Three types of analyses may be implemented in this phase of PA; time-based, cost-based and quality-based analyses. What types of analyses are used depends totally on the improvement objectives of the company.

Time has become a more and more important competitive factor and time-based analyses have become very powerful tools in process improvement. Time is an exact factor, and usually time-based data is well documented in companies. Time-based data gives a good picture how a process is functioning, and where time really adds value during the process.

Cost-based analyses are often difficult, because many costs are not chargeable to individual categories e.g. activities. This is particularly true of many overhead costs. Some form of allocation system (such as activity-based costing) must be used to employ cost-based analyses effectively. Many patterns are used, and companies vary appreciably in their format of cost-based analyses. Regardless of the difficulties involved, cost-based analyses are seen as a necessary element in the control and evaluation of processes.

Quality-based analyses help companies to identify their quality improvement needs in processes. The occurrence of different defects in time and quantity is listed and based on that information the cost of poor quality is calculated. Identifying cause-effect relations for defects is supported by the tool.

Improvement actions

As a result of the PA procedure the company should now have a clear picture what is happening in the assessed process. This consciousness is needed in the next phase, which is more problem solving oriented. The improvement work can direct its attention e.g. to:

- re-design of process structures;
- re-thinking of procedures;
- updating the performance measurement system;
- utilization of new technology;
- building and utilization of new competence;
- re-structuring company polices and organization.

The process assessment tool helps also to validate present measurement practices and to find new process oriented measures. These new measures do not compete with the overall performance management system. Instead, they must be integrated to the overall system in order to support the measurement of the daily operations.

To help people in selecting valid measures for processes the libarry of measures and drivers is introduced. It is a structured catalog of reasonable measures to support a new process oriented management paradigm. This catalogue is realized as a database in the PA tool, and possible measures can be found by using different search keys. The librarry serves thus as an idea bank when new process oriented measures are generated and validated.

4 SIMULATION GAMES

The rationale of simulation games in business process re-engineering

Simulation games are a powerful means to redesign process structures and to aid in their implementation. In a business process re-engineering project, simulation games can thus be used as a natural next step after the process assessment procedure. Process assessment gives an important input to the redesign phase: it helps in the selection of the target process, and it gives concrete measures and restructured objectives for the redesign task.

The virtue of social simulation games is that they bring together all those organizational variables that are crucial for process improvement: the tasks and activities, the formal and informal information flows, and individuals in their actual work roles. These participative games create the holistic understanding of the process that is needed in successful and efficient improvement work (Smeds 1994).

Simulation games can be used as a rapid process prototyping method in a participative manner. The results of simulation games in process restructuring have so far been excellent (Ruohomäki 1994, Smeds 1994, Smeds and Haho 1995a, 1995b). The new, game-designed processes are superior in both efficiency and human-orientedness, and the total lead-time of change could be considerably shortened.

Simulation games are applicable in almost every kind of processes. They can e.g. be used in redesigning the whole order to delivery chain or they can focus only on the production process.

The six main phases of the method are depicted in 'Figure 3'. The different phases in the framework are described in more detail in the text below.

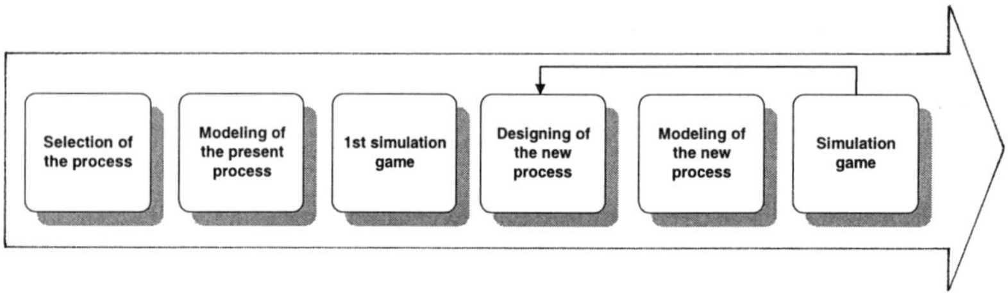

Figure 3 The main phases of the simulation game method.

Modeling of the present process

The modeling work can start after the target process for redesign has been selected (discussed in section 3). The modeling of the present process is a team effort and is done by using the wall chart technique. These modeling sessions are led by an external or an internal facilitator. The team involved is a mixture of employees that are familiar with the process.

The objectives for redesign direct the modeling work. The model can describe the general process flow, or it can be built on a more detailed level by using selected case examples. The case method is very useful e.g. in make-to-order processes. The simulation game itself is usually played using real case history material.

The basic model is a flow chart where the process is depicted through a set of activities and material and information flows. The model can include also information about actors, functions, description of activity and resources. The model is produced by the GAMEVIEW software, that is developed in TIME GUIDE project to support the simulation games.

First simulation game

The first simulation game visualizes to all participants the present state of the process. The game provides a common understanding of the whole chain with its strengths and weaknesses. It shows the interdependence between different departments and activities and stresses out the importance of cooperation and communication. During the game, concrete improvement proposals and ideas come up for further discussion.

Before the game, the participants and their roles have to be decided. The actors have an active role during the game. They play their actual real working roles. Questions about the problems in the process can be discussed together, as they rise. The observers follow the game and make notes of improvement ideas and questions that come up during the game. To observe a simulation game is often a very useful experience e.g. for managers and system designers. The number of the actors can be around 20, and an adequate number of observers is around five.

The game room should resemble in its lay-out the real departments or functions. The room should be equipped with different kinds of gaming material: a big process flow chart, documents used in the real process, raw materials and products (or scale models) handled in the process. The gaming material needs not to be very sophisticated. The most important issue is that there is something concrete to depict the flow of information and material in the process, not just the verbal explanation.

The basic idea of the simulation game is simple. The game follows the process flow that is depicted in the flowchart on the wall. The documents and the objects representing the raw material or the products are passed over from actor to actor like in the real process and each actor explains to the others their own tasks.

During the game, the other actors and the observers are free to ask about unclear things and make their own comments. The problems, improvement ideas and other comments that appear during the game, are documented on stickers by the actors and observers. These stickers are then attached to the flowchart on the wall.

The game is led by an external or internal facilitator who guides the game by making questions and comments. The facilitator directs the discussion in order to achieve the predefined objectives of the game.

After the game, a short debriefing workshop is organized where the problems and the development ideas that arise during the game are discussed and summarized.

Designing and modeling the new process

The experiences and documents from the first simulation game are used in smaller change project groups to design the new process structure. The design phase should not concentrate too much on resolving every single problem that came up in the first game. Rather, the whole process structure should be scrutinized and redesigned. When the whole process structure is reshaped, many of the minor problems that appeared in the first simulation can be eliminated.

The model of the new process must be more precise than the first model. Whereas in the first game the actors play in their own work roles, the second round is now based on the hypothetical future process. In other respects the preparation for the second game is a repetition of the first game.

The second and subsequent simulation games

The purpose of the second simulation game is to test the functionality of the designed process before implementation. The game provides also a learning experience to the actors as well as to the observers. The second game goes mainly like the first one. However, if the actors are not familiar with the designed process, the role of the facilitator is more critical. He or she must explain the new process structure and how it affects the actor's roles. The second game could be more structured than the first one. This means that the participants are not allowed to make so many questions and comments. The questions and comments are however written on the stickers and attached to the flow chart on the wall and summarized and discussed after the game.

The results of the second game are used as an input for further development of the process. Subsequent simulation games can be used to experiment different process alternatives before the final choice. Thereafter simulation games can be used for employee training before the actual implementation of the new process.

The GAMEVIEW software

GAMEVIEW is a software tool that is developed in TIME GUIDE to support the simulation games. It can be used for business process modeling, process visualization and process information gathering. The software produces automatically some of the material needed in the simulation game, and it also supports the documentation of the problems, improvement ideas and other comments that arise during the game. The models stored in the software can be analyzed using different kinds of views. These views can be either graphical or text based reports. The reports and the views of the software can e.g. be used as process documents for the company's quality manual.

5 CONCLUSIONS

The existing tools to support process re-engineering activities are still insufficient. The TIME GUIDE project aims to develop as a part of overall methodology tools and methods to evaluate the performance of processes and to design new process solutions. In 'Figure 1' we introduced the main phases of the process improvement cycle. Trough process assessment, an enterprise gets aware of the weak points in its processes and can define a new set of performance measures that supports process structures and process thinking. After the process assessment procedure, the company has a clear picture what is happening in the assessed process. This consciousness is needed in process redesign. Important inputs to the redesign phase are the concrete measures and restructured objectives, which provide the "red thread" for the restructuring.

Simulation games can be seen as a natural next step after the process assessment procedure in the business process improvement cycle. They allow a rapid participative process prototyping for business process redesign. The new, game-designed processes have turned out to be excellent in both efficiency and human-orientedness, and cost savings in implementation have been considerable.

6 REFERENCES

Barrett, J. (1994) Process Visualization Getting the Vision Right Is Key. Information System Management, Spring 1994, 14-23.

Bredrup, H. (1995) Productivity studies in Performance Management: A Business Process Benchmarking Approach. (editor: Rolstad=E5s, A.) Chapman & Hall, London, UK, pp. 17 - 43.

Colom, J. and Smeds, R. and Kleinhans, S. and Doumeningts, G. (1994) The role of benchmarking in the management and change process: some reflections from the TIME GUIDE project. In: Asbjörn Rostadås (ed.) Benchmarking - Theory and Practice, Chapman & Hall, London, 1995, 51-62.

Crosby, P. (1990) QES-Course material. Winter Park, Creative Factory, Inc. 146.

Harrington, H J (1991) Business Process Improvement. Quality Press, McGraw-Hill, Inc. New York, 267.

Kontio, J. (1994) Software Process Modeling: A Technology Review and Notation Analysis, Working document, Nokia Research Center, 50.

Moseng, B. and Bredrup, H. (1993) A methodology for industrial studies of productivity performance. Journal of Production Planning and Control, 4 (3).

Ruohomäki, V. (1994) Participative development and organizational learning with work flow game. In: Vartiainen, M. and Teikari, V. (eds.) Working papers of the third Otaniemi - Dresen Workshop, 16-17.12.1993, Report Series of Industrial Economics and Industrial Psychology Helsinki University of Technology, No 157, TKK Offset, Espoo, 63-75.

Smeds, R. (1994) Managing Change Towards Lean Enterprises. International Journal of Operations and Production Management, Vol. 14/3, 66-82.

Smeds, R. and Haho, P. (1995a) Tailored order-to-delivery process simulation game. In: Jens Riis (ed.) Simulation Games in Production Management. Chapman & Hall, London, 1995 (in press).

Smeds, R. and Haho, P. (1995b) Simulation games in business process re-engineering. In: Danny Saunders (ed.) The simulation and Gaming Yearbook Volume 3: Games and Simulations for Business. Kogan Page, London, 1995, 246-253 (in press).

7 BIOGRAPHY

Terho Laakso received his Lic. Tech in information processing science at Helsinki University of Technology in 1993. After his M.Sc.(Eng.) graduation in 1989 he worked as a system engineer and quality manager in Schneider Electric Oy for five years. Currently Terho Laakso is researcher at HUT and undertaking research on business process re-engineering in TIME GUIDE project.

Jussi Hakamäki, M.Sc.(Eng.) graduated from Helsinki University of Technology in 1993. He is currently working as a researcher at HUT in TIME GUIDE project. His special interest is the use of simulation games in business process improvement and re-engineering.

Kimmo Forsberg received his M.Sc.(Eng.) in industrial management at Tampere University of Technology in 1993. Since then he has worked as a researcher at HUT. His special interests are business process improvement and related issues of industrial buying behavior.

Riitta Smeds is Associate Professor at Helsinki University of Technology, Department of Industrial Management. Her current research interests concern enterprise evolution, technology strategy, management of change projects in enterprises and enterprise networks, business process re-engineering, management of innovations, implementation of new techno-organizational systems, participative modeling and simulation for business process redesign. She is currently leading the Finnish part of the two year EUREKA project TIME GUIDE "Guiding the evolution of industrial enterprises", as well as the EUREKA project SOFTMATCH "Social simulation method for the management of techno-organizational change".

30

Integrated Resource and Order Monitoring

H.-P. Wiendahl, J. Gläßner, M. Mittendorf
Institute for Production Systems (IFA), University of Hannover
Callinstr.36, 30167 Hannover, Germany
Telephone: + 49 511 762–2440
FAX: + 49 511 762–3814
E-mail: mit@thor.ifa.uni-hannover.de

Abstract

Growing global competition forces enterprises to improve their logistic quality to ensure customer satisfaction. Logistic quality requires good process capability as well as high process reliability. One approach for establishing a high logistic quality is the application of logistic-oriented and computer-aided monitoring systems, based on logistic models (e.g. the funnel model, developed at the IFA). This model describes and quantifies the dynamic interactions between logistic objectives.

To assist the creation of a company-specific balance between the contradictory logistic targets, monitoring systems have to provide an order-oriented and a resource-oriented view of logistic processes. With the regular processing and visualisation of logistically relevant information corresponding to these views, this integrated monitoring concept supports various logistic analyses and the derivation of improvement measures.

Keywords

Production Model, Monitoring, Logistic Controlling

1 INTRODUCTION

Competition forces the companies to strengthen their market power by improving their performance. Besides innovative solutions and high product quality, logistic efficiency is an important aspect for market success. The logistic efficiency can be described by delivery capability and delivery performance.

The delivery capability describes the ability of a company to promise a customer a certain desired deadline in consideration of the production situation. Delivery performance, on the other side, is the extent to which delivery dates promised at the placing of the order are met. If the desired delivery performance has been realized, process control exploits this logistic potential within the scope of resources planning, inventory management and operative control to realize a high logistic process reliability. Besides the mentioned performance indicators, to ensure market success, product prices and therefore the economic efficiency of all business processes must not be ignored.

Mainly by applying production planning and control, it is aimed to carry out the logistic objectives mentioned above. But as the systems used in companies often offer only insufficient support for these tasks the result is frequently unsatisfactory. Extensive analyses carried out by the IFA in various companies have shown that in most cases the deadlines are not met and there is often a lack of process reliability. Therefore many enterprises are very dissatisfied with the used PPC systems (Hautz, 1993). According to a survey carried out by GEIGER 77 % of 70 German medium-sized companies named PPC as the main source of problems in production (Glaser, 1991).

2 REQUIREMENTS FOR A SYSTEM FOR INTEGRATED ORDER AND PROCESS MONITORING

2.1 Integration into the planning and control process

A basic approach to improve this stuation is to provide logistic-oriented instruments for the controlling of the production processes. They should already consider interdependencies between the logistic objectives and facilitate statistically based statements about the aspired targets using key data and process graphics.

By using appropriate data processing and representation measures, a target-oriented process structuring and control can be derived along the logistic chain with these monitor systems, if

- the used key data correspond to the figures relating to quantity and time normally used in planning,
- the resource and order-related approach to the production process necessary for planning and control is supported and
- the supply of information is geared to the different questions concerning the strategic, materials planning and scheduling and operative level of planning.

Then the control of logistic processes can be transformed into a “closed-loop control” via the feedback element "monitor system" which is a main condition for a logistically regulated production process. To sum up, it can be said that monitor systems provide support for the following tasks:

- Target-oriented review and adaptation of the product and production structure.
- Deduction of realistic rated values for the planning of the order delivery date and the monitoring of order throughput.
- Continuous depiction of the actual status adapted to the information requirement of the various decision-makers.
- Analysis of deviations by rated-to-actual comparisons.
- Derivation of measures for process improvement.

2.2 Necessary monitoring approaches

To fulfil the above mentioned tasks, monitor systems have to support two approaches to the logistic procedures corresponding to the schedule and loading plan of a PPC system, these being an order-oriented and a resource-oriented approach.

The targets delivery performance and delivery time can be monitored by means of the **order-oriented** approach to the order throughput of a company. If the production is mainly customer-specific these targets determine directly the company's market success. The change in importance of the objectives as a function of the position of the point of customer decoupling and the procurement strategies belonging to it. Generally a part of the production process has to be realized on the basis of a production program, that is producing to stock, because normally the delivery time required by the market is shorter than the lead time needed for procurement, manufacturing and assembly of the whole product. Before the point of customer decoupling, the inventory as well as the utilization of the means of production come to the fore.

For companies in the capital goods industry, where the process of product development is largely customer-specific, order monitoring is of great importance. In case of customer-anonymous production the dimensionalization and continuous monitoring of the inventory and work systems – that is resource-oriented monitoring (see below) – come to the fore.

Depending on the level where the order monitoring is to support decision making, the system has to provide information with different levels of condensation:

- The rating of product structures according to the resulting logistic characteristics such as product structure lead time, point of variant formation, or the course of added value.
- The monitoring of the order network throughput.
- The monitoring of the throughput of individual procurement, manufacturing or assembly orders.

In addition to the order-oriented approach a **resource-oriented** monitoring for the order management is also necessary. Here the focus is not on the throughput of defined, mainly multi-stage, customer orders along the logistic chain, but rather on the logistic behavior of individual work centers within the procurement, manufacturing and assembly. In case of complex construction and operation scheduling processes, a development monitoring related to the work stations in non-manufacturing areas is necessary to support the order management.

The monitoring of work-system-related objectives like inventory, lead time, schedule performance and utilization reveals the quality of the individual logistic processes. If it is possible to control these key data via a closed-loop control, the market objectives delivery time and delivery performance can be influenced. Individual orders can be accelerated during throughput, for example by giving priority, but this has a negative effect on other orders, so that their key data deteriorate correspondingly. Thus there is a direct connection between the achievement of the objectives related to work system and order-related objectives. Only the control of the logistic procedures on every production process level guarantees the achievement of short delivery times and high delivery performance of the customer orders while at the same time considering the economic efficiency of the production process.

3 MODEL FOR A UNIVERSALLY APPLICABLE DEPICTION OF LOGISTIC PROCEDURES

The basis for the outlined monitoring concept must be a universally applicable process model that shows realistically the planned and actual order throughput and supports logistic processes from both the resource-related and the order-related points of view. In order to depict the interaction between the processing behavior of work systems and the throughput behavior of individual orders a logical linking of the resource-related and the order-related process model is required.

A further prerequisite is that all logistic objectives with their reciprocal relations can be depicted in the model as well as in a corresponding key data system. Finally a graphic and numerical representation of the logistic procedures based on common operational data has to be guaranteed.

3.1 Resource-oriented descriptive model

The funnel model developed at the IFA is a universally applicable model for the representation of resource-related logistic processes. Initially this model was developed to depict the throughput behavior of manufacturing systems but it is also suitable for depicting logistic processes in general (Wiendahl, 1987).

As shown in Figure 1, the basic idea is to consider the work systems involved in the order processing as a network of interlinked funnels, based on the material flow. To a work system described in this way, orders come and go with the waiting orders representing the work-in-process in the funnel. This process can be described quantitatively in the throughput diagram. As shown at the top right hand side of Figure 1, input and output of orders are plotted cumulatively as a function of time, beginning at a defined level of work-in-process. Here the value of the ordinate equals the size of the order, which is typically scaled in standard hours.

Depending on the definition of the system limits, the order throughput of an individual work station, a store or even a whole production sector can be represented. All main logistic objectives can be represented in the throughput diagram. The mean range corresponding to the lead time (defined as mean inventory divided by mean performance) and the mean inventory demonstrate, as shown in Figure 1, the vertical and horizontal distance respectively between the input and output curves. The mean performance can be determined by the gradient of the output curve. Schedule performance and utilization can also be shown, if the rated curves are depicted besides actual input and output.

The main purpose of the throughput diagram is to describe the dynamic process. It can be supplemented by a stationary description of the logistic behavior of work systems through logistic operating curves (Figure 1 at the bottom, right). Logistic operating curves condense various stationary operating states of a work system as a function of a parameter – the mean inventory – and show the functional relations between the logistic objectives of performance, range, lead time and interoperation time and the inventory (Nyhuis, 1991).

By using a logistic operating curve it can be shown that there are only small performance variations above a certain level of inventory. An interruption in work does not arise for there is a continuous supply of work. Below this value performance losses increasingly occur because of a temporary lack of work supply. Above a certain inventory level with almost

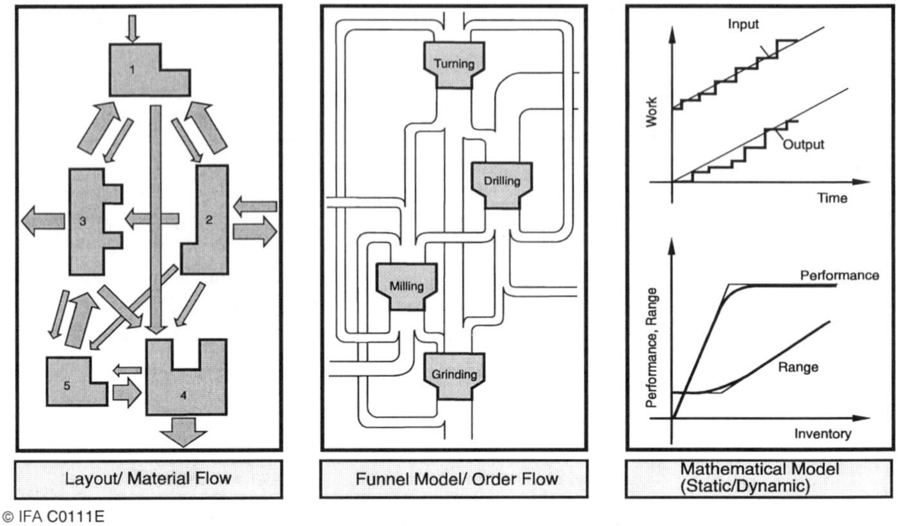

Figure 1 Resource-oriented modelling of logistic processes.

constant performance the lead time increases corresponding to the work-in-process. Below this value the lead time decreases with a decreasing inventory to a theoretical minimum which only depends on the mean order time and transit time (Nyhuis, 1991).

On the whole the basic form of a logistic operating curve is valid for all production systems. Beside the determination based on simulation studies, there is the possibility to determine these curves using a mathematical approximation, as described by Nyhuis (1991, 1994a and 1994b) and Wiendahl (1991b).

3.2 Order-oriented descriptive model

A basis for describing a throughput of customer orders, which can consist of a network of individual orders, is an assembly-oriented product structure. This structure determines the components, parts and procurement items with their corresponding production stages which are necessary for the making of a product (see Figure 2 at the top). Therefore it forms the basis for the structure of the order network that covers all the requirements necessary for the making of the desired product as a customer order. Now the order network can be represented as a time schedule under the course of the time (middle part of Figure 2). Every order is shown as a bar, the length of which represents the order's lead time. In this representation the sequence of orders results from the logical dependencies in the order network so it can be seen which manufacturing and procurement orders have to be completed before starting a particular assembly order.

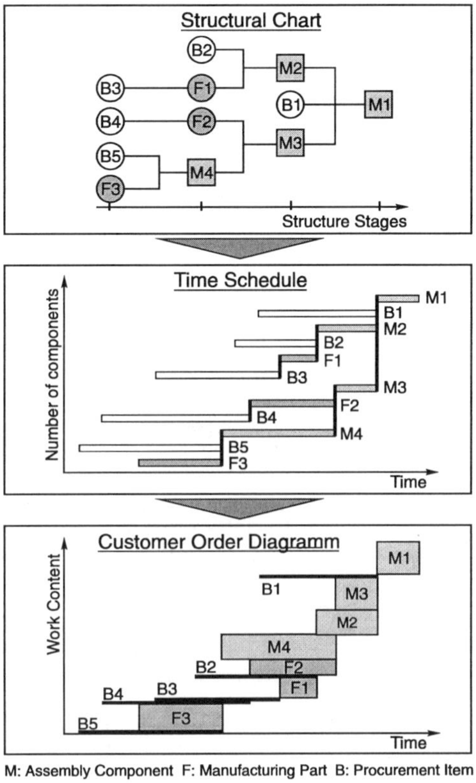

Figure 2 Derivation of the customer order diagram from the product structure.

As shown in the bottom part of Figure 2, a customer order throughput diagram is realized in such a way that the individual procurement, manufacturing and assembly orders of a customer order are classified by the completion date and plotted cumulatively as a function of time (Dombrowski, 1990). A throughput element is described as a rectangle. Its length represents the lead time and its height shows the size of the order. The cumulated work content of all working processes of manufacturing and assembly orders characterizes the size of the order. Since procurement items require virtually no capacity in the company concerned, they are merely represented as a bar in the diagram according to their respective replacement time. Thus the customer order throughput diagram illustrates the chronological procedure as well as the work content of all individual orders belonging to the customer order. Since the model has also to be suitable for the monitoring of individual manufacturing and assembly orders, as shown in Section 2.1, every throughput element of a customer order can be split up into its individual operations on the next lower hierarchic stage.

3.3 Interlinkage of both approaches

Since the operation throughput elements are the basis of the resource-related throughput diagram, a transferable representation of the logistic behavior of customer orders and work systems is possible. The relation between the description of work-system-specific logistic processes in the throughput diagram and the order-oriented depiction in the form of a customer order throughput diagram is shown in Figure 3 using as an example the throughput of 5 manufacturing orders through 4 work systems.

The throughput diagram for work system 2 (Figure 3 below) shows the input- and output-behavior of all the orders passing through this sector. Compared to this, the order throughput diagram represented on the left shows the throughput of order C with its 4 operations. The operation C2 at machine 2 is shown as a throughput element in both representations providing a connection between the order-oriented and the resource-oriented approach to logistic processes, so that the prerequisite for the integrated monitoring approach is given.

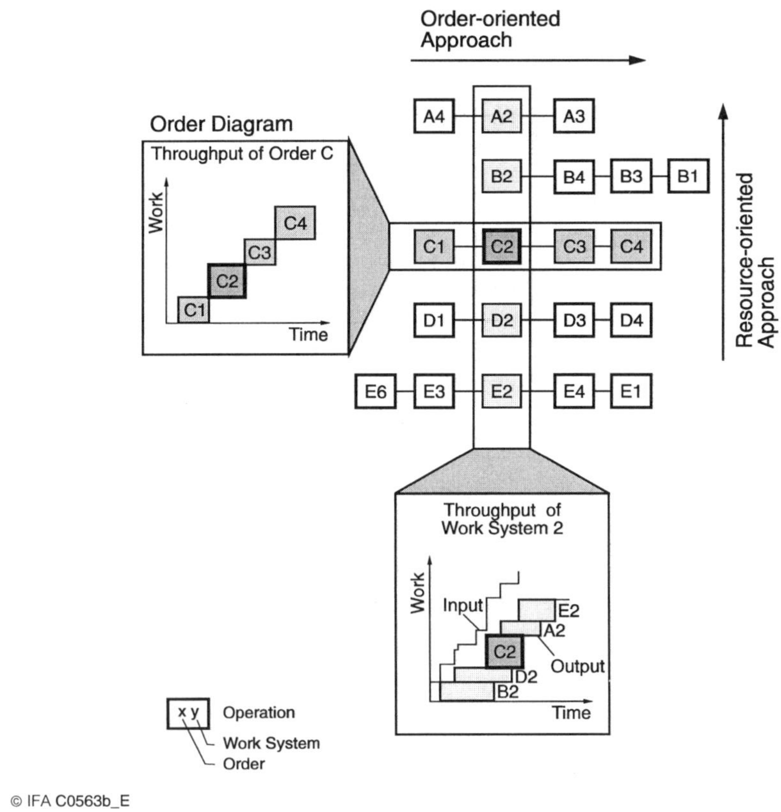

Figure 3 Interlinkage of the resource and the order-oriented approach to logistic processes.

4 STRUCTURE OF AN INTEGRATED ORDER AND PROCESS MONITORING

Figure 4 shows the classification aspects of an integrated order and process monitoring. It supports a resource-oriented and an order-oriented approach and enables the processing and representation of information adapted to the specific problems and planning horizons of the long, medium and short term planning stages (Wiendahl, 1992). Depending on the specific problem, the user is provided with flow charts, distributions and key data for the various approach levels.

For strategic decision making, analyses at long intervals are sufficient. They serve to check the logistic process capability, that is to examine to what extent the existing product and process structures allow the meeting of deadlines required by the market. As shown in the top of Figure 4, this can be realized on the resource side by demonstrating with logistic operating curves the marginal lead times that can be achieved while applying the existing or a new material flow structure, and considering a suitable utilization. The marginal lead times determined in that way can further be used to determine the shortest possible product lead time within the order-oriented approach. This information can be used, for example, to determine the point of customer decoupling (see Figure 2), to support a make-or-buy decision or to evaluate new products logistically. In this way aspects of a logistically suitable construction can be taken into account in the design and planning phase of the products.

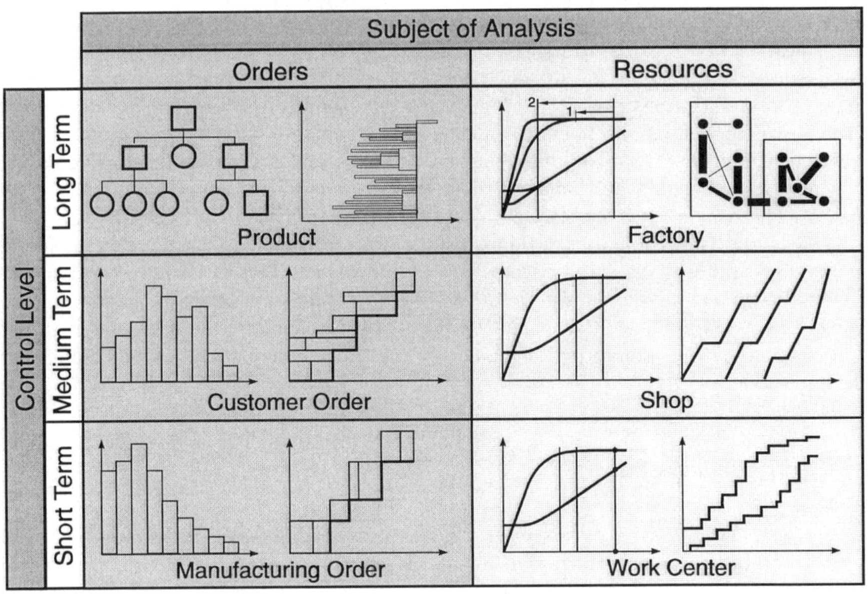

Figure 4 Classification aspects of an integrated resource and order monitoring.

Monitoring systems on the operative level serve as a feedback within the flow regulation by continuously monitoring the logistic process reliability. This enables a visualization and validation of planning parameters such as the scheduled lead time. Finally customer order throughput diagrams and evaluations for the analyses of the order structure serve to monitoring the throughput of order networks and in this way support a customer order control. Structure analyses and order diagrams based on individual operations are applied to plan the throughput and to survey short-term orders in a certain section. In the resource-oriented approach the state at the work systems can be visualized with logistic operating curves and throughput diagrams. In case of an unstatisfactory actual status the finding of corrective measures is supported.

The presented concept is to be realized in the form of individual systems which on the one hand are based on a universal approach and particularly on a consistent process model and which on the other hand meet the specific interests of the various departments of the company. The order monitoring is an exception, for it surveys the inter-departmental aspect of the throughput of complete order networks. In this way it supplements the specialised process monitoring.

5 SUMMARY

Logistic controlling is required to guarantee the logistic quality and efficiency of a company. It controls the process capability as well as the logistic process reliability. One way to realize these requirements is to apply logistic-oriented and computer-assisted monitoring systems which have to support both logistic approaches, order-related and resource-related. With the regular processing and visualisation of logistically relevant information corresponding to these views, this integrated monitoring concept supports various logistic analyses. On this basis the development and derivation of improvement measures and the structuring and control of processes and procedures are supported. Therefore this monitoring concept is a useful tool to enhance a company's customer orientation and market performance.

Besides prototypical solutions developed at the IFA (Wiendahl, 1995) there are monitoring systems to monitor logistic procedures already available. These systems (developed in different ways) are based on the described methods and applied in a great number of companies (e. g. (Holzhüter, 1990) and (Wiendahl, 1991)). For the time being the main fields of application are the monitoring of manufacturing procedures as well as procurement and inventory monitoring.

6 REFERENCES

Dombrowski, U. (1990) Logistische Produktanalyse als Ausgangsbasis für eine Reorganisation des gesamten Auftragsdurchlaufs. Wiendahl, H.-P. (eds.): Belastungsorientierte Fertigungssteuerung. Seminar of the Institut für Fabrikanlagen, University of Hannover, 10./11.10.1989 in Hannover and 21./22.3.1990 in München.

Glaser, H., Geiger, W., Rohde, V., (1991) *PPS, Grundlagen-Konzepte-Anwendungen*. Gabler, Wiesbaden.

Hautz, E. (1993) PPS und Logistik, zukünftig ein Widerspruch? In: H.-P. Wiendahl (Hrsg.): Conference *IFA-Kolloquium* 1993: Neue Wege der PPS. Institut für Fabrikanlagen, University of Hannover, gfmt-Verlags KG, München.

Holzhüter, E. (1990) Erfahrungen mit dem System KPSF zur belastungsorientierten Fertigungssteuerung in einem Maschinenbauunternehmen mittlerer Größe. In: H.-P. Wiendahl (Hrsg.): Belastungsorientierte Fertigungssteuerung. Dokumentation of seminar and conference of the Institut für Fabrikanlagen, University of Hannover, 21./22.3.1990, München.

Nyhuis, P. (1991) *Durchlauforientierte Losgrößenbestimmung*. Dissertation Universität Hannover 1991. Fortschritt-Bericht VDI, Reihe 2 Nr. 225, VDI-Verlag, Düsseldorf .

Nyhuis, P. (1994) Quantifizierung logistischer Rationalisierungspotentiale mit Betriebskennlinien. *ZfB Zeitschrift für Betriebswirtschaft*, Betriebswirtschaftlicher Verlag Dr. Th. Gabler GmbH, Wiesbaden, 64 (1994), H. 4, 443–464.

Nyhuis, P. (1994) Logistic operating curves - a comprehensive method of rating logistic potentials, Paper at conference "EURO XII / OR 36", University of Strathclyde, Glasgow.

Wiendahl, H.-P. (1987) The throughput diagramm - a universal model for illustration, control and supervision of logistic processes, *Annals of the CIRP*, Vol. 37/1/1987

Wiendahl, H.-P. (1990) PPS-Controlling - Grundgesetze einer logistikgerechten Produktionsplanung und Fertigungssteuerung. Paper at conference "7. Österreichischer Logistik-Dialog", Wien 1990.

Wiendahl, H.-P. (eds.) (1991) *Anwendung der belastungsorientierten Fertigungssteuerung*. Carl Hanser Verlag, München; Wien.

Wiendahl, H.-P., Kuprat, Th, (1991) Logistic Analysis of Production Processes by Operating Curves, *Annals of the CIRP*, Vol. 40/1/1991.

Wiendahl, H.-P., Ludwig, E., Ullmann, W. (1992) Produktions-Controlling - ein neuartiger Baustein zur Verbesserung der Fertigungsabläufe. Paper at conference "PPS im Wandel", 6./7. Okt. 1992, Zürich.

Wiendahl, H.P., Gläßner, J., Mittendorf, M. (1995) Integrated Resource and Order Monitoring, Paper at IFIP WG 5.7 working conference "Re-engineering the Enterprise", 20./21. April, 1995, Galway.

31

A System Solution to Concurrent Process Planning and Workshop Control

B. C. Schmidt
***IPH**-Institut für Integrierte Produktion Hannover gem. GmbH*
Hollerithallee 6, D-30419 Hannover, Germany
T.: +49 511 27976-42; F.: +49 511 27976-88;
email schmidt@iipr00.ifw.uni-hannover.de

Abstract:

This paper discusses current weak points in manufacturing planning and the need for more collaboration between process planning and scheduling. It then describes the approach and prototype developments of a new software system (COMPLAN) in which process planning and workshop scheduling are supported as collaborative tasks that operate on the same data base. The system uses net-shaped process plans with manufacturing alternatives to increase planning flexibility in the workshop. The system provides equally manual and automatic planning functions for process planning and scheduling and also provides an interactive interface to CAD.

Keywords
production management, concurrent engineering, process planning, workshop scheduling.

1 INTRODUCTION

„How to improve performance in industry" is the key question for which manufacturing engineers have to continuously find new and innovative answers. In recent years the efforts focused on two improvement areas.

Business Improvement, which is mainly achieved through organisational changes. These changes are inspired by new concepts or visions such as Lean Production, Business Process Re-engineering, Fractal Factory, Concurrent Engineering or Segmentation and Decentralisation.

Production Improvement, which is mainly achieved through technical and organisational changes that are driven by technology. In this area the efforts aim at continuously improving the processing of technical information (e.g. CAD/CAM, FEM) and logistical information (ERP/MRP), improving the employed manufacturing technology (High-Speed Machining, FMS). The advances are mostly characterised by making effective use of newly available technology such as lasers or supercomputers or by making use of newly developed methods such as fuzzy logic or genetic algorithms.

New concepts and visions for business improvement have received great attention in the last four years. The existing enormous potential for improvement by organisational changes in corporations had been neglected. Great successes were achieved and are still achieved today by Lean Production, Business Re-engineering and Decentralisation. But industrial competitiveness will always also be determined by the performance of manufacturing.

Especially, the successful implementation of concurrent engineering concepts will require the development of new underpinning technology and enabling information systems. The idea of concurrent engineering must not only be applied to product development but to the entire chain from development to regular production. It must not only aim at performing tasks in parallel and thus shortening the time-to-market, but it must focus on making the individual tasks in manufacturing also collaborative to thus achieve a better overall performance.

1.1 Weak points in manufacturing planning

Several weak points can be identified in the current state-of-the-art in manufacturing planning in industry. An efficient interface between CAD data and computer-aided process planning functions is missing. Also it is required to introduce generative process planning systems to industrial applications. Currently, automated process planning is only possible for a limited spectrum of workpieces. An introduction of automated generative planning modules with limited capabilities in industry requires that they are integrated with existing commercial CAPP systems. Process planning also has to provide more routing flexibility to workshop control. That requires on the other hand that information on logistical constraints is provided to process planning. The increased cost pressure no longer allows it to leave process plans unchanged for a long period of time. Revision and improvement of process plans will become a continuous task. Consequently, the functionality of existing process planning systems also needs to be extended to make CAPP systems a powerful tool for cost management and for the evaluation of make-or-buy decisions. Scheduling systems will have to improve reactivity and support decentralised organisation structures. To improve planning flexibility scheduling systems should interact better with process planning and make use of manufacturing alternatives. Finally, a continuous monitoring of workshop performance will be required as a first step towards a closed-loop workshop control.

Time and cost of manufacturing are significantly determined by the performance and degree collaboration of manufacturing planning, process planning and workshop scheduling. Today, process planning and workshop scheduling in industry are carried out as sequential, non-collaborative tasks. This has several disadvantages. The flexibility to react upon workshop disturbances is low. Sequential process planning and scheduling results in long through-put times. Generating process plans without considering actual workshop load and bottle-necks results in process plans which are not optimal for the situation in the workshop at the time of order execution. Moreover, the description of resources that are used by process planning and scheduling are often not consistent. The scheduler and shop floor control department can not call upon a process planner to assist whenever rescheduling also requires alternative manufacturing processes or routes to be searched for. This can be improved by making those activities partially concurrent or at least collaborative. The different information flows in conventional and concurrent process planning and scheduling are outlined in Figure 1.

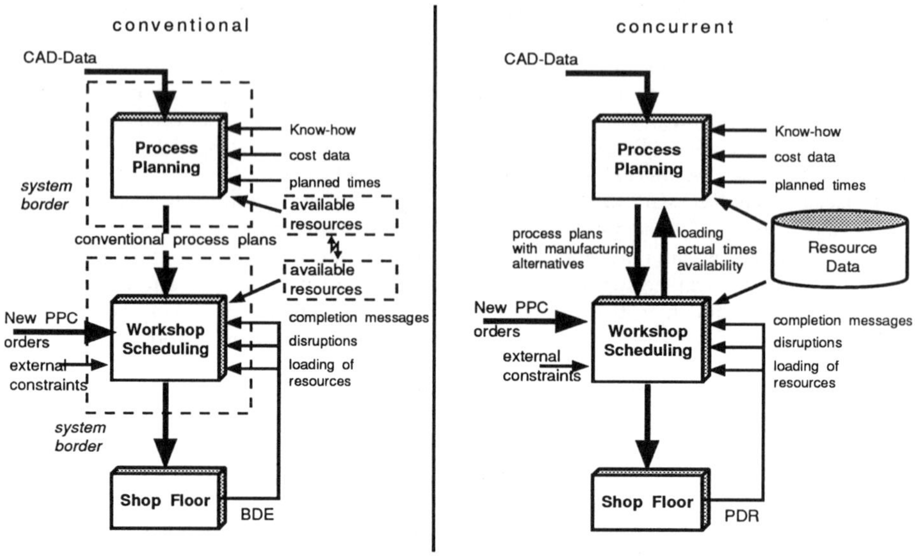

Figure 1 Flow of information between process planning and workshop scheduling

The integration or collaboration of scheduling and process planning has been the focus of extensive research efforts in recent years. A comprehensive overview of the world-wide efforts in this area is given by Elmaraghy (1993). An important research effort that has lead to a commercial system was performed in Twente (Houten 1991) where mainly the approach of dynamic, on-line process planning was pursued. In the COMPLAN project, only one order independent process plan is generated for each workpiece (Kreutzfeldt 1992). The flexibility for scheduling is improved by including manufacturing alternatives or possible changes in manufacturing sequence already in the process plan. The usually linear process plan sequence is enhanced to a net structure that contains the feasible operation steps. Several alternative routings or sequences of operations are combined in one Petri-net structure, so called non-linear process plans (NLPPs). The most important feature of the COMPLAN process planning system is its ability to handle so called non-linear process plans. An example for such a net-shaped process plan is given in Figure 2.

1.2 Approach

The target environment for the COMPLAN system is small batch production of complex products in a typical job shop environment. A certain complexity of the manufacturing process is required to be able to fully utilise the possible benefits of the system. Also, a certain level of computer support in design, order processing and workshop control should already exist in companies that aim at introducing the COMPLAN system. The following environment functions are required:

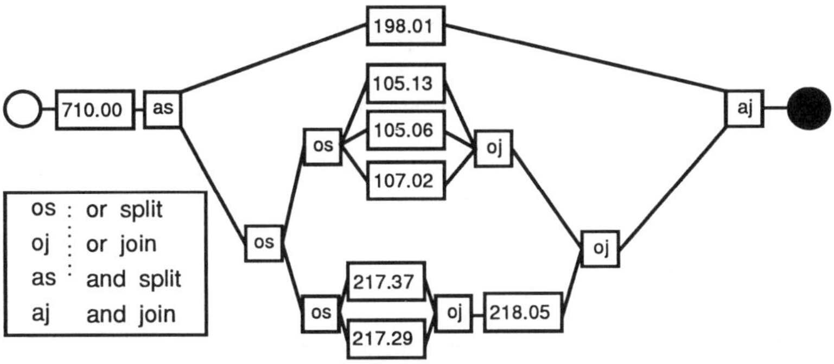

Mach nr.	Description	Setup time	Mfg. time
710.00	saw	5	10
198.01	gun drill machine	150	500
105.13	NC mill/drill/bore hor/vert 3 axes mach. centre	100	150
105.06	NC mill/drill/bore hor/vert 4 axes mach. centre	30	100
107.02	NC mill/drill/bore hor 4 axes mach. centre	50	110
217.37	mill machine	80	130
217.29	NC mill machine	100	90
218.05	NC drill/bore centre	100	50

Figure 2 An example NLPP

A PPC system or a MRP system that performs the explosion of the bill of materials, performs material requirements planning, lot sizing and assigns due dates for the manufacturing or assembly of parts. This manufacturing order information is passed to the COMPLAN system. Manufacturing order information must comprise at least an identification for the workpiece to be produced, the quantity and the due date for the workpiece.

A design department that provides workpiece information for each workpiece. This information must comprise all information needed to create a process plan. On the one hand, an interface is foreseen towards wire-frame CAD systems, using a neutral programming interface.

The workshop must be able to provide production registration data. As a minimum, the workshop feedback data must contain the actual start- and finish dates of work orders.

The project is carried out under the framework and with funding from the ESPRIT program of the European Commission. The consortium consists of seven partners, where IPH acts as co-ordinating partner. AIS and CCI are commercial software developers. IPH is an independent non-profit research and consulting organisation in the area of computer applications in manufacturing that co-operates with industry and the University of Hannover. WTCM is the collective research centre of the metalworking industry in Belgium. KUL division PMA and IFW of the University of Hannover are well established university institutes in the area of production technology. LVD is a medium sized enterprise that develops and produces machine tools for sheet metal forming and cutting. LVD employs about 800 people.

This consortium structure ensures that innovative ideas are implemented in state-of-the-art software that considers requirements and constraints of typical industrial application. The system is currently implemented on different UNIX workstation platforms with OSF/Motif user interfaces. Developments are currently performed in parallel on DEC, HP, IBM and SUN workstations. The system is mainly developed in C++. LISP is used within the Petri Net Process Planning Module (PN PPM). It is implemented on an ORACLE database system.

2 SYSTEM ARCHITECTURE

The COMPLAN system provides the framework where all modules work together on the basis of a concurrent approach. The system supports a flexible link between the process planner and the workshop scheduler, resulting in a better overall performance in manufacturing. The whole system is based on an integrated relational manufacturing database. The modules of the COMPLAN system can be grouped in four main functional units as represented in Figure 3. The module acronyms use in the diagram are explained in the following.

2.1 Process Planning (CPP)

The Complan Process Planner has to perform two main functions. It has to handle the description of workpieces in features (Workpiece Delineator) as basis for automated process planning and it has to provide process planning functions for different situations and requirements.

These functions are controlled by so-called Editor Control Modules (ECMs). These modules feature graphical editors that allow to manually enter data about either the workpiece (WP ECM) or the non-linear process plan (NLPP ECM). The ECMs cover the entire workpiece spectrum and the complete set of process planning data of a company.

Additionally to the ECMs, the COMPLAN Process Planner is extended with a set of „plug-and-play“ modules, allowing to obtain some of those data automatically or semi-automatically. These automatic modules are invoked and controlled from the editors.

The Workpiece Delineator (WPD) consists of a Workpiece Editor Control Module (WP ECM) allowing to manually enter workpiece data. To automatically obtain workpiece descriptions the WP ECM can call the CAD interface (CAD PPI) and the CODL file interface (CF PPI) modules that interpret CAD data.

Process planning functionality is provided by four modules: NLPP ECM, STS PPM, PN PPM, and AVOPLAN PPM. The NLPP Editor Control Module (NLPP ECM) offers the function to manually edit non-linear process plan data in a graphical editor. It also controls the State Space, the Petri Net and the AVOPLAN Process Planning Module. The STate Space Process Planning Module (STS PPM) and the Petri Net Process Planning Module (PN PPM) generate NLPPs automatically from feature based workpiece descriptions. The two modules apply different approaches to reach this goal. The AVOPLAN Process Planning Module (AVOPLAN PPM) generates process plans semi-automatically.

Getting the feature based workpiece description is the first step for automatic Process planning in COMPLAN. The Workpiece Editor Control Module (WP ECM) is able to edit manually a workpiece description and save it in the database. The WP ECM is also able to read the output of other Process Planning Interface (PPI) modules.

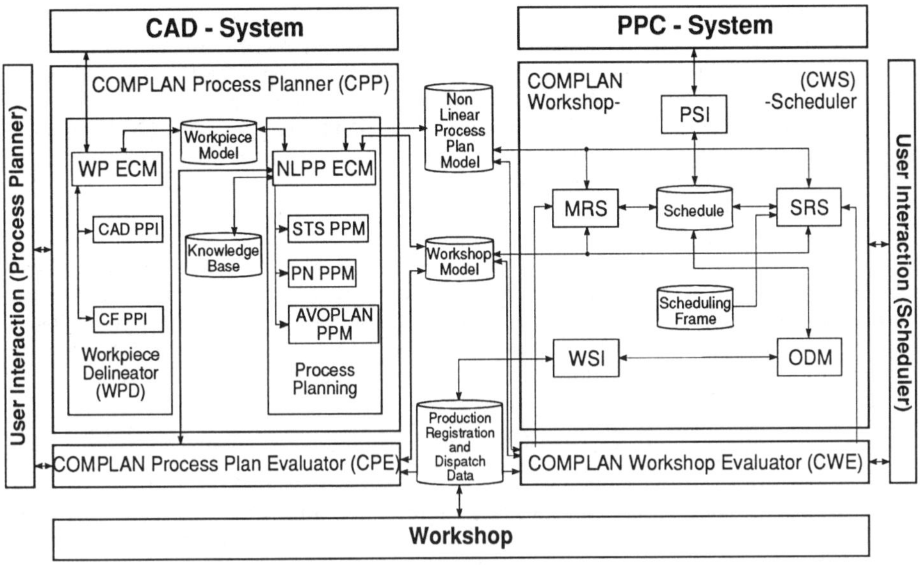

Figure 3 Architecture and Modules of the COMPLAN System

Using the CAD interface (CAD PPI), the feature-based description of a workpiece can be retrieved semi-automatically from 3-D wire frame CAD systems. This interface has been implemented for the *Unigraphics* system and for *Applicon Bravo*. For the ME 10 CAD system of LVD, a customised interface is being coded to overcome ME10 related problems. On the other hand, an interface is foreseen to feature-based CAD systems that write in standard file formats (so-called CODL-files).

Non-linear process plans (NLPPs) with manufacturing alternatives can be created or edited manually and be stored to the database with the graphical NLPP Editor Control Module.

Each of the automatic process planning modules can only be used for a limited spectrum of workpieces for which the feature catalogues and manufacturing constraints have previously been compiled. A workpiece testbed for different typical parts of LVD has been specified, and a feature catalogue has been compiled that specifies the process planning knowledge for each of the features used in those workpieces.

The PN PPM takes a feature based workpiece description as input and uses the concepts of Generic Petri nets, opportunistic process planning and constraint based search. The approach is described by Detand (1993) and Kruth et al. (1994). The current prototype supports all the steps of automatic process plan generation in a straightforward manner. The module is capable of generating fully automatically a NLPP for a simple LVD workpiece.

The State Space Process Planning Module uses a feature-based description of the workpiece that involves both geometrical and technological parameters. The current prototype supports an interactive generation of the operation steps of the NLPP. The STS PPM is linked to the ACIS modeller to allow geometric reasoning operations on the workpiece and features a graphical interface to display process planning information and a wire-frame model of the workpiece state.

The AVOPLAN PPM is based on and covers the full functionality embedded in the existing commercial AVOPLAN system which is being re-engineered for the use in the COMPLAN environment. It provides powerful time- and cost calculation methods, and the generation of a complete process planning data using variant or similarity planning methods.

It is invoked from the NLPP ECM and supports variant planning for some LVD workpiece families, manual process planning of all those parts that cannot be handled by the automatic modules, and time and cost calculation for an NLPP generated by any module

Other modules supporting process planning could be added to enhance the system with even more automated or user specific functions at a later time.

All communication between CPP modules is based on the Complan Universal Process Planning Language (CUPPL) which describes a context free grammar for NLPP, workpieces, commands and constraints. The Complan Communication Server (CCS) makes it possible to synchronise different processes on multiple CPUs. The link from Process Planning to Scheduling is achieved by the information in the common database where all modules access the same central NLPP tables.

2.2 Workshop Scheduling

Based on the manufacturing orders coming from an external PPC or MRP system the COMPLAN Workshop Scheduler (CWS) creates a schedule that is dispatched to the workshop. The CWS applies a two-level hierarchical approach to scheduling with two levels of scheduling detail and scheduling horizon. It is distinguished between workcenter and workstation level, where workcenter basically comprises a group of workstations with very similar machining capabilities. The architecture of the COMPLAN Workshop Scheduler is shown in Figure 4.

The Production Planning and Control System Interface (PSI) provides a connection to the required upstream PPC or MRP system. It receives data from the external system or reads it from external files and transfers the relevant information to the COMPLAN data base.

Scheduling is then performed by two modules. The Medium Range Scheduling module (MRS) has a wider time horizon for scheduling and operates on an aggregated level of detail. In the Short Range Scheduling module (SRS), individual orders will be allocated to specific workstations. The time horizon of scheduling is limited and the actual state of the workshop is taken into account for the compilation of the schedule.

In hierarchical scheduling several modules, mainly the MRS, SRS and ODM, work on the same data objects (work orders). Therefore, interaction between these modules has to be organised. Each module needs information on which scheduling actions have already been performed, which actions need to be done and what the actual state of an order in the shop is. This is necessary to avoid that orders are rescheduled when they are already dispatched to the shop floor where order preparation might already been started.

The MRS has a wider time horizon for scheduling and operates on an aggregated level of detail. On this aggregated level of workcenters, the system performs capacity planning and follows a load-oriented strategy for load levelling and order release. The applied algorithm follows the method of Load-Oriented Order Release that was developed by the Institute for Production Systems (IFA) at the University of Hannover (Wiendahl 1994). It has been enhanced to consider and investigate NLPP information for manufacturing alternatives at bottleneck resources. One of the objectives of Complan is the early recognition of possible capacity bottlenecks.

The output of the MRS are scheduled work orders with due dates and workcenters assigned to every operation in the process plan for the manufacturing of a workpiece. This Medium Range Schedule is the basis on which Short Range Scheduling is performed. Required input for Medium Range Scheduling are the manufacturing orders released by the external PPC or MRP system.

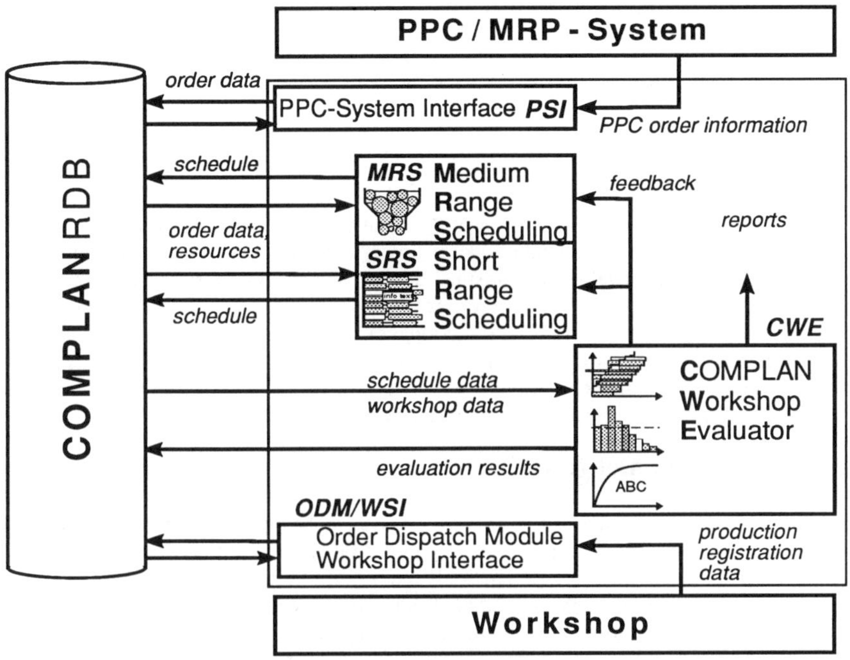

Figure 4 Architecture of the COMPLAN Workshop Scheduler

The current system mainly relies on user interaction for the selection of alternative routings and shall later be extended by automated strategies for selection of alternatives. The objective is to provide the SRS that operates with a shorter scheduling horizon with a rough scheduling basis to support and so to enable and improve more detailed scheduling decisions.

On the level of detailed scheduling (SRS), individual orders will be allocated to specific workstations. The SRS obtains the MR-schedule together with production registration data and returns the more detailed SR-schedule to the common data store. The returned SR-schedule updates the load profile that is used as a planning basis in the next MR scheduling run.

To automatically generate schedules, an automated planning modules uses simulation methods to execute Petri Nets. The Petri Nets are compiled from the non-linear process plan nets from all orders that have been marked for automated scheduling. The SRS also provides a Gantt-Chart as graphical user interface that will allow manual scheduling or changing of every order by user interaction. Additionally to commercially available detailed scheduling systems, the SRS will be able to access and display information on manufacturing alternatives. Schedules can be manipulated interactively on a graphical planning board.

The Order Dispatch Module (ODM) and the Workshop Interface (WSI) connect COMPLAN to the workshop. The ODM generates auxiliary orders. Both auxiliary orders and work orders are dispatched individually following specific workshop needs. The ODM works automatically and communicates via the database with other modules. The Workshop Interface (WSI) provides the required feedback information from the workshop to the Complan system. The WSI delivers scheduled orders to the workshop of LVD and records feedback data. In addition to the WSI a module simulating workshop feedback has been developed for testing and presentation purposes.

2.3 Process Plan Evaluator (CPE)

The evaluation of process plans (CPE) aims at the quality improvement of process planning by evaluating workshop performance data. The technical reliability of process planning data is improved by collecting and evaluating statistical data on actual process and setup times versus planned times. The quality of NLPPs with respect to logistical criteria or the actual situation in the shop is improved by forwarding production constraints such as e.g. unavailability or overload of certain resources to interactive and automatic process planning modules. Also a messaging system will be implemented where e.g. schedulers can attach notes to objects of a process plan that shall later be read and considered by the interactive process planners.

2.4 Production Evaluator (CWE)

As the demands on PPC increase, the amount of data to be processed and interpreted has grown significantly. Smaller lot sizes and throughput times along with shorter planning cycles also contribute to this. PPC systems can only be significantly improved when employing closed-loop control strategies that use shop floor data as feedback information to derive meaningful reference values as planning parameters for PPC.

These deficiencies are mainly due to insufficient skill in processing, aggregating and interpreting shop floor data. Another reason is the frequently poor quality of feedback information with respect to correctness, completeness and actuality. Production managers are supplied with insufficient information to cope with the ever increasing, complex decision making in manufacturing management.

The COMPLAN Workshop Evaluator (CWE) is designed to perform evaluation of shop floor activities and provide the user with aggregated information, textual as well as graphical. The CWE mainly will be used interactively to support decision making by the scheduler. On demand the evaluation results are stored in the database for later use. Certain restricted functionality of the CWE is also available to other modules to provide a convenient access to relevant shop floor data. Since the user rarely wants to overview the whole workshop all evaluations and visualisations are based on a pre-selection of orders, resources and time horizons. apart from tables with a user dependent selection of characteristic values three types of diagrams are generated on demand.

Throughput diagrams will give detailed information about the progress of order flow (in- and output), inventory, throughput time etc. and the related deviations of actual order data from the planned data respectively the scheduled times.

Bar charts to generate e.g. load diagrams for displaying resource related information like the capacity profile versus the load profile or to display frequency distributions of e.g. order sizes.

General 2D diagrams for user definable axis are e.g. to generate the characteristic operation curve which displays output versus inventory levels.

The evaluation aggregates workshop performance data and provides graphical representations for a quick overview on the workshop situation. It makes it possible to evaluate the quality of a schedule and determine the workshop performance. Evaluation results support scheduling decisions, the choice of the appropriate scheduling strategy and provide feedback to the process planning department.

3 INTEGRATION AT PILOT USER LVD

The project aims at the development of an integrated software prototype, of which the main modules shall be operational and tested at the pilot end-user's site after 33 months of project runtime in May 1995. The COMPLAN software will be installed in the workshop for mechanical manufacturing of discrete parts. The LVD shop floor is a typical example of a job shop for small batch production (batches from 1 to 10). To execute the operations, LVD utilises mainly universal machining centres and general purpose tools. The production flow in the LVD job shop is rather arbitrary, and the bottle-neck resources are dynamically moving around. The NLPP concept is especially suited for this kind of production environments.

The generation of NLPPs (by the COMPLAN CAPP system) can be done automatically using the Petri Net or State Space Planning modules, semi-automatically using AVOPLAN variant and similarity planning, or manually in a graphical NLPP editor. The automated functions will only be worked out for a limited scope of workpieces. For hydraulic building blocks, that are an important part family at LVD, the knowledge base and manufacturing constraints have already been compiled and tested.

The modular structure of the COMPLAN system enables an integration of the system step by step in the existing data processing environment as well as a „sophisticated" integration in future systems. In the start-up phase the first modules - mainly for the manual activities - will be installed and the COMPLAN data base will be set up. The scheduling modules (CWS) require the interfaces to PPC for the reception of manufacturing orders and to the data collection system for the feed back from the shop floor. Both interfaces can be realised by connection to the *OnLine Process Control* system (OLPC), which was developed by LVD for DNC functions and the collection of shop floor data.

4 EFFECTS ON BUSINESSES

Computer-aided process planning systems offer the last major and almost unused potential for a significant shortening of the throughput time from part design to part manufacturing. The growing demand for effective CAPP systems is caused by high effort in manufacturing planning required by complex CNC-controlled manufacturing systems. The envisaged COMPLAN system would significantly reduce process planning time and effort. It will also provide additional flexibility for the workshop by effectively making manufacturing alternatives available. The project will provide the missing link between CAD and CAPP and will so contribute an important element towards a flexible and fully computer-aided process chain from design to NC-programming.

The current practice of rushed and improvised changes in the manufacturing routing endangers product quality and transparency of manufacturing costs. COMPLAN will support quality assurance efforts by providing process planning support to workshop operators. Together with an evaluation tool this will allow more rational decision making on the shop floor and will strengthen the competitiveness of small batch manufacturing of discrete part which is predominant in many SMEs.

5 REFERENCES

Detand, J. (1993) A Computer Aided Process Planning System Generating Non-Linear Process Plans, PhD Thesis, KU Leuven 1993.

Elmaraghy, Hoda A. et al. (1993) Evolution and Future Perspectives of CAPP. Annals of the CIRP Vol 42/2/1993.

Houten, F.J.A.M. van (1991) PART, a computer-aided process planning system. PhD Thesis, University of Twente, Enschede 1991

Kreutzfeldt, J. and Schmidt, B.(1992) Integrated Process Planning and Workshop Scheduling, Proceedings of the 8th Int. IFIP WG5.3 Conference PROLAMAT 92, Tokio, June 23-26, 1992

Kruth, J.-P.; Detand, J.; Van Zeir, G.; Kempenaers, J; Pinte, J. (1994) Opportunistic Process Planning: A knowledge based technique for CAPP applications. 14th ASME Int. Computers in Engineering (CIE94) Conf.; Minneapolis, September 11-14 1994

Lenderink, A. (1994) The integration of process and production planning in small batch manufacturing. PhD Thesis, University of Twente, Enschede 1994

Wiendahl, H.-P.(1994) Load Oriented Manufacturing Control, Springer, Berlin London 1994

6 BIOGRAPHY

Bernd C. Schmidt studied production technology at the University of Hannover, Germany and finished with a Master Degree (Dipl.-Ing.) in 1990. In 1987/1988 he spent a year as guest worker at the Automated Production Technology Division of the National Institute of Standards and Technology in Gaithersburg, Maryland, USA. In 1990 he joined the logistics department of the former CIM-Fabrik Hannover as assistant to Prof. Wiendahl. When CFH changed its name to ***IPH***-Institut für Integrierte Produktion Hannover in 1992 he became head of the logistics department, which today employs 8 research and project engineers. His main task is currently the co-ordination of the COMPLAN project.

INDEX OF CONTRIBUTORS

KEYWORD INDEX